W9-DES-228

On the Edge of Earth

On the Edge of Earth

The Future of American Space Power

Steven Lambakis

The University Press of Kentucky

Scholarly publisher for the Commonwealth, serving Bellarmine University, Berea College, Centre College of Kentucky, Eastern Kentucky University, The Filson Historical Society, Georgetown College, Kentucky Historical Society, Kentucky State University, Morehead State University, Murray State University, Northern Kentucky University, Transylvania University, University of Kentucky, University of Louisville, and Western Kentucky University.

Title page: This near-infrared photograph of the Earth was taken by the Galileo spacecraft at 6:07 a.m. PST on Dec. 11, 1990, at a range of about 1.32 million miles. The camera used light with a wavelength of 1 micron, which easily penetrates atmospheric hazes and enhances the brightness of land surfaces. South America is prominent near the center; at the top, the East Coast of the United States, including Florida, is visible. The West Coast of Africa is visible on the horizon at the right. Photo from the National Aeronautics and Space Administration archives.

Editorial and Sales Offices: The University Press of Kentucky
663 South Limestone Street, Lexington, Kentucky 40508-4008
www.kentuckypress.com

12 11 10 09 08 6 5 4 3 2

Library of Congress Cataloging-in-Publication Data

Lambakis, Steven James.
On the edge of Earth : the future of American space power /
Steven Lambakis
p. cm.
Inclues bibliographical references.
ISBN-10: 0-8131-2198-1 (cloth : alk. paper)
1. Astronautics and state—United States. 2. Astronautics—United States—Planning. 3. National security—United States. I. Title.
TL789.8.U5 L35 2001 00-012288
387.8'0973—dc21
ISBN-13 : 978-0-8131-2198-7 (cloth : alk. paper)

This book is printed on acid-free recycled paper meeting the requirements of the American National Standard for Permanence in Paper for Printed Library Materials.

Manufactured in the United States of America.

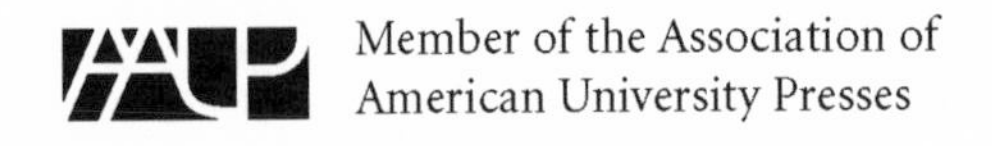
Member of the Association of American University Presses

Contents

For Tracie,

always and forever my source of joy,

and for Matthew, Lindsey, and Alexander,

for whom the future waits

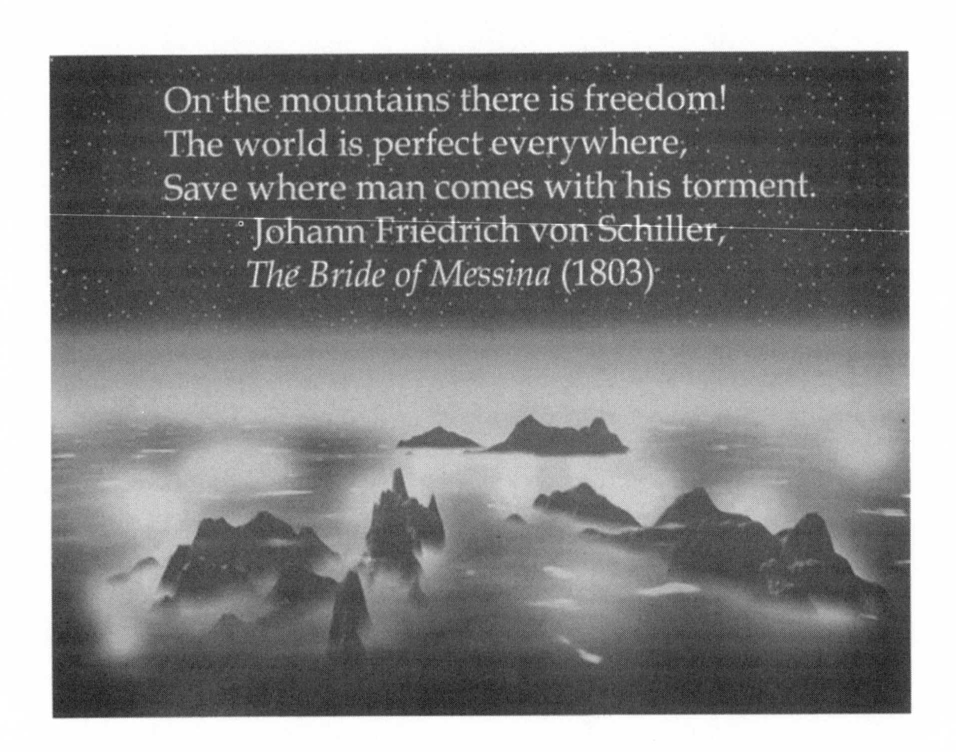
On the mountains there is freedom!
The world is perfect everywhere,
Save where man comes with his torment.
Johann Friedrich von Schiller,
The Bride of Messina (1803)

Acknowledgments

This book would not have been possible without the assistance of the Smith Richardson Foundation. My appreciation to the people of Smith Richardson who saw a compelling need to produce this book and who elected to throw their trust and support behind me is beyond measure. The foundation's generous sponsorship allowed me to devote considerable time over a period of two years to explore the subject matter of this book.

For their special counsel, oversight, and criticism, I would like to recognize three individuals. Dr. William R. Graham, President of National Security Research, Inc. and former Science Advisor to President Reagan, offered historical insight and invaluable policy and technical advice. Our conversations were an immense help. I also am deeply indebted to Dr. Colin S. Gray, Professor of International Politics and Strategic Studies at the University of Reading in the United Kingdom. Dr. Gray not only is a keen critic at the levels of policy and strategy, but his historical knowledge and expertise in military and defense affairs proved to be important guideposts for me. Finally, I am grateful for the perceptive counsel of Mr. Richard W. Scott Jr., who at various stages in his career worked in the defense, civil, and commercial space sectors in the United States. I found his experience in policy, military, and technical matters to be very helpful. I also would like to acknowledge the special contributions he made to the development of chapters 4 and 5.

While I have followed and analyzed the space power subject since the early 1990s, writing this book was a tremendous learning experience for me. My conversations and interviews with numerous individuals during 1998 and 1999 enhanced the quality of this work. There are several people I would like to thank, many of whom took time out of very busy schedules to meet with me or to review and comment on portions of this manuscript. I am ever appreciative for their insights and helpful comments. In alphabetical order, they are: Colonel David Anhalt, USAF (Assistant to the Director, Office of Net Assessments, U.S. Defense Department); Commander Thomas Berg, USN (U.S. Space Command liaison to the U.S. Ballistic Missile Defense Organization); Mr. Marc Berkowitz (Space Policy Director for the Assistant Secretary of Defense for Command, Control, Communications and Intelligence in the U.S. Defense Department); Major Brad Broemmel, USAF (Air Force Space Command staff); Ambassador Henry Cooper (Chairman of High Frontier); Mr. Charles Cunningham (Deputy Assistant Secretary of Defense for Intelligence, the U.S. Defense Department); Major General Robert Dickman, USAF (U.S. National Space Architect); Colonel Wayne Dillingham, USAF (Legal Advisor, U.S. Space Command); Brigadier General Stephen Ferrell, USA (Director of Strategy and Plans, U.S. Space Command); Colonel Bill Fruland, USAF (Directorate for Operations, U.S. Space Command); Dr. Robert Fugate (Senior Scientist and Technical Advisor, U.S. Air Force Research Laboratory); Mr. Wayne Glass (Defense Legislative Assistant for U.S. Senator Jeff Bingaman); Dr. Daniel Goure (Center for Strategic and Security Studies); Mr. Brian Greene (staff member on the U.S. House Armed Services Committee); Mr. Dana Krupa (Assistant to U.S. Senator Jeff Bingaman); Mr. Christopher Lay (SAIC); Lieutenant General Lester L. Lyles, USAF (Director of the U.S. Ballistic Missile Defense Organization); Mr. Andrew Marshall (Director, Office of Net Assessments, U.S. Defense Department); Lieutenant Colonel Sam McCraw, USAF (Air Force Space Command staff); Lieutenant Colonel Cynthia McKinley, USAF (Commander, 21st Space Wing, Air Force Space Command); Mr. E. Clayton Mowry (President, Satellite Industry Association); Mr. James Muncy (Defense Legislative Assistant for Congressman Dana Rohrabacher); Dr. Mitch Nikolich (Senior Analyst, National Security Research, Inc.); Mr. Pete Norris (President, SPIN-2); Major Perry Nouis, USAF (Public Affairs, U.S. Space Command); Colonel James Painter, USAF (Chief, Policy and International Affairs Division, U.S. Space Command); Lieutenant Colonel Anthony Russo, USAF (Space Warfare Center, Air Force Space Command); General Bernard Schriever, USAF Retired (pioneering program manager for the U.S. Atlas ICBM Program); Mr. Andrei Shoumikin (National Institute for Public Policy); Mr. Gil Siegert (Space Ventures); Captain Don Slaton (U.S. Joint Chiefs of Staff, Space Division); Ms. Marcia Smith (Congressional Research Service); Mr. Robert Snyder (Executive Director, Ballistic Missile Defense Organization); Mr. Eric Sterner (Legislative Assistant for Congressman James Sensenbrenner); Dr. Thomas Ward (Director, Threat and Countermeasures, Ballistic Missile Defense Organization); Rear Admiral Richard West, USN (Deputy Director, Ballistic Missile Defense Organization); Brigadier General (sel) Simon "Pete" Worden, USAF (Deputy Chief of Staff for Air and Space Operations).

I would like to thank Mr. Bernard Victory from the National Institute for Public Policy for his hard work on the appendix. I also owe debts of gratitude to Dr. Keith Payne, President of the National Institute, and the National Institute's Mr. John Kohout III, both of whom counseled me from time to time over the past couple of years, and to Ms. Amy Joseph and Ms. Karen Lynd of the National Institute for Public Policy, who provided reliable administrative support.

I also am profoundly grateful for the precision and professionalism of Karin Kaufman, whose copyediting skills put a high gloss on this work.

My debts to others in the writing of this book extend back very far. There are two individuals in particular I would like to recognize for their encouragement, teaching, and guidance. I will long be grateful to Mr. Paul Peterson, who teaches American government and political philosophy at Coastal Carolina University in South Carolina. With lessons in the works of Aristotle, Plato, and *The Federalist* that were as engaging as they were provocative, Paul lit within me an ambition to learn, not only about political things, but also about the world around me. His intervention changed my educational outlook and challenged me to think hard about fundamental political questions and, at the same time, to reevaluate my understanding of the good life. His inspiration and teachings gave me the courage to follow a career path that, thus far, has been very rewarding.

My career as a defense analyst got off to the best possible start when I joined the National Institute for Public Policy, a non-profit public education organization founded by Drs. Colin Gray and Keith Payne in 1981 and specializing in long-term national security policy and strategy analysis. I have worked with Colin Gray off and on now for ten years, and I would be remiss if I did not recognize that his works in the areas of space power, strategy, and military affairs were the impetus behind many of the ideas presented in this book. We truly do stand on the shoulders of others, and Colin has provided me a firm base from which to work and grow.

Lastly, the demands of writing mean that many long hours must be spent apart from family. For the loving and supportive setting she provided me during the writing of this book, I am eternally obliged to my bride, Tracie.

Introduction

On the Edge of Earth: The Future of American Space Power addresses major shortcomings in the defense space policy of the United States and offers recommendations to those who make, influence, and study national security decisions. Its comprehensiveness, references to American political traditions, attention to military and space history, and focus on policy-level considerations distinguish it from other works on the subject.

Space is a vital force in the life of the United States, the country's future bound to the development of capabilities for exploiting this environment. Indeed, America's security today depends mightily on its space power. Yet Americans do not spin in orbit alone, nor are potential enemies oblivious to the contributions of satellites to military victory. Equally ambitious and resourceful foreign governments, enabled by the proliferation of space technology and operational expertise, have compelled Washington to take its foreign affairs and security concerns to space. Today the United States is secure in space by default, not because there is a deliberate policy framework and well-resourced, organized, and strategically guided military force to guard national space interests. Security cannot be stable when it exists by accident. History supports the belief that hostile foreign governments and nongovernmental entities will endeavor to impair America's space capabilities or use satellites to their own advantage. How prepared is the country for this day?

What in some respects may be called America's second Manifest Destiny, beginning in earnest some forty years ago, has sparked a great revolution in the international security environment. Man's activities in space have transformed the dark arena surrounding Earth into a traversable medium having increasing influence over strategic interests and defense planning. The universal application of space power and the speed and singular contributions of satellite operations help the country cope with many traditional security matters in ways that are unique to the space age, a development that has encouraged the evolution of new war-fighting concepts and doctrines.

Yet in defiance of these consequential changes in the security realm, highlighted by the growing involvement of foreign governments and commercial entities in space, the nation's leaders have offered few convincing options for defending our interests in this outlying region. As surely as that "geographic" environment touches and affects the survival and livelihood of U.S. citizens, responsible officials must endeavor to influence and control the activities that take place there. In the end, however, defects within U.S. defense space policy pose the greatest obstacles to assuring national safety in the outermost frontier. Official rhetoric about the importance of space to security has been deceptively forceful. Rather than preparing adequately to meet new dangers to national security, policy makers have spent the past forty years arbitrarily, and with astonishing irregularity, constraining military activities in the realm above the atmosphere without clear public justification.

An essential yet difficult task of statesmanship as it relates to providing for national defense is to educate the populace and those in positions of leadership to care about the national future "out there." Statesmanship can do this by dispelling false ideas and shutting out fanciful political and strategic options. Leaders must strive to clarify public opinion and offer compelling choices befitting the national situation. *On the Edge of Earth* will show in this regard that not all opinions are created equal, that common sense and prudence compel us to elect some opinions as superior to others. I believe that the American people must know that, contrary to what elite opinion makers often suggest, the country's vulnerability in space does not enhance its overall security. In the absence of timely moves to counter it, this deception will endure until the shock of the first armed clash on the fringes of Earth's atmosphere jars Washington from its sleep. The reader, moreover, will learn that satellite orbits neither traverse a pristine "heaven" nor carve out a sanctuary, a longed-for place to escape earthly discord. Notions holding that the uses of military power are antithetical to peace only cloud public understanding and hinder meaningful political discourse and defense planning.

As U.S. investments and interests in space grow, citizens and government leaders must seek positive affirmation of the country's decision forty years ago to rely on its space power for national security. To preserve peace by preparing for war on the highest "mountain" man has ever climbed is in every respect an act of prudence driven by necessity. America's military power, after all, makes peace possible. Citizens of the United States and foreign countries can travel on the land, sail at sea, and soar through the air in relative security today precisely because of, not in spite of, America's armed presence in these environments. *On the Edge of Earth* provides a basis for believing that it can be no different in space.

PART 1 ▸ THE VITAL FORCE

To call something "vital" is to underscore its indispensability. Vital implies life and that which is necessary to survival. It also suggests an alternative state: mortality and the absence of that which is necessary to sustain life. Although man's (or a nation's) physical survival does not hinge on activities in space, our lives are nevertheless tethered, sometimes imperceptibly so, to objects orbiting Earth. Space sustains a modern economy, impacts international relationships and diplomacy, and, most important, contributes profoundly to the evolution of modern military forces.

The reader should take away from part 1 of this book three broad points. First, the space age is an integral feature of the modern condition. As the ages of agriculture and industry profoundly affected the commercial and power relationships of societies, so the robots routinely set in motion outside the atmosphere influence the national and global distributions of power, help determine the courses of science and understanding in the world, and powerfully impact the creation of national and personal wealth. Indeed, commercial reliance on space and commercial space operations boldly underscore the national commitment to exploiting and cultivating Earth's magnificent artificial "rings."

Second, the growing number of space powers and spacefaring nations has internationalized space. Diplomacy and actions of strategic importance increasingly revolve around activities in the cold, dark sphere enveloping Earth. Given the United States' resourceful and powerful position in the world, it will continue to influence the near-term course of human activity in space and lead other nations in the creation of new techniques (including military, legal, and commercial ones) for adapting space to the realities of world security and economics. By extension, we ought to expect international expressions of hostility to manifest themselves in space. Wherever man goes he is shadowed by his torment.

Third, the reader will appreciate that the strength of the United States' security rides confidently on its space power. The ascending military importance of space, today centered on information gathering and handling operations to assist national decision makers and war fighters, is undeniable. With its enhancements of the national early warning, communications, and reconnaissance functions, the satellite is one of the largest, most important contributors to national security over the past forty years. Since the 1991 Persian Gulf War, space power has brought information critical to tactics and operations directly to the combatants. Space activities, in other words, are changing the way states deter and fight one another.

Part 1 is not for the fainthearted. It covers a lot of material, including information of historical and technical interest. When appropriate, facts of economic or commercial import are presented, but in keeping with the defense focus of this book, most of the material is crafted to help the reader identify the significant military and policy issues at hand. Indeed, one will find that military space issues are becoming, if anything, more complicated. Part 1, therefore, introduces the reader to many of the key and sometimes emotion-laden policy issues of today and tomorrow, including interpretations of the "peaceful uses" of space, the requirement for space control, the impact of military space activities on foreign relations, the rise of space commerce, and the prodigious appropriations and organizational challenges now confronting the United States.

1
In Space Is Our Trust

How and Why Does Space Impact the United States?

Space is America's passion. As an endless and virtually unexplored frontier, accessible only once we have availed ourselves of our collective technological and engineering genius, space also is a truly American passion. Satellites, the flowers of our obsession, have spawned a global social revolution, affecting how we think and go about our daily business, entertain ourselves, and relate at home and abroad to our family, friends, and business associates. Space, in fact, has affected in a fundamental way how we function governmentally—even how we fight wars and guard the peace.

Feeding off the computer revolution, robots orbiting overhead bestow upon us a historically unique means to connect to one another globally and deliver information with uncanny speed and reliability. Satellites are largely responsible for the exponential growth of information products and services.[1] The space bridges constructed over the years are particularly desirable to Americans, whose family members commonly are spread across the United States and even around the world.

Satellites, we shall see, are a boon for business. Indeed, space has become a prosperous business in its own right. From the Fortune 500 companies to the local "mom-and-pop" stores, unless you've got a hook-up to space, your business operations antedate the 1960s and clearly are not geared for the high-tech, competitive economy of the twenty-first century.

Although integral to our everyday lives, space operations are transparent. We take for granted having at our fingertips devices that send and receive voice and video data in any combination to and from practically anywhere in the world. We routinely view breathtaking overhead perspectives of cities and nature's geological formations from hundreds or even thousands of miles above the planet. We accept as commonplace the possibilities of worldwide navigation. There are untold national security, civil, and commercial implications in all of this, a reality that relentlessly and beyond any force of will that we possess intertwines the prosperity and security of the United States ever more tightly with activities in outer space.

Only within the last hundred years or so have scientists and engineers attempted space travel. Manned and unmanned spacecraft have skirted the edges of Earth along the *high*ways that crisscross the skies only for slightly more than four decades. We have found over these forty years that space allows us to do better, in some cases a lot better, many of the things we have always done well. In many instances, satellites have permitted us to do things we never would have imagined only a generation or two ago.

The United States places its hope for the future, its care for its people, its confidence, its interest, and its trust in space precisely because space sustains and expands national industry, economy, science, and security. With more that $100 billion invested in space as of 1999, the incentive to gain access to Earth orbits and protect national space property and activities will continue to expand.[2] While we Americans certainly could live well enough without satellites (not a really viable option today in any case), thus enervated we could not thrive, nor prosper, nor live in our accustomed states of comfort and safety.

How Did We Reach Space?

Now as in the beginning, space enthralls us. The cold, hard vacuum enveloping Mother Earth beckons us, as would an oracle, to approach it for answers to our most profound and fanciful curiosities. Largely dark (though peppered with explosions of radiant light), shapeless (though characterized by magnificent celestial formations), transparent (yet ultimately nebulous to our intellects), without measurable depth or limited substance (its distances best measured in time or the speed of light—186,000 miles per second), timeless (at least to humans accustomed to day-night cycles), a container of all material things (though with no definable edges), void of recognizable direction (no north, south, east, or west to speak of), the living force (and, with planet-destroying asteroids, dying suns, and black holes, the reason for our ultimate destruction), outer space is the greatest of all enigmas.[3]

Throughout human history, intense theological wonderment, scientific investigation and exploration, philosophical reflections, imaginative flights of fancy, and practical national missions have allowed our minds to dance around the mysteries of this vast realm. Indeed, our interrogations and observations of space began with basic and eternal questions dealing with the existence of God and the creation of the universe[4] and our desire to learn about the characteristics and properties of neighboring celestial bodies. Yet the search for answers also extends to more temporal concerns, including how to derive commercial utility from space and accomplish research of practical importance to life on Earth.

Pre-twentieth-century efforts to reach out into space necessarily were confined to the reasoning powers of the mind and imagination. Millennia have passed since Babylonian, Chinese, Egyptian, Indian, and Greek astronomers observed the heavens to perfect concepts of time, predict spellbinding eclipses, and hypothesize about the shape of the earth. Literature has acted as a vessel from long ago, carrying through the ages many of man's earliest thoughts about space. Plutarch, Lucian of Samosata, Bishop Francis Godwin, Cyrano de Bergerac, Jules Verne, and H.G. Wells gave us memorable and influential works of fiction and speculation. New concepts about our solar system appeared after the astronomical revolution led by Nicholas Copernicus, Johannes Kepler, and Galileo Galilei.

Prior to the dawn of the space age in the middle of the twentieth century, theologians, philosophers, and storytellers molded space legacy, and only within the past one hundred years or so have concerted efforts been made to realize dreams more than two millennia old. Theoretical and technological advances in rocketry in the late nineteenth and early twentieth centuries, coupled with major strides in the understanding of engineering challenges posed by space flight, cleared the way for a giant move beyond science fiction into the realm of applied space science. The engineers, technologists, and scientists of the early space age gave us the "legs" to climb into space and stay there.

For the purposes of this book we will concern ourselves only with a minuscule portion of that inky depth surrounding our planet—an exosphere called circumterrestrial space. Sandwiched between the exosphere and terra firma is an atmosphere made up of regions familiar to us as the troposphere (the lowest ten miles of air), the stratosphere (a place where only a few aircraft will venture), and, from thirty to fifty miles out, the mesosphere. Boundaries for circumterrestrial space (to the extent they can be "charted") begin roughly at the top of the atmosphere, or sixty miles above Earth. At this point, the effects of aerodynamic drag and atmospheric friction are sufficiently forgiving to permit at least one orbit around Earth by a small spacecraft. At an altitude of fifty thousand miles, we may draw the uppermost boundary of this "geographic" region.(The appendix reviews the types and characteristics of circumterrestrial orbits.)[5]

Spacecraft sent into this region must survive and cope in a truly hostile environment. Solar winds and flares, cosmic rays, unrelenting gravity, both torrid and frigid temperatures, electromagnetic forces, and two now-famous radiation belts discovered by James Van Allen greet satellites, shuttles, and stations designed for human habitation once they have burned through the atmosphere to reach orbit.

It was Van Allen who designed the *Explorer 1* satellite, the first satellite launched by the United States, to examine some of these features of space.[6]

The conditions in circumterrestrial space are potentially lethal to man and destructive to machine. As visitors to this region, therefore, we face more than the physically daunting challenge of getting there. Once we or the robotic extensions of ourselves arrive, survival demands the utmost in ingenuity and resourcefulness—and, in the case of manned adventures, a healthy endowment of courage, emotional balance, mental discipline, and faith.

Yet the first challenge remains one of escaping Earth's gravity with enough velocity to either enter an orbital path or break away entirely from the earth system. The science of rocketry overcame that physical barrier. The Chinese employed early rocketry in warfare using the explosive force of black powder as early as the thirteenth century. After Roger Bacon, an English monk, brought improved explosive powder to the shores of Europe in the mid-thirteenth century, longer-range projectiles made their appearance in warfare among the European powers. As early as 1591, Johannes Schmidlap produced drawings of rockets with stages, an idea put into practice with only partial success in seventeenth-century Poland. Experiments in pyrotechnics took place in Europe and Asia until the invention of rocket artillery in the early 1800s. The rocket launcher, soon perfected and honed into a powerful and effective weapon, became the progenitor of a new science—rocketry.

Technological and engineering advances of the twentieth century, led by the Russian Konstantin Tsiolkovsky, the American Robert Goddard, and German citizen Hermann Oberth (Oberth was born in Hermannstadt, Transylvania, in 1894), proved rocketry to be the long-sought-after solution for traveling beyond the atmosphere. One might also include in this group of pioneers Robert Esnault-Pelterie of France, who gave aviation and astronautics a scientific foundation.

Tsiolkovsky's research and originality gave birth to the new science, although his thoughts and theories were not shared with other scientists for many years. The space fantasies of Jules Verne sowed the seeds for Tsiolkovsky's ideas on rocket-powered space flight. Verne, Tsiolkovsky wrote, "directed my thought along certain channels, then came a desire, and after that, the work of the mind." Published in 1903, the year the Wright Brothers accomplished the first powered airplane flight at Kitty Hawk, North Carolina, Tsiolkovsky's *Exploration of Cosmic Space with Reactive Devices* developed many familiar space-flight concepts, such as rocket staging, and laid bare fundamental mathematical principles needed to understand launch dynamics and master orbital physics.[7]

Independently of Tsiolkovsky's work, Robert Goddard labored hard to give life to remarkably similar theories and analytical reasoning by designing, building, and launching the first high-altitude rockets. Indeed, this Clark University professor of physics strongly influenced the emergence of a new discipline called astronautics, or astronautical engineering, a combination of physics, engineering, chemistry, and astronomy. The works of Jules Verne and H.G. Wells also had a profound effect on Goddard, though Verne, noted Goddard, had taken some disturbing liberties with facts about space travel.[8]

Through his studies and experimentation, Goddard helped transform what was in essence the pseudoscience of rocketry into a respectable discipline of systematized knowledge, and he made a name for himself not only in the United States but also internationally. Fearing the ridicule applied by a misinformed public to the pioneers of air flight, Goddard avoided the limelight and significant involvement in the newly formed rocket societies. He rarely advertised his real objective—that of designing a rocket to reach the Moon—preferring instead to refer publicly to his "high-altitude atmospheric research." Nevertheless, Goddard's plan to send a rocket to the Moon and signal its arrival with an enormous powder flash was the reason he was known in some circles as a crackpot—the "Moon Man."

Goddard published in 1919 the test results from the early period of experimentation with rockets in *A Method of Reaching Extreme Altitudes,* the first scientific treatment of space travel. Most notably, Goddard established the utility of the combustion chamber and nozzle in a rocket's design. He also saw the need to discard individual rocket stages as they expelled the last of their propellant.[9] Modern rockets continue to exploit each of these engineering innovations.

Trials involving liquid-fuel rockets, which never flew above nine thousand feet, helped him improve the steady flow of the propellant to the combustion chamber, an important requirement for rapid ascent. Indeed, a rocket needs to travel about seventeen thousand miles per hour to fly above the atmosphere and fall into an orbit around Earth. A speed of twenty-five thousand miles per hour breaks the rocket free of the earth system for a journey to the Moon and beyond. One persistent problem standing in Goddard's way was rocket instability in flight, which he partially solved through the application of gyroscopes and gimbaled engines. In 1936, he published his findings in a second Smithsonian report, entitled *Liquid Propellant Rocket Development.* By the 1940s, his work in this area was so advanced that German rocket scientist Wernher von Braun remarked that "Goddard's experiments in liquid fuel saved us years of work, and enabled us to perfect the V-2 years before it would have been possible."[10]

Elaborating on Issac Newton's Third Law of Motion, that "for every action there is an opposite and equal reaction," men such as Goddard and Tsiolkovsky, and their disciples in rocketry, helped disprove the popular notion that rockets would not work in space because there was no "air" to push against. Their labors demonstrated that the rocket, in fact, was a self-contained unit capable of producing a reaction within its own body to cause movement in the opposite direction of the thrust, even in the vacuum of space.

It remained for mathematician Hermann Oberth to persuade people that space travel, a flight among the stars, was a practical idea. Oberth laid out a vision and a theoretical technological path for catapulting man out of the earth system. Such was the inspiration for his famous 1923 work, *The Rocket into Interplanetary Space.* Oberth's fertile ideas about space travel captured the imaginations of the public and the scientific community. He was a significant force behind the prestigious German Rocket Society and responsible for inspiring the international space travel movement of the 1920s and 1930s.[11] He designed spacesuits, considered the problems of eating and handling bodily wastes in space, contemplated space walks, and proposed shuttles and space stations to facilitate interplanetary travel. In 1929 Oberth published *Ways to Spaceflight.* His work was significant because he advanced the theory that, using the high energy potential of liquid propellants, it was feasible to put large, manned craft into space. Oberth, who lived to see man venture into space, conceded later in life that the problems of space travel had proved to be more complicated than he originally thought.[12]

While rocket engineers and scientists in the United States and Russia relied on private financing for their work, the German government was the first to give significant support to rocket research. The origin of the famous V-2, *die Vergeltungswaffe,* or "weapon of retribution," may be traced to Adolph Hitler's attempt to circumvent the provisions of the post–World War I Versailles Treaty restricting armaments, munitions, and war material. This treaty, signed in 1919, sought to ban artillery from Germany's arsenal, but it was silent about the future development of rockets, which at that time were not perceived as weapons. After observing the activities of the German rocketeers, German ballistics expert and visionary Col. Karl Becker believed rockets could be turned into super-long-range bombs and provide the Third Reich a novel source of military power without violating the letter of the treaty, which Hitler publicly repudiated in any case in March 1935.[13]

Beginning in the mid-1930s, Gen. Walter Dornberger sponsored the work of von Braun's rocket team on the V-2 and supervised operations at Peenemünde on the Baltic coast, the site from which the Germans launched as many as 1,402 V-2s against targets in London and southern England. Other cities on the Continent suffered damage and casualties from V-2 strikes, including Antwerp and Liège in Belgium and Paris, France. In one of his many reports from London during the war, Columbia Broadcasting System's Edward R. Murrow astutely observed that "the significance of this demonstration of German skill and ingenuity lies in the fact that it makes complete nonsense out of strategic frontiers, mountains and river barriers."[14]

Further experimentation during the late 1940s and early 1950s brought rockets to the point where they generated sufficient thrust to slip the bonds of Earth, enough at least to insert an object into orbit. The 1950s witnessed the birth of a family of familiar rocket systems and intercontinental ballistic missiles (ICBMs), including Atlas, Thor (later Delta), and Titan, all of which could deliver hydrogen warheads. While rocket and missile technology matured in these early years, so, with the help of emerging computer technologies, did related communications, navigation, and control systems.

The 1950s and 1960s also were notable for the intense political and strategic rivalry that existed between the United States and the Soviet Union. Cold war competition propelled East and West headlong into a race to field the first intercontinental ballistic missiles and to become the world's first space power. Moscow's embrace of the space revolution overshadowed the relatively staid enthusiasm for space among U.S. officials before the launch of *Sputnik* in 1957. The Soviet Union approved its ICBM program the year Stalin died, 1953, and Soviet scientists that same year announced the feasibility of sending a spacecraft to the Moon and deploying artificial satellites around Earth.[15]

So when, in 1955, the International Geophysical Year (IGY) Committee announced that for eighteen months beginning on July 1, 1957, the nations of the world would take aim at space to study the physical properties of Earth, it was no surprise that the Soviet Union plunged into a program to be the first to send a satellite into orbit. At the time, Moscow declared it would deploy an automatic laboratory for scientific research in space. Yet space represented much more to Soviet leaders than a laboratory. Since becoming the first nation to endorse the goal of space travel in 1924, Soviet space objectives had included concepts for enhancing national prestige and security and exploiting the military applications of rockets and satellites.[16]

Having made no prior policy decisions regarding the use of space, the IGY provided the rationale for Washington to enter the space age. The United States publicized in 1955, on the day before the Soviet announcement, its own plans to launch a small scientific spacecraft. President Dwight Eisenhower recognized that considerable prestige would accrue to the first country to successfully launch a satellite, but, he decided, the United States would have to make its mark in space using nonmilitary, scientific, or "peaceful" technologies and systems. Consistent with scientific methodology and practices, all space experiments would be open and their results published.

During the early postwar years, German expatriate Wernher von Braun made significant progress in rocketry by launching numerous military rockets at White Sands Proving Ground in New Mexico. Indeed, largely because of von Braun's early work, by October 1957 the United States excelled in the areas of rocket guidance, solid-fuel technology, and warhead design. By nearly every account, von Braun, had he been given the go-ahead, could have put the United States in space months ahead of the Soviet Union.

The president nevertheless deferred to his political conscience, rather than to expediency, to protect a public image. Eisenhower preferred to emphasize the "civilian" aspects of the U.S. space program and refused to use von Braun's army-sponsored team to make America's first jump into space. In his quest to set the United States' space program onto a nonmilitary path, Eisenhower and his advisors ironically turned to the Naval Research Laboratory, or NRL, to develop Project Vanguard's Viking booster to represent the United States in the IGY. The NRL's Viking sounding rocket was launched successfully twelve times between 1949 and 1954. Yet Vanguard could hardly be called a civilian program. Not only was there the obvious relationship with the U.S. Navy, but naval officials boasted that the information provided by the Viking tests would feed directly into the design of long-range missiles.[17]

Eisenhower's decision, in conjunction with his determination to initiate a secret military program to construct a reconnaissance satellite, set out two distinct avenues for development—one public and scientific, the other secret and military.[18] Eisenhower's early posturing on this issue may well have harmed the U.S. military establishment by darkening the country's mood regarding the possible military uses of space, despite the reality that military research and development, investments, and programs permitted the United States to be an early space entrant in 1958 and to become the space power it is today. Nearly every facet of the United States' present-day involvement in space has origins in the nation's military.

Preliminary analysis of the political impact of the Soviet space achievement highlighted a number of disturbing concerns. After October 4, 1957, the Soviet Union appeared in the eyes of some as an equal with the United States.[19] *Sputnik* bolstered Soviet claims of scientific and technological superiority, enhancing Soviet propaganda and improving Moscow's hand as it competed in the cold war. *Sputnik* also threw veils of doubt over the minds of America's allies and friends about the ability of the United States to lead and protect the free world.

The NRL-developed rocket was simply not as technologically advanced as the U.S. Army's Redstone military missile developed by the more experienced von Braun team. When on November 3, 1957, the Soviet Union launched its second satellite, this time with a live dog named Laika as a passenger (Laika, of course, did not survive the journey), Eisenhower was motivated to bring the United States swiftly into the space age, and he requested that the army ready its Jupiter-C rocket for a possible launch. The Vanguard program experienced another aborted attempt to reach orbit in December 1957. Politically, the United States could no longer afford to keep the army-developed technologies on the shelf. In the end,

Eisenhower turned from the navy to the army to launch the first U.S. satellite into orbit, the ten-and-a-half-pound *Explorer 1,* on January 31, 1958, a mere ninety-one days after von Braun received the authority to proceed.[20]

What Can a World-Circling Spacecraft Do for Us?

The 184-pound ball called *Sputnik* was the first artificial Earth satellite. It used a one-watt transmitter to generate signals throughout its ninety-two-day life-span, sufficiently strong to be heard by amateur radio operators. Indeed, the message was loud and clear to all who listened: space holds a key to the future greatness of nations. Yet the idea of placing an object in orbit around Earth was not new to the twentieth century. Sir Isaac Newton, while working out his theory of gravitation, understood that artificial satellites were theoretically possible. Scientists and engineers had to await advances in rocketry before testing the English mathematician's theory. For then as now, the rocket remains the only practical means for hurtling an object fast enough to attain orbit.

Although the United States did not have a space program until the U.S. Air Force began development of the first reconnaissance satellite in 1954, official interest in the practical uses of space grew after World War II. The U.S. Army Air Force sponsored several feasibility studies during the early postwar years. One of the more notable studies was a 1946 report credited with undertaking the first thorough review of the feasibility of designing a man-made satellite using the technology of that time and known methods of engineering and propulsion. In *Preliminary Design of an Experimental World-Circling Spaceship,* a report by the RAND Corporation, one author hypothesized, "If a vehicle can be accelerated to a speed of about 17,000 m.p.h. and aimed properly, it will revolve on a great circle path above the earth's atmosphere as a new satellite. The centrifugal force will just balance the pull of gravity. Such a vehicle [deployed at an altitude of about 300 miles] will make a complete circuit of the earth in approximately 1-1/2 hours."[21] Satellites would "undoubtedly prove," according to another contributor, to be militarily significant. The authors of this study understood that the deployment of a "spaceship" would have profound strategic and worldwide implications, noting,

> We can see no more clearly all the utility and implications of spaceships than the Wright brothers could see fleets of B-29's [*sic*] bombing Japan and air transports circling the globe. . . .
>
> The achievement of a satellite craft by the United States would inflame the imagination of mankind, and would probably produce repercussions in the world comparable to the explosion of the atom bomb.[22]

From the defense planner's perspective of that time, not only would satellites be invulnerable to the threats posed by artillery and radars against airborne assets, but there might also be a role for "high velocity missiles," or "space-missiles." In this prophetic report we see early interest in deploying orbiting weapons and platforms to improve reconnaissance, communications, targeting, bomb damage assessment capabilities, and the ability to monitor weather conditions over territory controlled by the enemy.[23]

Expanding Our Neighborhoods through Communication

Until the age of satellites, wires, cables, and electronics easily took care of man's communication needs. Transcontinental telephone links using undersea cable enabled the exchange of information overseas. Although radio links existed, they were not reliable relays for messages across vast oceans. In 1945, Arthur C. Clarke, the legendary British science fiction writer who gave us *2001: A Space Odyssey,* started a communications revolution. In his paper "Extraterrestrial Relays," Clarke proposed using an artificial satellite to overcome the limitations of a ground-based, tower-to-tower microwave radio system. Wrote Clarke, "A true broadcast service, giving constant field strength at all times over the whole globe, would be invaluable, not to say indispensable, in a world society."[24]

Terrestrial microwave towers have some notable limitations. First, a series of towers must be constructed. Second, because radio waves must travel on a line-of-sight path, the towers cannot be spaced too far apart, making them useless for transoceanic communications. Furthermore, costly towers are not likely to be constructed in remote regions of the world, effectively denying radio-wave communications to some populations.[25]

Clarke proposed placing three satellites at an altitude of 22,300 miles in geosynchronous Earth orbit (GEO). Geosynchronous orbit, also known as "Clarke orbit," means that a satellite deployed at that altitude above the equator moves at the same speed and direction as the rotating Earth. Under these conditions, the satellite appears to remain stationary. Three satellites spaced 120 degrees apart over the equator provide nearly complete coverage of Earth's surface. By receiving signals from one point on Earth and repeating and relaying them to another, satellites act as space towers and may be linked by radio to one another and to ground stations. Nevertheless, transmission time delay (which produces annoying echoes in the receiver's ear) and the challenge of placing a satellite at such a high altitude led scientists to look beyond the development of GEO satellites to low-Earth-orbit (LEO) alternatives.

In 1960, the United States successfully launched its first communications satellite, a one-hundred-foot diameter aluminum-coated Mylar balloon called *Echo 1.* A giant reflector visible to the naked eye, *Echo 1* was used to bounce radio and television signals across the continent, and from a transmitter in New Jersey to a receiver in France. This passive communication satellite was inefficient, however, leading scientists to experiment with active, real-time systems. In 1962 *Telstar* became the first satellite to receive, amplify, and immediately retransmit signals, making it possible for live television and real-time telephone conversations across the Atlantic Ocean.

Yet *Telstar* and other experimental satellites, such as *Relay*, were disappointing. Deployed in medium Earth orbit (MEO), they moved from horizon to horizon too quickly, taking with them any strong signal connection within a few minutes. Two hours later, they would reappear, only to race away once again. One solution was to develop a constellation of satellites networked with ground stations so that a communications tower would be in sight at any time of the day. Several technological and financial challenges prevented this idea from coming into being until the late 1990s, when Motorola's Iridium deployed the first low-altitude satellites for mobile communications systems.

The "Clarke orbit" once again appeared to offer the best solution. With the Delta rocket, a derivative of the Thor intermediate-range ballistic/antisatellite missile developed by the air force, the National Aeronautics and Space Administration (NASA) held a reasonably reliable solution for the problem of long-haul transportation into space. Technological advances eventually fixed the bothersome echo. A practical GEO communications system, however, had to await the support of a space-minded president (John F. Kennedy) and a congressional compromise that resulted in the Communications Satellite Act of 1962.[26] This law established the Communications Satellite Corporation, or COMSAT, a regulated private company chartered to develop with all due speed a global satellite communications system.

After reviewing a few alternatives, COMSAT executives settled on an investment in a series of Syncom satellites developed by Hughes Aircraft. The successful deployment and operation of *Syncom 2* on July 26, 1963 (*Syncom 1*, though launched successfully, experienced malfunctions in orbit), confirmed COMSAT's faith in a geosynchronous satellite communications system. As a sign of the international political implications of global satellite communications, President Kennedy used *Syncom 2* to speak by telephone directly with Nigeria's prime minister.[27]

Today, the services provided by communications satellites are woven into the fabric of our lives. They were, and are, the true catalyst for globalization, or the worldwide melding together of different financial and economic systems. As channels for relaying voice calls, computer data, television signals, facsimile messages, and the digital projection of movies from studios direct to theaters,[28] satellites have become pillars of a global telecommunications infrastructure. Communications satellites not only expanded the number of phone calls that could be handled at one time (when compared to the capacity of microwave radio signals) but also made business cheaper for broadcasters, governments, and corporations. Underscoring the growing faith among seafarers in satellite communications, on February 1, 1999, cargo vessels over three hundred tons and oceangoing passenger ships officially discarded the use of Morse code as a means of communication.[29]

The myriad uses of communications satellites are staggering. Space-based communications, though a nearly invisible force in our everyday lives, is a major and even indispensable linchpin in the most significant economic, governmental, and military activities of the country.[30]

The Skies Have Eyes

Elaborating on possible military benefits of space, Wernher von Braun noted in 1945 that "the whole of the Earth's surface could be continually observed from . . . a rocket [or satellite]. The [rocket] crew could be equipped with very powerful telescopes and be able to observe even small objects, such as ships, icebergs, troop movements, construction work, etc."[31] Indeed, space offers a vantage point comparable to none other on Earth. The satellite's field of view is far wider than that offered by aircraft cameras, making it possible to view the entire Earth in a twenty-four-hour period, assuming the satellite traveled around the rotating Earth in a low orbit over the poles. In fact, weather satellites based in geosynchronous orbit provide time-sequenced snapshots of an entire hemisphere.

Retrieval of the film package from space, however, posed a major problem. Von Braun's statement about observing Earth from space assumed a crew of spacemen, in part because there was no technology for getting raw and processed imaging data from space back to Earth.[32] Satellites, theoretically, could radio findings to Mother Earth. But how might this be accomplished without a "man in the loop"? Proposals to transmit film images to Earth or develop a film-return satellite initially were infeasible for technical and logistical reasons. Engineers entertained the idea that television signals could be transmitted from satellites to return images of selected territories, but enthusiasm for this method faded quickly when experimenters determined that the poor quality pictures received were worthless for many military endeavors—further evidence that the armed forces were a major force in the development of satellite technologies.

With the invention in 1956 of ablative material, which allowed a satellite payload to reenter Earth's atmosphere without burning up, the satellite film-return design became feasible. Concepts emerged for deorbiting a film payload and retrieving it upon return to Earth in the air or on the ground. Other critical progress in this area included advances in spin-stabilized camera designs and a hundredfold improvement over World War II camera technology. These developments led to the establishment of Corona, the United States' first covert satellite reconnaissance program.[33]

Intelligence purposes drove Corona. Eisenhower blessed the development of a reconnaissance satellite once U.S. defense and foreign policy officials determined that *Sputnik* had done the United States a favor by establishing the legal concept of freedom of international space. The United States launched into the highly classified program, which was under the direction of the Central Intelligence Agency (CIA), using the cover story that it was preparing to orbit a series of satellites to conduct biomedical experiments. In fact, the Discoverer satellites produced the first photographic intelligence from space. The recovery of the *Discoverer 13* capsule, the first such recovery of an object from space, inaugurated a string of successes for this program. *Discoverer 14*, launched in August 1960, sent back the first satellite pictures of an airfield in the Soviet Union, an advance base for Soviet bombers.

Prior to the arrival of observation satellites, the United States relied on spies, reconnaissance flights along the periphery of the Soviet Union, radars based in Europe, the Middle East, and Alaska, and communications intercepts to gain insight on the development of Soviet strategic forces. Yet a significant knowledge gap remained. With the success of *Discoverer 14*, it became possible to view routinely denied areas in order to find ICBM sites, bomber fields, and other installations and activities of military significance. Good intelligence, or course, will not only uncover military threats but also reveal weaknesses in foreign militaries and help ensure that resources are not spent preparing for the wrong danger.

The Corona program, which ended in 1972, helped dispel the notion that American nuclear forces were inferior to those deployed by the Soviets. These early reconnaissance satellites, however, had their limitations. The inflexibility of the Discoverer satellites, their infrequent overpasses and film-drop design, made them practically useless during the 1962 Cuban Missile Crisis, when Moscow had deployed only a handful of intermediate-range missiles on Cuban soil. Other national security programs developed in the wake of Corona included SAMOS (satellite and missile observation system), Close Look, and the not-so-secret Key Hole (KH) series of "spy" satellites.[34] KH-9, the first in the new series, also known as *Big Bird*, retained the old film-based retrieval system. However, the satellite was too cumbersome for operational use. The time increments between picture deliveries was significant, and raw data was often not available for days. Once the data had been retrieved, it still had to be analyzed before it could be disseminated. These shortcomings made KH-9 an irrelevant asset during the 1973 Arab-Israeli War, although these early reconnaissance satellites responded well to the original charter of the National Reconnaissance Office (NRO) to provide intelligence on strategic threats.[35]

Satellite remote-sensing technologies also sparked a meteorological revolution and today provide information increasingly relied upon by weather forecasters. Visible light and infrared sensors help track meteorological events on Earth and allow weather watchers to peer into remote locations (the oceans, deserts, and poles) for a glimpse of weather patterns that may affect surrounding regions. Images of Earth's regions help produce accurate and detailed maps and act as guides in the quest for valuable Earth resources and record vegetation levels, monitor disappearing rain forests, and measure soil moisture content. Imagery satellites also warn of crop blight or even uncover lost worlds, which is what happened in the 1980s when observation satellites discovered roads covered by sand dunes belonging to Ad, an ancient Middle Eastern metropolis referenced in the Koran.[36]

The Satellite Will Be Our Guide

The Sun, Moon, stars, and other celestial objects, such as Jupiter's satellites, long have served as reference points to aid and expedite man's travels on Earth. The Sun helps a traveler to determine east and west, and by evaluating the direction of shadows cast from objects on Earth, he can know north and south. At night, the North Star remains a faithful guide. Observations of the "Pole Star" and calculations using such instruments as quadrants, astrolabes, cross-staffs, and sextants, which factor in direction and the relationship of the Pole Star to the horizon, have helped seamen determine with a fair degree of accuracy their latitude on Earth. Latitude measurements became important when seafarers sailed within large seas such as the Mediterranean, but especially when they ventured outside this familiar body of water and into the vast oceans, at which point it became critical to know how far north or south they had traveled.

Longitude (or the measurement of an east-west position on Earth) was a more challenging matter. With Earth's rotation, the stars and Sun are always in motion around it, a reality that frustrated for centuries all efforts to use stars to determine east-west position. The Royal Observatory at Greenwich, England, in the late seventeenth century offered a long-sought-after solution. The key involved knowing in particular the fixed stars and understanding the motions of heavenly bodies. In the interest of establishing a standard set of charts that record longitudes (different countries at one time based their charts on different meridians), the United States proposed in 1884, and it eventually was accepted, that all nations use Greenwich as the prime meridian for the purpose of calculating longitude.

For more than six hundred years, the science of navigation evolved in concert with developments in astronomy, mathematics, and cartography and benefited from the ingenuity of successive instrument makers. Visual triangulation, dead reckoning, and the hand-held sextant became essential navigation tools. Travelers in search of their bearings relative to the north position still derive benefit from the twelfth-century invention of the magnetic compass, though these devices are often troublesome and unreliable. The dangers of collision motivated early pioneers to find a way to maintain appropriate distances between ships and determine their positions relative to known points on land. Early navigation techniques, however, left pilots at the mercy of the weather, as storms usually spoiled positioning calculations and cloudy nights concealed any information the heavens might hold. Thinking in the area of navigation expanded and advanced markedly with the demands of aviation. Radio-beam and inertial navigation systems, though incapable of providing highly precise location and direction information, offered ways of arriving reasonably close to a preselected destination point.

The advent of the digital computer, the means for establishing a reliable and continuous flow of information, paved the way for the next big revolution—navigation satellites. Signals from space give us information we need to calculate a satellite's position and help us to determine locations and bearings at nearly any point on the globe. Satellites significantly enhance navigation capabilities because they not only provide consistent, reliable, global coverage, but also faithfully transmit through fog and darkness.

The U.S. Navy launched the first experimental Transit system navigation satellite in 1959. Transit's prime mission was to assist submarine captains to determine their position prior to launching ballistic missiles. Transit satellites utilized what is known as the Doppler shift, which recognizes that a satellite's signal frequency will seem higher as it approaches the location of the ground receiver and lower as it passes overhead and away from the receiver. Transit's defect was that it required very expensive and cumbersome electronic ground equipment. It also had operational drawbacks. Accurate position readings required that the low-Earth-orbit Transit satellite make at least two passes, necessitating a wait of at least one hundred minutes. Even under the most ideal circumstances, Transit signal measurements had a margin of error of about one meter, making it unreliable for land surveyors.[37]

During the early phases of the Transit program, the air force began to think about a more sophisticated use of satellites that involved measuring the distance to satellites, or ranging. The U.S. Air Force contracted with Rockwell International in 1973 to design the NAVSTAR (navigation system using timing and ranging) GPS satellite. First launched in 1978, GPS (global positioning system) satellites use a highly accurate atomic clock to provide precise information to anyone, anywhere, and at any time of the day regarding latitude, longitude, altitude, travel velocity and direction, and time.

The nominal constellation of twenty-four GPS satellites, with four satellites deployed at twelve thousand miles above Earth on six separate planes, provide this information more frequently than Transit satellites. Each satellite transmits a unique signal to ground receivers, and the receiver in turn generates a set of codes identical to those transmitted by the satellite. The calculation of time delay determines the receiver's distance from the satellite. By capturing the signals of three or, ideally, four satellites, a user's receiver measures the time of arrival of each signal, provides a "pseudo-range" value for each satellite (clocks in the receivers are not likely to be in synch with the satellite's more accurate clock—and with GPS *timing is everything*), uses the fourth satellite to synchronize the time, and then processes the data to provide positioning information.[38] This satellite navigation system has been so successful, celestial navigation classes at the U.S. Naval Academy may now be in peril.[39]

As we continue the discussion below and in the chapters to follow, it will become increasingly evident that satellites have countless uses that we take for granted. We can expect to see new space applications for civil, commercial, and military purposes emerge in the future with a frequency so regular and a consequence so great that the country's launchers, satellites, space planes, and space stations increasingly will be considered vital national assets.

How Does Space Serve the Public Good?

It is not appropriate to consider in these pages NASA's future. Suffice to say that questions about its long-term goals, high-fixed costs of projects with long operational lives, and recent management woes have posed daunting challenges to an organization that, as the repository of the greatest body of space expertise in the world, was and remains a source of pride for the nation.[40] The reader need only be aware that NASA evolved as an institution endowed with the purpose of managing a space program important to the life and progress of the United States. Indeed, it was not too long after *Explorer 1* fell into orbit around our planet that our national leadership recognized that a viable and energetic space program could perform a supreme service to the public good.

Washington established NASA, as it has other civil agencies, to pursue objectives that are not readily supplied by the marketplace—in other words, if the government did not engage in these activities, the odds against them happening at all would be very high. The U.S. civil space program has evolved from highly visible programs such as Mercury, Gemini, and Apollo to programs setting in motion prolonged operations, such as the use of the space shuttle, the operation of the Hubble Space Telescope, and NASA's multifaceted deep-space exploration initiatives.

NASA's primary mission is to perform basic science and exploration and enhance understanding of our planet and the universe. It has produced technologies that account in part for America's commercial competitiveness in a number of areas, not just space. Civil space programs have contributed to education on a national level and enable contacts with our "neighbors" in foreign lands, enhancing to some degree global understanding. NASA also has made important contributions to national security, including remote sensing and communications, and its scientific achievements continue to inspire national confidence.[41] Civil space programs, moreover, stimulate scientific and technical employment. Investment in space exploration initiatives means investment in high technology and improvement in national competitiveness.

Through manned and unmanned scientific and astronomical observations and measurements, civil space programs have broadened our knowledge base significantly. "Before-and-after" examples of space successes are plentiful. Before our physical entry into space, the nature of the medium between Earth and Sun was in question; after, it was proved that there was a continuous solar wind of ionized gas from the Sun that continued past Earth to points outside the solar system. Before, there was no knowledge of magnetic fields of other planets; after, the Pioneer and Voyager missions confirmed giant magnetic fields on Mercury, Jupiter, and Saturn. Before, scientists assumed that the moons of other planets in our solar system were of the same origin; after, encounters with Jupiter revealed that a planet's moons may be very different from one another.[42] To some scientists in this field, missions of discovery in space not only must be continued to advance the cause of science, but failure to pursue this kind of exploration in the future could cost the United States its stature as a world civilization.[43]

Numerous satellites have been launched into orbit to send back information on space physics and make astounding astronomical observations. Space physics satellites, such as the NASA Explorer series (1958–75), observe ultraviolet, infrared, and X-ray radiation outside the solar system and perform near-Earth astrophysics research experiments. The Hubble Space Telescope, perhaps one of NASA's most familiar satellites, allows astronomers to observe distant stars and galaxies and make scientific measurements without the blurring effects of Earth's atmosphere.[44] Hubble's "photo shots" routinely make the pages of popular magazines and newspapers, leaving people around the world to marvel time and again at the magnificence of outer space.

Through the development of new technologies and growth in commercial markets, civil space contributes to the U.S. economy and the general prosperity of the nation. Advances in the conversion of solar power to electricity and the exploitation of other energy sources in space will have numerous applications to enterprises on Earth and in space. Materials processing in space exploits the zero-gravity, near-perfect vacuum of space to manufacture pharmaceuticals, electronic components, optical equipment, and metal alloys with far greater precision and with fewer defects than is possible on Earth. This work could help advance research on known terrestrial manufacturing processes and discover new processes, although there is no evidence to date that we will break into this field anytime soon.

NASA has developed satellites to enhance our practical understanding of life on Earth—to study natural resources and to monitor weather and ocean conditions, the movement of Earth's continents and earthquakes (using geodetic Earth-orbiting satellites), and sources of pollution. The U.S. Department of Commerce, through the National Oceanic and Atmospheric Administration (NOAA), has the lead responsibility for managing remote-sensing satellite operations and archiving imagery. Variously operated by NASA, NOAA, the Department of Defense (DoD), and the U.S. Geological Survey (USGS), the NASA-developed multispectral, moderate-resolution Landsat remote-sensing satellites have provided a continuous record of land-surface data since 1972.[45]

The Department of Agriculture uses Landsat data to forecast crop production worldwide. Other civilian applications of national remote-sensing satellite programs, to include advanced spy satellites, include monitoring of world food, water, and energy supplies and mineral resources, forestry and rangeland management, fish and wildlife management, water resources management, and geological mapping.[46] NIMA, or the National Imagery and Mapping Agency, in cooperation with the CIA, has made available to NOAA very high resolution pictures.[47] Advances in computer processing and development of a mapping software known as Geographic Information System have aided the civilian mapping projects performed by the U.S. Geological Survey. The Army Corps of Engineers refers to USGS imaging data from space prior to the construction of dams, levies, river channels, roads, and pipelines.

Since the 1960 launch of the first series of weather satellites called TIROS (television and infrared observation satellite), weather forecasters have been able to demonstrate for television viewers cloud patterns and the movement of storm fronts using pictures from space that show the location of jet streams, ice flows, snow-covered regions, and the approach of tropical storms and hurricanes. Whereas most U.S. imagery satellites use a polar orbit to fly over regions of interest, the GOES (geostationary operational environmental satellite) series satellites are able to take pictures continuously of an entire hemisphere.

Raw weather data from GOES satellites are transmitted to Wallops Island, Virginia, for processing and distribution to data users, including NOAA's National Weather Service.[48] The National Weather Service provides storm and flood warnings as well as weather forecasts using information supplied by several satellites, including the Defense Meteorological Satellite Program (DMSP) satellites operated by the U.S. Air Force. NOAA's polar-orbiting operational environmental satellite (POES) system and the GOES system (including the latest GOES-Next satellites) measure cloud cover and motions, understand hurricane activity, and register temperatures on Earth's surface.[49] The Federal Emergency Management Agency uses satellite data to help cope with national crises wrought by severe storms, floods, earthquakes, tornadoes, fires, and volcanic eruptions. Remote sensors make initial assessments of the damaged area, which may form the basis for recovery planning and confirm the best evacuation or emergency supply routes.

From a geographic standpoint, routine and continuous study of the world's oceans is possible only from space. For years, satellites such as NASA's SEASAT enabled detailed mapping of ocean surface atmospheric fronts and hurricanes and allowed scientists to measure ocean temperatures (actually estimated by measuring the height of the ocean surface) and understand better the general circulation of the ocean, ice-flow activity, and oceanic biological productivity.

Finally, the United States is the only country today capable of routinely delivering a broad range of payloads into space, including manned and unmanned vessels and spacecraft intended for travel well beyond the earth system. There are, in fact, several "families" of unmanned expendable launch vehicles (ELVs) built around the ballistic missile technologies developed by the armed forces. The oldest is the Titan line. Titan 1 first launched successfully in February 1959. Today, the Titan 4 rocket is the most powerful unmanned space launcher in the inventory, capable of lifting thirty-nine thousand pounds three hundred miles high. (The Saturn 5, capable of lifting two hundred thousand pounds to orbit, was abandoned in 1973 after taking Americans to the Moon and placing the U.S. *Skylab* in orbit.) The Atlas family of rockets boosted several probes to the Moon and to other planets in our solar system. Later generations of Atlas deployed the U.S. Navy communications and GOES weather satellites. The Delta family of rockets, first operational in 1960, has carried numerous science and astronomy payloads satellite payloads to all orbits.[50]

The space shuttle, or the Space Transportation System (STS), is a semi-reusable, aircraft-like vessel with a cargo bay that routinely takes forty thousand to forty-five thousand pounds into LEO carrying a crew of up to seven people on missions lasting days and even weeks. At $400 million to $600 million per flight, the STS, first launched in April 1981, carries scientific and military payloads and will be used to ferry materials required for the construction of the *International Space Station.* The shuttle not only allows NASA to continue its tradition of manned space flights, as a partially reusable launch vehicle, it is perhaps a harbinger of a new way of conducting the space launch business.

NASA has a number of other experimental programs. The DC-X single-stage-to-orbit flight-test vehicle became the first to demonstrate successfully a vertical landing and turn-around time to make possible the next launch in a matter of hours, not days or weeks. DC-X test operations went a long way to showing that low-cost rockets could be flight-tested like aircraft. The DC-X worked using existing technologies, whereas the remaining "X-vehicles" (the X-33, X-34, X-43, and X-37) concentrate heavily on the development of radically new technologies.[51] NASA is working with Lockheed Martin to develop the X-33, a 53 percent working scale model of the VentureStar reusable launch vehicle (RLV). Throughout the lifetime of the program, NASA's X-33 project was confronted by mounting technical problems.[52] In March 2001, given excessive cost growth and technical uncertainties, NASA scrubbed the X-33 project, which was intended to revolutionize the space launch industry by, among other things, reducing the cost per pound to orbit to a fraction of what it is today (more than ten thousand dollars).[53]

Were the X-33 program to succeed, its technological achievements would have "found their roots" by contributing to the development of the country's first operational military space plane. The X-33 program was one more example of a "civilian" program that has had to borrow heavily from past military programs. The X-33's thermal protection system, one of the more critical technologies and structural features being developed by Lockheed Martin for inclusion in the VentureStar spacecraft, may be traced back to the 1980s' U.S. Air Force classified "black" programs to develop a military space plane.[54] The "dirty little secret" is that U.S. military purposes and the visionary work done by the armed forces and the military laboratories are largely responsible for the country's current position of leadership in space.

Why Is Business Looking Up?

Not only does the United States anticipate significant economic benefits from space, but space goods and services supplied by an expanding domestic commercial space market will empower national civil and security activities. The 1990s have been a prosperous decade for commercial space services in many areas, including communications, remote sensing, navigation, and launch. The space industry, which grossed $65.9 billion in 1998, can be divided into three categories: manufacturing, operational and service support, and satellite service providers.[55]

Manufacturing

Space manufacturers are producing spacecraft in record numbers and, in some instances, at a pace worthy of Henry Ford. Motorola has done for satellite manufacturing what Henry Ford did for automobiles in 1908 with his invention of the production line. Sixteen-hundred-pound Iridium satellites were produced on an assembly line one every 4.5 days.[56] Satellite manufacturers worldwide, led by companies such as Hughes, Lockheed Martin, SS/L, Matra, and Aerospatiale, pulled in more than $17 billion in revenues in 1998, the commercial and government sectors evenly splitting demand. With the telecommunications industry in the vanguard, almost 250 geostationary satellites were launched during the 1990s, as compared to 69 in the 1980s.[57]

Expendable commercial space launch vehicle manufacturers and service providers, domestic and international, are stretched to meet surging demand for lift to orbit. U.S.-built Delta 3, the improved Atlas, the evolved expendable launch vehicle boosters, Boeing's Sea Launch vehicle (an improved Ukrainian Zenit 2), and France's Ariane 5 are all expected to dominate the future launch market. A string of launch failures in 1998 and 1999 involving Titan 4s, Delta 3s, China's Long March, the Zenit, Lockheed Martin's Athena and failures attributable to upper stages have caused delays in the satellite service industry (although 1998 and 1999 still went down on the record books as the busiest in history) and underscore just how risky this business remains.[58]

A 1999 report by the Federal Aviation Administration and industry projected that there will more than fifty commercial launches worldwide annually over the next decade, with roughly twenty-six launches annually devoted to lifting satellites to nongeostationary orbits.[59] The findings of this report later were tempered by a new caution generated by bankruptcies in the mobile telecommunications industry, which led some analysts to conclude that the number of commercial launches in the coming years would not be as great as anticipated.

There are also a number of small, private ventures, such as Kistler Aerospace, Rotary Rocket Company, Kelly Aerospace, Pioneer Astronautics, Space Access LLC, and Starcraft Boosters that have laid plans to develop and then market reusable launch rockets, space planes or space helicopters using rocket-driven rotary blades. In July 1999, Rotary Rocket's Atmospheric Test Vehicle flew for just under five minutes up to a height of eight feet, a sign that concerted efforts are being made to develop a reusable launcher. Other proposed transatmospheric vehicles will leave Earth's surface under their own power, use rocket-staging, deliver a payload to orbit, and then return to Earth for refueling and follow-on missions. Several of these start-up companies plan to spend about $6 billion each to develop a commercial vehicle, although the likely affect of this novel mode of transportation on the marketplace is still in question.[60]

Current RLV research aims to get costs down, to allow launch vehicles to fly frequently, require less maintenance or support, and last for many years—to discover technologies to ferry a satellite into orbit for less than one thousand dollars a pound.[61] When you consider that the major expenses of a space company are the cost of the satellite, launch services, and insurance, the development of more efficient, cost-effective space launch systems could reduce dramatically the cost of deploying and operating a satellite constellation.

The largest industry manufacturing segment deals with ground equipment. While satellites and space launch vehicles routinely grab the headlines, they are useless without the associated terrestrial receiver, transmitter, processing, and dissemination equipment. The areas of ground-equipment manufacturing, testing, and data analysis operations together provide the greatest employment and bring the most revenue to the industry. Revenues generated across the space ground-segment industry are expected to climb above $31 billion per year by 2001. Earth stations (for transmitting, receiving, and processing satellite signals), consumer electronic equipment (including television satellite dishes and GPS navigation equipment), and computer software and hardware needed to plan and launch satellites and monitor their status (or "health") make up a significant portion of the ever expanding ground space industry.[62]

Operational and Service Support

Ground stations, equipment, and skilled technical support personnel permit smooth operations throughout the life of a satellite mission. Operational support includes satellite control and telemetry operations, utilization of planning software and component testing facilities, and operations at launch-vehicle spaceports. Successful satellite operations also require commercial legal, financial, and consulting support from business and technical experts. With an increase in space business opportunities there has been, not too surprisingly, steady growth in a financial sector to assist in mergers, acquisitions, debt financing, and raising venture capital. Legal services have sprouted to engage in contract preparation, lobbying, and licensing of satellites and launch vehicles. Professional legal, financial, and media publishing services have averaged more than $2.5 billion annually over the past few years. The space insurance market in 1997 pulled in over $1 billion in revenues for the first time, $464 million of that being profits, although the industry faced more than $1.4 billion in claims (a figure that does not include policies written to protect satellite owners from launch delays) during 1998. Service providers rely on insurance to offset financial penalties stemming from launch failures or on-orbit failures caused by faulty hardware or software, solar flares, orbital debris, or meteoroid impact.[63]

Satellite Service Providers

The space industry is made up of a multitude of service providers.[64] The strongest, most diversified service segment is telecommunications. As late as the mid-1990s, space-based telecommunications comprised primarily fixed voice and video-broadcast services offered by international and private communications networks, cable television programming, telemedicine, and tele-education. Today the fixed satellite services market, relaying data through spacecraft parked in geostationary orbit, still dominates telecommunications and undergoes steady growth (a growth rate of more than 30 percent is expected beyond 2001). This service industry pulled in $6 billion in 1998. The ability of these satellites to provide services to wide geographic areas and sparsely populated regions make them very attractive to investors.

Transponders carried on these satellites number in the hundreds and are owned by such companies as PanAmSat, Intelsat, Asiasat, and Loral. Each transponder pulls in between $1.5 and $4 million per year. The fixed telecommunication market has expanded to accommodate the growing demand for broadband systems, or systems having large bandwidth, to deliver high-speed Internet access, interactive video, and video on demand. Bandwidth is like a beverage straw, the larger the straw, the more liquid can be drawn through it. The larger the bandwidth, the more data can be passed through a particular channel. The larger the bandwidth, the more services companies can provide. It is in this area that high-capacity ground-based fiber-optic cable services compete very strongly with satellites, although the two services in many ways complement each other.[65]

Most of the world's population is without telephone service. Inexpensive, global, mobile satellite systems can provide regionwide and worldwide connectivity—an especially significant development for countries such as China, Brazil, and India that currently do not have extensive national infrastructures in place. Mobile satellite systems use multiple spacecraft in low Earth orbit, the orbits used by the pioneers of satellite communications back in the late 1950s and early 1960s, to ensure reliable point-to-point (handset-to-handset) global service. The eighty-four-nation Inmarsat maritime satellite consortium, which for decades brought mobile communications capabilities to oceangoing vessels, stripped itself of its treaty-organization status in order expand services as a private company. It remains to be seen whether many of the new satellite ventures will survive crushing financial challenges.[66]

Several companies, known as the "Little LEOs," make the capabilities of landline providers of nonvoice communications services faster, cheaper, and more extensive. Using satellites at altitudes of 500 to 1500 kilometers, these companies handle two-way limited-data transmission throughout the world. Little LEO system providers will facilitate automatic reading and data acquisition from gas meters, oil wells, and scientific data equipment, assist mobile computing, paging, and e-mail usage for business and mobile users, help monitor locations of trucks and rail cars, and enable quick processing of credit and automatic teller machine (ATM) cards.[67]

Led by the Teledesic venture of Bill Gates, Craig McCaw, and Boeing, a Lockheed Martin–led (Italy's Telecom Italia Group and TRW are partners) satellite service known as Astrolink, and a consortium called Skybridge (Alcatel Space of Paris, France is the prime investor), broadband services will take the Internet to space, facilitate desktop video-teleconferencing, enhance direct-to-home video and electronic messaging, improve telemedicine and electronic transaction processing, and provide important channels for news gathering and dissemination. As envisioned by its creators, Teledesic would make it possible to produce inexpensive children's watches with tiny location devices, put a global navigation and security system in all new cars, and allow farmers to analyze crops from laptops in the fields. Teledesic plans to orbit 288 or fewer satellites deployed 825 miles above Earth. While there are a number of technical, marketing, and regulatory issues yet to be resolved, analysts expect that more than $52 billion will be invested in this service sector by 2005, and that by 2010 it will be generating more than $77 billion a year in revenues.[68]

Television is also hitching a ride in space. Television broadcast satellites are delivery sources for hundreds of audio, radio, and data networks. Global television broadcasters—Cable News Network (CNN) in the United States, BBC World Service Television, and France's TV5—rely on large-scale satellite networks to link the entire globe. Direct-to-home TV services have emerged as the satellite industry's growth engine. A growing number of viewers are receiving television signals for movies, sports, and other TV programming from direct-to-home services. DirectTV (which debuted in 1994), Echostar, and Primestar provided more than two hundred channels of sharp digital video and CD-quality audio to more than 8.5 million subscribers in early 1998. The primary downfall of this service today is its inability to bring local channels to its customers, but this is a regulatory limitation (an effort to shelter broadcast and cable companies from serious competition), not a technical shortcoming.[69]

Telecommunication services have evolved over the course of the last forty years to embed further the satellite into our culture, politics, and economy. Banks, brokerage firms, and other financial institutions use satellites to relay financial information; satellites constitute the backbone of an ever expanding (and increasingly "essential") ATM network. Satellite communications improve store operations and customer service at major retail stores such as Wal-Mart, Kmart, and JC Penney by making possible instant credit card and check authorizations, efficient inventory management, and instant reporting on prices and sales.

If current trends hold, three-fourths of all satellites launched by 2007 will fly in LEO and make up part of our future telecommunications infrastructure. Industry analysts forecast that between 1999 and 2008, around 1,017 satellites valued at more than $50 billion will be launched.[70] Mobile phone systems will grow to about 30 million users by 2005–7. Interactive broadband satellite systems may have 150 to 200 million subscribers by 2008. The space telecommunications market may rise to $41 billion by 2001, and to $96 billion by 2006. This is a serious industry.

Though relatively small when compared to telecommunications service providers, the remote-sensing industry clearly has a future, with one of its more prominent customers being agriculture. Since 1994, more than $1 billion has been invested industrywide in remote sensing.[71] The Commerce Department predicted in 1998 that this industry would have an annual growth rate of 15 to 20 percent. Higher-resolution spacecraft and new computer hardware and software analysis capabilities have spurred growth and capital investment. There is an expanding market for electro-optical systems using direct downlink to ground stations, although companies such as SPIN-2 offer film-based optical systems using film canisters dropped into the atmosphere to be retrieved and processed on Earth.[72]

Competitors of film-based operations will use electro-optical systems to collect images in more timely fashion and downlink their data directly to ground stations. Companies competing in this area include Space Imaging (U.S.), SPOT Image (France), Radarsat (Canada), and IRS (India). SPOT (ten-meter resolution panchromatic) has been a leading seller of commercial imagery. Newly launched companies offering images at one-meter resolution within two to three days of an order include EarthWatch (Quickbird satellites), OrbImage, Space Imaging (Ikonos satellites), West Indian Space (U.S.-Israel), and Israel's Eros. New high-resolution spacecraft such as Ikonos are able to image pipelines and roads, trucks and large equipment, making its one-meter resolution images, which are touted by company spokesman to provide more information than aerial images and to be effective instruments for urban planning, map making, and natural disaster assessment. Ikonos satellites are capable of overflying a target site every three days, or every one and a half days if the customer can settle for one-and-a-half-meter imagery. Estimates of the worldwide satellite imagery market range anywhere from $500 million to $2 billion a year.[73]

The Department of Defense, and defense officials of foreign countries, are also potential customers of high- and low-resolution commercial remote-sensing data. NIMA looks to the commercial sector to obtain one-meter resolution imagery for ground processing. The National Reconnaissance Office, responsible for U.S. satellite intelligence activities, announced in 1999 its intention to invest $1 billion over five years to use one-meter resolution commercial satellite imagery.[74]

Geographic Information System (GIS) software allows analysts to combine satellite imagery data with layered information from other sources (including position information from GPS) to provide a geographic view of such things as elevation maps, population demographics, utility maps, political boundaries, zoning classification for gas and electricity, and water supply and waste management.

By integrating large, spatially referenced sets of data, satellite imagery can help provide a rich, visual geographic context that will help people make complex decisions and use the information more productively. GIS data services generate the bulk of the revenues in the remote-sensing industry.[75]

The final major space service provider is the U.S. Department of Defense, as the owner and operator of the GPS satellites providing navigational services. DoD probably never will be compensated for an investment of more than $12 billion (operating costs run about $500 million a year), which is expected to lead to GPS services worth more than $50 billion by 2005. Commercial and civilian use of differential GPS, or D-GPS, has further expanded possible applications. In light of the multifaceted commercial uses of GPS, Washington developed two new frequencies that will provide significant improvements in precision and reliability to civil and commercial users.[76] Navigating the airways is now a matter of punching a predetermined course into a GPS processor and letting the plane find its way, figuratively speaking. The precise timing mechanism of this satellite systems is used to transfer funds securely and facilitate credit payments at the gas pumps. Indeed, the number of possible commercial applications is astounding when one considers that this technology originated with the creation of atomic clocks used for studying physics.

A taxi company in Singapore developed an automated GPS-based location and dispatching system that allows customers to book rides from wireless terminals. This system reportedly shortens response time, increases security, lowers fuel consumption, and decreases overhead. Similarly, the regional transportation authority in Paris, France, is equipping all four thousand of its buses and vehicles with GPS terminals to improve fleet management. GPS, of course, may also be used to improve personal services. Punch in an address on the console of a late-model Honda Acura RL and a computer-generated voice will tell you where to turn and let you know when you are off your precalculated course. Golfers may use GPS receivers to get an accurate fix on distance to the green.[77]

Analysts expect the market to expand further with the development of smaller GPS receivers and the integration of GPS technology into multifunctional products (mobile computers and cellular phones, for example). Smaller and more affordable receivers will proliferate with the growth in the number of manufacturers. The market for GPS will improve as coverage become constant, continuous, and reliable, as assurances are received that access will never be substantially disrupted by government operators.

Commercial use of space is expanding at such a pace that from 1998 to 2007, more than one thousand satellites are projected to be launched. This represents a total investment of more than half a trillion dollars.[78] According to Merrill Lynch, U.S. commercial satellite revenues will grow from $38 billion in 1997 to $171 billion by 2007.[79] Ten years from now we are likely to be doing the same types of things in space—communications, navigation, and imaging. Only the technology is getting smaller and more capable and is being tailored to individual use. Improvements will come in speed, mobility, and bandwidth. This means that the real commercial revolution in the years ahead may well come in the form of more personalized services.[80]

How Is Space Revolutionizing National Defense?

Our dependence on space for national security is a matter of daily fact and official record. Washington's defense space budget in recent years has been roughly $13 billion a year (with another $13 billion spent on civil space programs). Not only is space a key intelligence-gathering arena, it also has become so integral to the operation of the U.S. armed forces that our defense leadership can state that "space power has become as important to the nation as land, sea, and air power."[81] In chapter 3 we will examine this claim in greater detail. For now, let us simply consider very broadly why satellites today are indispensable agents for future military victory.

Granted, the importance of satellites to a particular mission hinges rather decisively on the availability and applicability of friendly land, air, or sea assets. Clearly, reconnaissance may be performed by aircraft, and microwave towers or fiber-optic cables in the theater of battle may meet communications requirements. If these alternatives to satellites are available, and if battlefield conditions and strategic circumstances permit the use of terrestrial capabilities, alternatives to satellites most certainly can, and will, be exploited, probably even in conjunction with space capabilities.

Likewise, satellites may not qualify as "critical" mission assets where space can only play a very limited role, such as in a jungle or urban warfare campaign. Satellite images may not accurately portray what lies beneath the foliage, nor are views from space likely to reveal snipers laying in wait around a particular corner within a particular building entrance way. Satellite communications and navigation, on the other hand, may be more readily exploited under these conditions. In the grand scheme of things, satellites lose nothing to sea or air or land power assets, the utility of which also depends mightily on geography and the military and political situations at hand. Sea power has little to offer a commander of ground forces in a land-locked nation such as Afghanistan, yet naval vessels were critical to British success in the retaking the Falkland Islands in 1982.

So while one may make the case that space is not always important in warfare, it is not a very helpful observation. More noteworthy is that the American military today is so functionally reliant on satellites that, should it ever be completely or even partially shut out of space, it is not unreasonable to argue that it would relinquish significant efficiency, which, under some circumstances, might paralyze it. We look to space to help accomplish routine and vital peacetime and wartime military missions. This is the case simply because the skillful exploitation of space translates into greater efficiency in communications, navigation, and remote sensing, and it allows the armed services to reach out to any point on the globe in a reliable and timely manner. The space frontier over the last forty years has come to represent the ultimate "high ground" or overhead "flank" for protecting U.S. interests.[82]

Exploiting a Vantage Unlike Any Other

The Central Intelligence Agency, created in 1947 out of the wartime Office of Strategic Services, historically has displayed an insatiable appetite for strategic reconnaissance. Prior to the space age, intelligence and defense officials looked to high-altitude balloons, or high-flying aircraft such as the U-2 and the SR-71, to fly through restricted air spaces to search for any information that might betray a foreign government's intentions or military capabilities.[83] Since Corona, reconnaissance satellites have matured as reliable and productive sources of military, economic, and political intelligence.[84]

Sometimes referred to as National Technical Means, or NTM, there are essentially three categories of intelligence-gathering space assets.[85] Imagery intelligence (IMINT) systems use visible light, infrared, or radar to produce digitized "photos" of objects on Earth. The KH-11, launched in 1976, used charge-coupled devices (CCDs) to deliver data to ground stations and became the first satellite to adopt this revolutionary rethinking of what it meant to take "pictures" from space. The United States reportedly operates a very sophisticated $750 million to $1 billion reconnaissance satellite known as Advanced KH-11 (sometimes referred to as KH-12), which is said to be capable of viewing objects six inches in length from several hundred miles above Earth, sharp enough to identify ballistic missiles.[86] The Advanced KH-11 also has infrared sensors that function at night or in haze and smoke to penetrate camouflage and isolate objects having different temperatures. It also, reportedly, has a maneuvering capability that allows it to cross targets at different photographic angles.[87]

Another way to view Earth from space is to use radar satellites to produce photolike images after processing on the ground. Rather than passively collecting visible light or infrared signals, bus-size satellites in the Lacrosse series actively transmit and receive energy.[88] The advantage of using powerful beams of microwave energy to record and measure the timing of reflections back from Earth is one of "seeing" objects through cloud cover, foliage, and darkness. Radar sensors also can detect some objects buried under sand. Reconnaissance satellites passed over the Iraqi theater more than a dozen times daily during Desert Storm in 1991, providing real- or near-real-time intelligence on tactical targets, airfield damage, and Republican Guard emplacements.

The second category is signals intelligence (SIGINT). The National Security Agency, or NSA, collects and analyzes SIGINT in the United States. The United States deploys SIGINT satellites in all orbits—geosynchronous orbits to pick up UHF (ultra-high frequency) and VHF (very high frequency) communications, and low to medium Earth orbits to collect electronic signals from air-defense and early warning radars. Electronic intelligence (ELINT) satellites are believed to have monitored antiballistic missile (ABM) radar emissions around Moscow, while others have gathered telemetry from Soviet missile tests and verified compliance with arms control treaties. The Magnum series of advanced ELINT satellites reportedly scan for signals across the complete radio spectrum. Late-model SIGINT satellites, such as the *Vortex 2,* are said to be so sensitive that they can sweep a thousand mile area, capture tens of thousands of conversations on hundreds of frequencies, and locate an area so small that an analyst could conclude that a message came from the White House or the U.S. Capitol.[89]

Measurement and signals intelligence (MASINT) assets include the early warning spacecraft that carry infrared sensors in geosynchronous and very high elliptical orbits to detect rocket plumes and evaluate missile launches. In the 1991 Persian Gulf War, the Defense Support Program (DSP) satellites served as the primary Scud launch detection system and helped paint a picture of the battlefield during the Kosovo operations. Nuclear detonation detection sensors on GPS navigation satellites stand poised twenty-four hours a day to detect and report on electromagnetic pulse emissions, a fairly certain sign of a nuclear event in Earth's atmosphere or near-space. While satellites can never fully replace human intelligence sources, space-based reconnaissance assets using multispectral, hyperspectral, and ultraspectral sensors help analysts detect and identify a growing number of "signatures" (including nuclear, biological, and chemical weapons production activities) on Earth that aid policy makers and strategic and operational planning.[90]

What might satellites reveal? Information relayed through space and down to Earth is processed, integrated, analyzed, and disseminated from ground stations.[91] Expert interpreters of intelligence data may be able to learn of foreign research and development plans for next-generation weapons or weapons of mass destruction, identify development of new transportation, urban, or communications infrastructures in closed societies such as North Korea or China, evaluate preparations to execute underground nuclear tests, and detect nonroutine activities at railroad and shipyards and airfields. America's intelligence satellites also reportedly were able to monitor the phone conversations of terrorist Osama Bin Laden, assisting in the prevention of several attacks on embassies around the world.[92]

Spacecraft may help reveal information on activities of operational importance in wartime, such as the massing of troops and military equipment, the deployment of mobile theater ballistic missiles, locations of possible targets to be attacked in wartime, the launch of ballistic missiles, and, of course, the weather. Knowledge of weather and cloud cover is an important consideration for use of laser and infrared guided weapons.[93] War-planning uses may require overhead assets to evaluate the enemy order of battle, observe the traffic and daily activities at the defense headquarters in the enemy's capital, or provide battle damage assessments from previous combat operations.[94] Information supplied by mapping satellites of Earth's terrain, together with digital information provided by GPS positioning satellites, can guide cruise missiles to enemy targets having known coordinates. In the future, satellites using synthetic aperture radar will provide high-resolution images that may be plugged into cruise missiles to provide very accurate guidance.[95]

The strategic uses of space assets throughout the cold war really only applied to military activities during peacetime (e.g., monitoring arms agreements, tracking enemy naval movements). Wartime uses of space focused on the communications and intelligence needs of the National Command Authorities, not on the operational needs of the soldiers, airmen, and sailors. Since the end of the Persian Gulf War, the nation's intelligence and defense officials have sought to derive greater tactical benefits from space to better meet the needs of the war fighters.[96] Today, NIMA strives to organize the enormous amount of classified information that comes into the agency from various sources in order to provide tailored information on demand to commanders and the fighting men in the field.[97]

Command, Control, and Communications

Satellites perform an expanding number of peacetime and wartime missions, not the least of which is keeping national policy leaders and National Command Authorities—the president, the secretary of defense, or their duly deputized alternates or successors—informed and in a position to communicate with and relay decisions to commanders and U.S. or allied officials around the world. Satellites assist political and military leaders in the command, control, communications, and intelligence gathering functions, or C^3I. C^3I ensures that commanders in national headquarters and on the battlefield can direct military operations with higher levels of confidence and enables decision makers at all levels to think, plan, and act with greater precision and speed. Effective C^3I can make military units more agile and assist in the integration, coordination, and control of armed operations involving more than one service and more than one country.

With the integration of computers into military operations, the demand for bandwidth—getting information reliably and concisely to the fighting men and women on the ground, in the air, and at sea—has risen considerably. When the fighting must be done abroad, a good communications infrastructure established in the theater of operations cannot be assured. Determined to stay on the cutting edge of satellite communications, U.S. armed forces are preparing to take that infrastructure along, and the Department of Defense is examining commercial mobile satellite possibilities.[98]

Information passed through satellites may range from tactical intelligence to top-priority orders handed down by the president. The U.S. Air Force operates a series of Defense Satellite Communication System (DSCS) spacecraft to provide the bulk of the Pentagon's long-haul, high-priority, wideband satellite communications (voice and data), facilitate the exchange of wartime information between defense officials, battlefield commanders, and distribute data from early warning satellites.[99] During Desert Storm, coalition satellites carried 85 percent of the inter- and intra-theater communications load, more than 700,000 telephone calls and 152,000 messages per day.[100] MILSTAR satellites and new advanced wideband and advanced extremely high frequency (EHF) systems operated by the air force will route all sensitive military tactical and strategic message traffic and conversations worldwide.[101]

The U.S. Navy UHF follow-on, or UFO, satellites provide high-bandwidth communications for stationary and mobile users. UFO satellites will carry EHF and global broadcast service (GBS) transponders to deliver high-data-rate, one-way broadcasts of a large volume of information. The GBS, which operates like civilian direct broadcast television satellites and exploits commercial direct broadcast technologies, will coordinate C^{3}I systems and surveillance and reconnaissance data to give U.S. and allied commanders a common picture of the battlefield, which will allow them to coordinate operations anywhere in the world.[102] While the U.S. armed forces will turn to commercial satellite communications alternatives to facilitate routine operations, there will always be a requirement for dedicated, protected, secure, and reliable communications.

Navigation and Positioning

The global positioning system satellites developed by the Department of Defense have a master control station at Schriever Air Force Base in Colorado Springs, Colorado, and additional monitoring stations located in Hawaii, Diego Garcia, Ascension Island, and Kwajalein in the Marshall Islands. GPS provides two service levels, one of which is encoded for use only by U.S. armed forces. The DoD built in separate codes using "selective availability," which alters slightly the timing of the satellites' atomic clocks in order to transmit a signal via the C/A code to ordinary users that is not as accurate as the P-code used by the military. The C/A code provides a fix on a position that is accurate to within one hundred meters 95 percent of the time. The military receivers use the P-code to work out more refined positions by rectifying the timing errors, which provides accuracies of sixteen meters 50 percent of the time and twenty-five meters 95 percent of the time.[103]

During the Persian Gulf War, the desert terrain was an ideal operating environment for very mobile U.S. armed forces, which were trained during the cold war to move rapidly across the hilly grounds in central Europe. The lack of landmarks in the desert did not hinder fighting units receiving a continuous stream of location and direction data from GPS satellites. These space-based navigation aids made possible the deep penetration into Iraq and breathed confidence into the spirits of American soldiers and commanders, who also exploited overhead imagery and reliable communications to move vigorously with relatively little loss of life against inferior military opponents.[104] GPS also allowed coalition troops to be resupplied with great precision and reduced the risk to special operations forces who rescued downed aircrews in enemy-held territory.[105] In Desert Storm and Operation Allied Force (1999), GPS provided midcourse guidance for the standoff land-attack missile, the GPS-aided munitions, and the Tomahawk Land-Attack cruise missile, improving weapon accuracy and reducing the need for pilots to visually acquire a target.

There is no commercial alternative to this extraordinarily reliable navigation service. One U.S. official quite accurately observed that the Defense Department "erected the single, biggest barrier to entry to anyone who wants to get into the space-based navigation business. How do you beat somebody's service that's that good and offered for nothing?"[106] In the future, GPS signals will enhance several nonmilitary applications, such as aircraft navigation and landings. They will also make precision location information available to potential enemies of the United States, who may use it to orchestrate troop movements or enhance missile accuracy.

Access to Space

A true space power must have unimpeded access to space and be able to operate in and from space. Assured access to space demands capabilities for transporting mission assets to, through, and from space, operating on-orbit assets, and servicing and recovering "ailing" satellites.

There are two primary U.S. launch sites, one on each coast. The oldest (first used for launches in 1950) is the Cape Canaveral Air Station and is collocated with the Eastern Range (operated by the U.S. Air Force 45th Space Wing) adjacent to NASA's Kennedy Space Center in Florida. The Western Range (operated by the

U.S. Air Force 30th Space Wing), collocated at Vandenberg Air Force Base, California, is ideally situated to launch satellites in a southerly direction for insertion into polar orbit. DoD uses its two primary launch facilities to deliver payloads aboard Taurus, Delta, Atlas, and Titan expendable launch vehicles. Military payloads also may be launched aboard the Pegasus air-launched ELV and the space shuttle. The Centaur upper stage and the inertial upper stage take payloads from low Earth orbit to a higher altitude.[107]

NASA has the lead in the development of a reusable launch vehicle, so the Pentagon is focusing on the development of next-generation expendable launch vehicles. The evolved expendable launch vehicle (EELV), derived from existing Delta and Atlas rockets, is expected to reduce launch costs by 25 to 50 percent, which stand today anywhere between $50 and $100 million for medium-lift, and roughly $200 to $350 million for heavy-lift launches. With the Pentagon subsidizing EELV development by Boeing and Lockheed Martin by 50 percent, it is hoped that a common core of medium- and heavy-lift rockets will be available in abundance to the Defense Department and that U.S. launch companies can be more competitive against heavily subsidized foreign systems.[108]

The inadequacy of current launch facilities is a significant concern among space watchers. Not only are the equipment and infrastructure in need of general repair and major upgrading, but Cape Canaveral and Vandenberg today are not up to the task of launching the lion's share of some twelve hundred commercial spacecraft projected to be operational by 2008. Pressure is on the U.S. Air Force, which operates the ranges, to modernize them by replacing 1960s communications and telemetry technology and eliminating manual tasks. The launch pads also need to be configured to smoothly handle different launch vehicles used to lift commercial and defense payloads into orbit.[109]

What Is the Bottom Line?

Both public and private sectors in the United States are committing significant energy and resources to space operations, ensuring the country's growing dependence on reliable and unrestricted access to Earth's orbits. The space industry is a rigorous and demanding one, commanding highly competent skilled labor and excellence in education, requiring the most advanced technologies, handsomely rewarding innovation, and striving for the most efficient, reliable, and safe performance in its systems. The underpinnings of an experienced spacefaring nation consist of its operational, engineering, and technical skills, its technological sophistication, and a collective mind to apply these tools in order to reap the myriad advantages space offers. These qualities are all invaluable national resources and comprise the prerequisites of a truly competent space power. A great nation, a formidable power such as the United States, should seek to thrive in all of the geographic environments that touch it and affect it, including space.

Given these conditions, it may be argued that the United States' leadership must be prepared not only to pursue national economic, security, and political interests in space, but also to protect them. The importance of satellites to military forces cannot be exaggerated. Space assets have enabled a U.S. military strapped by level or declining defense budgets to accomplish more with less. Loss of military bases overseas and increasing force deployments to regions lacking modern communications and logistical infrastructures mean that accessibility to such theaters in times of crisis or war may only come reliably through space.[110] Space has become, and will remain, a vital part of the national security infrastructure—there are at present no political, strategic, military, economic, or technological indicators to the contrary.

The commitments of the country speak volumes. The United States is committed to nurturing the nation's scientific and engineering communities in order to maintain its global leadership in science, exploration, and technological innovation. It is politically committed to staking the country's security and, perhaps, economic future on space. The United States remains militarily committed to establishing armed forces that exploit space-based information capabilities and is strategically committed to retaining capabilities to project military power globally. Our present commitments will shape our national future.

2
Space and International Security Affairs

What Role Does Space Play in International Relations?

The systems supporting the exercise of military power help distinguish peace and war in any age.[1] Much as we could not recognize twentieth-century world politics absent a consideration of nuclear weapons or the rise of air power, explaining the decline of the western Roman Empire in the first centuries of the first millennium would make less sense without recognizing the central role cavalry played in the hands of Rome's enemies. Similarly, space systems have made their mark on international relations. Future historians will be remiss should they attempt to explain diplomacy, warfare, and peace from the late twentieth century onward without factoring in the role of satellites. Not only will the cogs in our giant space information machine (the reconnaissance, communications, and navigation satellites) continue to have geostrategic consequences, but because space is more than a medium through which we simply pass our information, the twenty-first century is apt to see for the first time space combat and the development of a new discipline in warfare.

As political units with primary command over the world's resources, the global distribution of power, and the allegiance of citizens, states remain the star players (as they have since the Peace of Westphalia in 1648), the emergence of nongovernmental organizations and transnational threats notwithstanding. Now as in the past, the material and human resources a state devotes to diplomatic-strategic action largely determine its status in the world. How or even whether a state participates in the ongoing space revolution will lend to or detract from its resources and influence. The relentless spread of space capabilities and the rise of spacefaring nations and space powers are factors woven into existing patterns of international competition. From this point forward, a proper assessment of the military balances in the world must incorporate a state's information infrastructure and space prowess.

This chapter considers some of these emerging complications by examining the impact space systems have on foreign relations. Among the diplomatic lessons of the last forty years, reconnaissance satellites deserve particular attention, if only because they are powerful tools of persuasion and media manipulation. We also cannot ignore the modern penchant to use international legal agreements to ensure stability and order in the world, for we have striven, with mixed results, to develop a body of laws and regulations to govern space activities. Space promises to be an exciting part of foreign affairs in this century.

What Are the Significant Global Trends?

International power, the role of technology, and wealth-producing activities are three areas that have been fundamentally affected by the events of the preceding decade. Global trends within these areas will continue to influence activities in the space medium.

A New Distribution of International Power

The 1990s may be described as a period during which the United States rose to a position of undisputed world leadership and stature. The bipolar international environment shaped by the United States and the Soviet Union evolved into a new strategic context in which the United States became the preeminent political, economic, and military power, unlikely to be contested by a single challenger for at least the next twenty years.[2] These dimensions of power, to include perhaps technological and cultural considerations, confer upon it a global clout relegating it to a class by itself.[3]

Historically, states have worked to improve their military postures or fashion alliances to "balance" the power of a state possessing overweening might. Yet this has not happened as of this writing. "No nation has flung down the gauntlet," writes Josef Joffe. "None has unleashed an arms race, none has tried to engineer a hostile coalition. The United States faces neither an existential enemy nor the threat of encirclement as far as the eye can see."[4] To be sure, Europe and China have increased their regional influence and are economically stronger than Russia. Russia remains the state with the most territory and a formidable nuclear power, although its political and economic futures cast a dark uncertainty over all of Eurasia.

Power may be defined as the capacity to do, make, or destroy, while international power traditionally is understood to be the capacity of a nation to impose its will on other nations.[5] To be sure, the United States practices hegemony, but it is a benign sort. Through the exercise of so-called soft power the United States works its will in the international system through the promotion of democracy and free markets rather than military compulsion or hard-core imperialism.[6] Others have observed that the United States "irks and domineers, but it does not conquer. It tries to call the shots and bend the rules, but it does not go to war for land and glory."[7] The sole remaining global power is neither inclined to coerce others forcibly (short of responding to aggression) nor pursue a policy of aggrandizement.

Democracy and free-market values are taking hold in many states, developments America's leadership believes will lead to "new opportunities to promote peace, prosperity, and enhanced cooperation among nations."[8] Western ideas such as representative government, individualism, the rule of law, and social pluralism, along with the spread of Western popular culture (pop music, movies, clothing), accompany American ideas and institutions abroad, all of which bolster U.S. global influence. To be sure, civilizations comprising different cultures and societies and regimes driven by ideas and objectives opposed to characteristically American ones will persist in the world and remain a source of international friction.

Our world today is also one in which many states, including members of the defunct Warsaw Pact, view the United States as the security partner of choice.[9] The United States leads the greatest, most active, and most successful security alliance in the world today, the North Atlantic Treaty Organization (NATO). Indeed, NATO, now more than half a century old, continues to be a dominant player in European security politics and has grown to nineteen states with the addition of former communist nations.

The United States has other notably strong partnerships with the Republic of Korea (where more than thirty-seven thousand U.S. troops are still stationed in a defensive posture to contain North Korea), Japan, Israel, and, in the wake of the 1991 Persian Gulf war, Arab gulf states such as Kuwait and Saudi Arabia. Major powers such as Russia and China cooperate with the United States on a number of security issues. That said, Russia and China are not entirely accommodating, and, in some cases, elements within those two regimes undertake activities contrary to U.S. security interests (one could point to the sales of weapons of mass destruction, or WMD, and missile technologies and expertise by Russian and Chinese officials and entities to states and subgovernmental entities hostile toward the United States). Besides a handful of bilateral arrangements between states opposing the expansion of U.S. influence in Eurasia (most notably, Russia and Iran or Russia and India), there is no other significantly influential international security alliance.

The international security realities of our age remain dynamic, with states and groups of states vying to join regional or interstate arrangements in order to secure their political, security, and economic interests. Indeed, the geopolitical landscapes of Europe, Asia, and Africa may not be settled for decades. The failure of regimes and political disintegration in Europe's Balkans region and Africa have been particular sources of instability. Countries from these troubled continents contend with traditional causes of war stemming from economic, ideological, religious, racial, or ethnic strife.[10]

The spread of advanced and dangerous weapons technologies also distinguishes today's international environment. The proliferation of WMD, including explosive nuclear, biological, and chemical devices, and ballistic and cruise missile technologies and systems dominates the security dialogue among nations, especially among the United States and its allies. We live in a "second nuclear age," characterized by more active "rogue" states, a proliferation of local crises, and the spread of WMD and missile technologies.[11] Less technologically developed regimes have very limited resources and seek expansion of their regional power and wealth.[12]

Finally, nonstate actors today have influence, if not over the distribution of international power in the world, then over the foreign and security policies of many states. Terrorists, organized criminal enterprises, illegal traffickers in weapons and WMD materials, and drug cartels as well as multinational corporations and other nongovernmental organizations have more than marginally influenced the global "community."[13] These forces will continue to pose sporadic transnational challenges in the world.[14]

A Technological Revolution

Information capabilities, made possible by giant leaps in solid-state electronics and computer microchips, are easily and cheaply mass produced and made available to individuals around the world. The doubling of computer processing power every eighteen months for the last thirty years, coupled with the rapid expansion of very low cost communications capabilities (costs have declined exponentially since the 1980s), are the two factors most responsible for the current information revolution.[15] Progress in this area is constant. According to estimates by the U.S. Commerce Department, information technologies have accounted for more than 25 percent of domestic economic growth in a recent five-year period. By 2010, a single microchip might hold as many as a billion transistors.[16]

One characteristic of the technology age is that the global information network transmutes many national or international problems into global ones. Information technology helps overcome many political, military, and economic challenges posed by time and distance. The cheap flow of information has expanded the number of global contacts, as individuals increasingly feel empowered to communicate more freely with one another, unrestricted by geographic distance or political boundaries. One may make the case that the evolution of information and communication technologies is altering the global landscape as we know it, allowing nonstate actors to exercise considerable influence in the world and become new focuses of power.[17] We ought to recognize, however, that the demise of the nation-state has been predicted since Immanuel Kant's 1795 *Perpetual Peace* essay.[18]

Information technologies are expected to play a key role in opening up politically closed societies.[19] The information revolution was a powerful enough force to encourage Eastern European countries to rebel against Moscow's control near the end of the cold war. Rebels watched the 1988 Czechoslovakia "velvet revolution" unfold before them on the Cable News Network—Soviet satellites, ironically, carried the CNN programming.[20] High-speed, high-volume telecommunications capabilities, as we already have seen, allow parties around the world to be networked globally, with vast, interactive databases at their disposal.

The information technologies now spreading with lightning speed around the world will have a variety of offensive and defensive military applications. The commercial development of information technology, according to several senior American experts on the 1997 National Defense Panel, is so "widespread, accessible, and cheap that it promises to create both opportunities and risks for our nation."[21] Information, properly processed, will grant significant diplomatic and defense advantages to any nation. The foundations of international power remain a nation's economy, military power, land, resources, and population, yet important advantages will accrue to the country possessing the educational and institutional flexibility that allows it to handle information efficiently.

The exploitation of information and recent technological innovations form the crux of the much-ballyhooed Revolution in Military Affairs (RMA). RMAs and military technology revolutions occur periodically in history, usually stimulated by surges in the areas of technology and engineering (RMAs may also be cultural, ideological, financial, conceptual, political, or administrative in nature).[22] To qualify as an RMA, the changes have to lead to marked improvements in effectiveness of military tactics and organization and bring about significant changes in the means and methods of fighting wars. Sometimes the change is so fundamental that dominant weapons and military forces central to a previous strategic age fade away in relevance. Past examples of military revolution include the appearance of mechanized armored forces between the world wars, which markedly reduced military interest in horse cavalry, and the displacement of battleships caused by ships capable of projecting air power from the sea.

Proponents of the latest RMA see "smart weapons" (weapons with rapid onboard processors, advanced guidance systems, and sensors) changing the face of the battlefield. Featuring new levels of speed, accuracy, and range, and capabilities allowing them to operate with minimal input from humans, these weapons exploit informational dimensions, giving their users supreme advantages over "industrial age" (or "Second Wave") opponents.[23]

One may doubt the claim that military affairs are transforming at all. At the strategic level, for example, the advanced weapons systems touted as the harbingers of the new RMA may be only marginally useful against many of the newly emerging transnational threats. Cruise missiles and smart laser bombs guided by global positioning system satellites are significantly less efficient in jungle warfare or military campaigns against drug trafficking. One may also question the advantage of such weapons in a battle between near-equals in strength, where "friction" on the battlefield gums up information flow and the "fog of war" makes the modern, information-intense battlefield an especially confusing arena to tactical and theater commanders. Bearing in mind these caveats, it is also true that a commander who can see through the fog will have better footing in battle than one who cannot.[24]

Although there is an ongoing debate about today's Revolution in Military Affairs, or over just what that RMA actually entails, there are certain developments not open to debate, including the growing relevance of space.[25] Recent technological developments have increased the potency of conventional arms. Conventional weapons are becoming more lethal, precise, and discriminating, and therefore are more likely than nuclear weapons to be used in war. Advances in military weapons and platforms make possible long-range conventional-strike operations. According to some, conventional space weapons may one day cut out nuclear weapons or air-, land-, and sea-conventional variants from certain tactical scenarios.[26] Many anticipate that information-age war will pose minimal risk to troops and civilians, an important benefit in this age of heightened sensitivity in the United States to battlefield casualties. If there is an information-based RMA today, the United States, with its technological and educational infrastructures, appears to be in the best position to exploit it.

Global Economic Interdependence

New information systems, satellites included, have globalized financial markets and generally broadened international transactions, including goods, services, and international investments. With the continuous transfer of technology worldwide, the global economy will expand and deepen as more countries are able to participate in more meaningful ways. Private sectors have paved the way for more open trade between nations, with net gains for almost all nations. The volume of world merchandise exports expanded by almost 10 percent in 1997 to $5.455 trillion, the second highest rate recorded in more than twenty years. Naturally, the purveyors of advanced technologies, the countries of North America, Western Europe, and East Asia, have come to dominate international trade and investment. "The new economic powerhouses," according to Citibank/Citicorp chairman and former CEO Walter Wriston, "are masters not of huge material resources, but of ideas and technology."[27]

The result of these developments is greater global economic interdependence. International finance, trade, and investment considerations necessarily affect domestic economic policies. According to one observer of international economics, "National sovereignty is progressively constrained as a practical matter by a deepening network of economic policy commitments, trade dependencies, and large volumes of private capital flows."[28] Progress in telecommunications has supported the development of private capital markets that cause more than a trillion dollars each day to change hands.

As the global economy grows, it becomes more intertwined with security issues, raising the national importance of technological and commercial considerations. Some might suggest that the new international economic realities reduce governmental incentives to resort to military force, as the stakes states have in maintaining a stable and peaceful trading order are high.[29] At the same time, this global economic interdependence raises the sensitivity levels of all states to economic developments abroad, a potentially destabilizing factor. In either case, it must be recognized that the United States is part of a global society, meaning that the economic problems of states (criminal and drug activities, recessions and depressions, the ups and downs of labor markets, natural resource depletion, technology and weapons transfer) know no boundaries.

What Is a Spacefaring Nation, and What Is a Space Power?

In this day and age, it is necessary to broaden our understanding of spacefaring nations and space powers. In the past, a country generally qualified as a spacefaring nation, and by extension a space power, if it had a manned space program. Minimally, at least, a spacefaring country had to possess a fairly robust launch infrastructure and indigenous capabilities to manufacture and operate space systems. Historically, only two nations, the United States and Russia, possessed the full range of small and large launch vehicles and capabilities to deploy manned and robotic spacecraft above Earth's atmosphere. Indeed, the technical difficulties and financial burdens associated with the full range of space activities remain prohibitive for most countries.

The focus on *national* space capabilities remains a fundamental determinant when considering whether a state is spacefaring. Today, the number of countries capable of inserting spacecraft in orbit has grown markedly. Capabilities for launching payloads into orbit and the competence and skills useful for designing, manufacturing, and operating satellites are spreading. Moreover, it is possible in this age to buy turnkey satellite systems, bypassing in some measure the earlier requirement to possess an extensive domestic space manufacturing and operating infrastructure.

It appears reasonable, then, to assume that the attributes of a spacefaring nation include ownership and control over something functional in orbit, regardless of whether that something was boosted into space domestically. Satellite payloads are tangible signs of a state's reliance on, or mastery of, the space environment. Additional levels of space expertise in areas of launch and data receiving and processing, for example, indicate greater national commitment to the spacefaring ideal. In so far as we would not consider a state's purchasing of air- or sea-shipping services as the basis for its qualifying as either a seafaring or air-faring nation, neither should we consider a nation that simply buys or leases space services to be a spacefaring nation.

In considering the question of what is a space power, we must look even harder at the evolution, and increasingly international character of, the global commercial infrastructure. Strategic partnerships and private corporations have carried us to the point where national ownership is no longer the only criterion in our definition. It may be more useful, therefore, to regard a space power to be any entity that has the capacity to utilize effectively the space medium for commercial or national security purposes, with some pieces of its space operations coming from dedicated national satellites and others belonging to the private sector and/or government-initiated commercial activities. The baseline measure of space power will be a country's ability to integrate space capabilities with other national activities and manage the rapid and immense flow of information. Clever space powers will be those that can effectively utilize the combinations of all the space services and elements available to it. Superior space powers will own and confidently apply significant space capabilities and possess, as part of their national infrastructures, the requisite skills to exploit them fully.

With the development of a global commercial infrastructure, does a country have to invest significantly in space infrastructure (such as that currently owned by the United States or Russia) to be a space power? Although there clearly will be gradations of space power that hinge upon domestic infrastructures, the short answer to this question is undoubtedly no. Any country will be able to buy and use what it needs by employing commercial services or engaging in partnerships with other states. How well it uses space and integrates it into its military operations will help determine the overall hierarchy among space powers.

How Do Satellites Empower Other Nations?

For forty years space activities were closely associated with the military and strategic rivalry manifested between two superpowers. Serving the national security needs of the United States and the Soviet Union were nuclear-tipped intercontinental ballistic missiles (which must travel through space to reach their targets) and satellites for performing important intelligence, communications, and warning as well as navigation and mapping missions. Space also became an important battleground in the propaganda war, as both sides pursued big-ticket, high-profile space projects to demonstrate superior political ideas and international leadership in science, technology, and engineering. During this same period, space commerce was in its infancy, and space science and civil programs never really dominated national space funding in either country, the exceptions being, of course, costly exhibition programs such as Apollo and Skylab.

Competition, international competition not excepted, reflects a passion to affirm one's superiority. Throughout the cold war, Washington and Moscow understood and acted as though one day space naturally would become an arena for the performance of important intelligence and military functions. Space was at first simply a newly accessible geographic environment abutting the more familiar air environment. In peacetime and wartime, national military forces for more than four decades now have undertaken the large-scale acquisition and deployment of reconnaissance, mapping, navigation, and communication space systems as well as produced associated strategic concepts and plans.

As the United States and the Soviet Union occupied the first circle of space activities during this period, a few other countries managed to enter the space age with investments of their own (relying heavily, however, on U.S. and Russian expertise and financing). Europe (led by the United Kingdom and France), Japan, Canada, China, and India understood that there were important political, security, and economic benefits tied to space. These second-tier countries generally understood that without the ability to access space and operate satellites, a state might easily fall out of step with other advanced nations.

Although the military space programs of many second-tier countries did not compare in size and sophistication to those of the superpowers, their civilian, commercial, and scientific programs produced technologies and systems having possible security applications. Scarce investment resources mean that international space cooperation among many governments and foreign firms are an important avenue for developing *national* capabilities—infrastructures, technologies, and expertise.

Since the early 1990s, worldwide interest in interstate space cooperation has risen dramatically along with opportunities for multinational commercial space ventures. There is a desire among many U.S. allies (France, Germany, Israel, and Japan are examples) to develop autonomous space capabilities in order to reduce their reliance upon the United States, although the prohibitive cost of any ambitious space project will compel governments to consider a cooperative strategy to share the costs. For example, during the 1970s, the United States had a monopoly on space launches from the Western Hemisphere. European nations viewed their dependence as intolerable and were energized to develop their own launch vehicle (the Ariane booster) following a dispute in the mid-1970s over the launching of European communications satellites.[30]

Technology transfer has made possible the global distribution of space technologies. Even though countries may not have a dedicated military space program, the technologies involved in commercial, scientific, or civil space programs have military applications. Consider how Antonio Rodota, the general director of the European Space Agency (ESA), believed that traditionally separate space functions have melded together, further blending uses of space for national security as well as civil and commercial purposes: "The ESA has the mission of dealing only with peaceful applications. But today the military applications lie in a border area of technology. And the role of the military is moving more and more away from war to peacekeeping functions. Some of the previous military applications are being used today for peaceful purposes."[31]

In 1987, there were 850 space and missile launches worldwide. The United States and the Soviet Union accounted for all but 150 of those. By 1989, worldwide missile launches doubled to 1,700, lesser powers having conducted 1,000 of them.[32] Countries possessing domestic space launch capabilities, in order of first achievements, are Russia (1957), the United States (1958), France (1965 in Algiers), Japan (1970), China (1970), Great Britain (1971 in Australia), India (1980), and Israel (1988). Italy was the third nation to launch a payload into orbit in 1964 using a U.S. Scout rocket from a platform in the Indian Ocean, and the European Space Agency launched its first satellite from Kourou, French Guiana, in 1979. Iraq's failed launch of the Tammouz rocket in 1989 exposed for the first time its secret space activities. Brazil, North Korea, South Africa, and Pakistan also one day soon may place a satellite in orbit.

In August 1998, North Korea tried but failed to boost a spacecraft through the atmosphere. The flight, which dropped a booster in Japanese waters and propelled its third stage nearly four thousand miles into the Pacific Ocean, sparked international concern that North Korea might one day use its technology to strike targets on U.S. territory. North Korea, by conducting space launches and proving their accuracy, is also advertising itself as a top international seller of ballistic missiles and space launch vehicles.[33]

Several states (including the United Kingdom, France, Russia, India, Germany, France, China, Canada, Indonesia, and Japan) have domestic satellite communication systems, and others, such as Pakistan, Taiwan, Nigeria, and Iran, have plans to acquire them. Aside from using national systems or turning to commercial providers to meet telecommunications needs, countries may also take advantage of international consortiums such as the European Telecommunications Satellite Organization (Eutelsat), the International Telecommunications Satellite System (Intelsat, a commercial cooperative of 125 nations providing routine communication and television services and distributing Armed Forces Radio), the International Maritime Satellite Organization (Inmarsat), and the Arab Satellite Communication Organization (Arabsat, formed by the Arab League in 1976 and first serving Arab cities in 1985) to communicate inter- or intraregionally. Nations with expansive territories, such as India and Brazil, or territories comprising thousands of islands, such as Indonesia, will be able to exploit mobile commercial satellite communications to facilitate economic development and consolidate power.

The United States also is not alone when it comes to reporting on weather conditions from space, as Russia, the European states, Japan, and China have launched their own weather satellites. Since 1962 Russia has operated a range of remote-sensing satellites to perform military reconnaissance, oceanographic, earth-science, and meteorological missions. Today, discounting Russia, the Indians have the largest remote-sensing satellite program of any foreign country.[34] During the 1970s, India built satellites that were lifted into orbit by the Soviet Union to observe the Indian subcontinent and surrounding oceans, and those satellites have allowed it to become a worldwide competitor and a major supplier of multispectral imaging data.

Some U.S. allies are motivated to deploy their own eyes and ears in space in order to overcome a reliance on the United States. Although the United States shares many strategic concerns with its European allies, strategic priorities sometimes do not coincide. The director of the Western European Union (WEU) Satellite Center, Bernard Molard, insisted that it would be imprudent for Europe to rely on others during wars or crises. The Satellite Center, he observed, provides European governments with "an independent look at crises, environmental catastrophes, movements of refugees and the status of treaty compliance. It is of strategic importance for Europe to have this capacity—not in competition with NATO, but as a complement to NATO. Nowhere is it written that the United States will always share Europe's strategic priorities."[35]

In step with this philosophy, Europe's defense and foreign ministers agreed in 1998 to develop a space-observation system using optical and radar sensors for civilian and defense missions. France has shown the most active independent streak in space and other defense areas. It launched its first civilian-led satellite pour

Table 1. Foreign Space Investments, 1997

Austria	$44 million	Italy	$762 million
Belgium	$169 million	Japan	$2.001 billion
Brazil	$96 million	Netherlands	$116 million
Canada	$185 million	Norway	$30 million
European Space Agency	$3.119 billion	Spain	$156 million
Finland	$24 million	Sweden	$88 million
France	$1.544 billion	Switzerland	$78 million
Germany	$712 million	United Kingdom	$320 million
India	$303 million	United States	$12.730 billion

observation de l'terre (SPOT) spacecraft in 1986 (with satellite-development help from the United States). Requirements for a dedicated military satellite led Paris to fund the development and deployment of the *Helios 1A*, launched from Kourou in July 1995, to provide high-quality images of military sites of interest.[36] Chapter 5 will give a fuller accounting of national capabilities in this area.

Countries also have access to precise navigation signals that may be used to enhance military operations as well as civil and commercial activities. The space-based components of the U.S. GPS system and Russian Global Navigation Satellite System, GLONASS, are not proliferating. However, the ground-based receivers are widely available and inexpensive.[37] The availability of receivers means that every foreign military will have access to technologies that could enable them to provide navigation support to the war fighters and deliver timely positioning data to advanced weapon systems.

A growing number of countries recognize the national importance of space. Table 1 presents a sampling of 1997 national investments in space activities and the development of space technologies, many of which are multi-use and may be used for military missions.[38]

There are also a number of other countries with small space programs and budgets, including Bulgaria, Chile, Greece, Hungary, Indonesia, Iran, Iraq, Luxembourg, Mexico, New Zealand, Nigeria, Poland, Portugal, South Africa, Syria, Thailand, and Turkey.

The proliferation of multilateral space ventures in the public and private sectors is gradually multinationalizing space. The spread of space technologies and services to nearly all nations is at the root of this phenomenon. The availability of higher-resolution multispectral imagery has grown, and the response time in delivering imagery from systems such as Landsat is now a matter of hours (assuming a country has a ground receiving station) rather than days or weeks. Indeed, military access to space—whether it be through commercial networks, dedicated military assets, universally available systems, or assets controlled by multinational institutions—is today more affordable and easy. The armed forces of other nations are incorporating space into their military doctrine and operations and are doing so more rapidly than previously possible by not having to retrace the development steps of the United States and Russia.[39]

How Do Nations Cooperate?

The world's spacefaring nations do not, by and large, work in isolation. The disappearance of the international security structure shaped by the cold war has opened the door to space cooperation among a large number of countries. Major motivations for international cooperation include gaining access to foreign capabilities and technologies. Cooperation can increase the effectiveness of national programs by minimizing duplication, pooling resources for large projects, supporting foreign policy objectives, strengthening the bonds of alliance and friendship, and bolstering national prestige. Proponents of a particular space program within a country might embrace international cooperation as a means for ensuring the program's survival—for once they are agreed to, given the diplomatic sensitivities involved, cooperative programs are very hard to cancel despite strong domestic reasons for doing so. Declining national budgets are a major restraint on the space ambitions of states and a key reason international cooperation is thriving. Nevertheless, space investments are on the rise.[40] Collaboration usually centers on pursuit of common interests (although individual national strategies incorporate those interests differently) and a desire to maximize scarce financial and technical resources.[41]

The global interlinking of human affairs naturally has led to multilateral approaches for exploiting commercial and economic activities and for addressing emerging threats. International space activities have tended to solidify the foundations of political, military, and economic power among participating nations—hence there continues to be strong interest in working in and through the space environment. Examples of international cooperation may include the exchange of data and research scientists, technology development, the execution of experimental trials, co-development of a satellite system, the provision of a satellite payload by one country to be flown by another, and even the coordinated use of systems.

National motivations to obtain an economic payoff from this activity at the government level may be expected to boost further the creation of national and international frameworks to govern interstate space activities. The establishment of trade "rules of the road" internationally, or the development of national policies to prevent the imposition of unnecessary barriers to trade, are important regulators of this process. Although governments will be hip-deep in many of the space activities taking place around the world, there will also be a great deal of collaborative money-making activity at the industry level.

Another important area of cooperation is defense. Although states historically have pursued security objectives unilaterally, certain factors may compel them to consider cooperative military endeavors with allies and friends. Sharing of costs, having access to foreign technologies, and ensuring interoperability during crises and conflicts are primary motivations for international cooperation. A small state may be unable to afford participation at the national level in the development of a military satellite unless it can do so through a collaborative arrangement that divides financial responsibilities.

Additionally, if the Persian Gulf War and the 1999 air campaign against Belgrade are indicators, we ought to expect allies and friends to cooperate more aggressively in coalition military operations—a fact that, in turn, will compel greater peacetime international cooperation in military space operations and production. Although not a perfect example of an equitable cooperative relationship, the United States has shared early warning data of worldwide ballistic missile launches with its allies and friends, especially as systems to defend against theater ballistic missiles are deployed and integrated into combined operations. Already the United States has arrangements to provide launch-point and projected-impact-point information to its erstwhile allies, the United Kingdom, Canada, Australia, South Korea, and Israel.[42] It would be, after all, cost-prohibitive for U.S. allies to have their own global complex of space- and ground-based early warning systems.

Cooperation, even with one's potential enemy, can be of paramount concern to national security. Russia's security strategy strongly emphasizes its nuclear retaliatory force, hence its requirement for reliable early warning information. Russia's general decline, its decaying ballistic-missile early warning satellites, and aging nuclear forces has caused considerable consternation in the United States. Its current constellation of early warning satellites completely missed the 1998 North Korean missile launch. Russia's coverage of U.S. missile fields has significant gaps, and its satellites cannot reliably detect ballistic missiles launched from large areas of Earth's oceans.

This general state of disrepair was complicated further still in the late 1990s by the so-called Y2K problem, that end-of-the-millennium software bug that many feared would shut down key national computer systems in Russia. A microchip meltdown might have caused severe disruption or failure of computers controlling critical military assets, such as Russia's early warning network and nuclear weapons command and control. These developments compelled U.S. policy makers in 1998 to begin working on arrangements to share with Russia U.S. early warning data collected from its Defense Support Program satellites and ground-based radars in order to prevent Russian miscalculations of missile launches around the world. Past miscues highlight the need for clear communication in this area. In January 1995, for example, Russian operators reported a NASA research rocket launched out of Norway carrying sensors to study the northern lights as a ballistic missile attack. Although nothing came of this incident, an alert was sent to Moscow and nuclear handlers activated President Boris Yeltsin's nuclear-command suitcase.[43]

The United States also engages in ongoing bilateral discussions with the United Kingdom, and it cooperates in the defense space area with a number of other allies, including Italy, Germany, Canada, and Australia. In 1993 the United States signed a Memorandum of Understanding with France to undertake cooperative defense space projects. Other cooperative activities with France, namely, sharing military space intelligence and co-developing satellites, are under consideration.[44] Future cooperative space partnerships may involve Israel, Japan, NATO, Indonesia, and Thailand. In the spirit of international cooperation in more scientific areas, the United States also has hosted aboard the space shuttle a number of foreign experiments and astronauts from the European Space Agency, France, Germany, Italy, Japan, and Russia.

In 1998, the United States agreed to work with European nations to operate an integrated weather satellite system. The so-called Joint Polar System, with operations beginning in 2003, will consist of one satellite supplied by the National Oceanic and Atmospheric Administration (using the NOAA-N series) and one provided by the European Meteorological Satellite, or Eumetsat (from the Meteorological Operational, or Metop, series). The satellites will ensure compatibility through the use of common instruments. The U.S. satellite will provide afternoon coverage, and the European satellite will make morning observations.[45] Although there were some disagreements over data distribution on both sides, this venture is a splendid example of cooperation among allies and may be a sign of things to come.

Nations will band together to work collaboratively should the projects prove to be done more cost-effectively on an international level (cost-effective from the point of view of less wealthy countries—such international collaboration generally does not award a cost savings to the United States) or there are diplomatic gains to be made through collaboration. The political dimension of a cooperative relationship at times will overshadow the economic dimension. Although there may be tremendous commercial benefits, questions about the impact of technology transfers on national security will determine the level of cooperation with other states. U.S. satellite makers, for example, have limited their contacts with communist China because of sensitivities in Washington over the transfer of technology that may help China improve the performance of its ballistic missiles.[46] In response, China has looked to Europe for access to satellite technologies.[47]

But politics do change. One of the more intriguing cooperative relationships, especially in the areas of space science and exploration, has been between the United States and Russia. Although Washington and Moscow collaborated to a limited extent in space during the cold war, cooperative activities (again, largely paid for by the United States) have expanded following the demise of the Soviet Union in order to help Russia make a successful transition to democracy, develop a market economy, and reduce military production. Besides giving work to Russian scientists and engineers, U.S. purchases of space-related goods and services also provided much-needed hard currency for the Russian economy. Other cooperative activities with Russia have taken place in telecommunications, science and technology projects, and by using Russian engines to enhance the performance of U.S. Atlas 3A launchers.[48]

The largest international cooperative body is the European Space Agency, which involves fourteen countries and is the primary avenue to space activities for a number of those countries. The agency's programs include involvement with the Ariane rocket program and work on the *International Space Station,* a program that also involves the United States, Russia, Japan, Canada, and Brazil.[49]

Numerous bilateral cooperative arrangements exist to facilitate data and technology exchange and ease the financial burden associated with space programs. India has cooperated with Russia in space since 1963, while its cooperative relationship with the United States has endured ups and downs.[50] Russia has had a strong cooperative relationship with France since 1966, having collaborated in several space research projects and experiments. The breakdown of the Intercosmos cooperative space alliance among communist nations (which included Eastern European nations, Cuba, and Vietnam sharing in scientific, technical, and manned-space-flight activities) compelled Russia in the 1990s to search out more dynamic cooperative relationships with other countries, chiefly to offer launch services from cosmodromes in Russia and Kazakhstan.[51] In 1997, Brazil signed cooperative space agreements with France, Russia, and the United States. Brazil also has a longstanding relationship with China to develop Earth resources CBERS (China-Brazil earth research satellite) satellites and scientific microsatellites.

Finally, a number of cooperative international commercial partnerships have arisen. A new breed of corporation, WorldSpace, which holds a license from the Federal Communications Commission (FCC), is owned by WorldSpace Corporation. WorldSpace Corporation activities (including activities of WorldSpace International Network, which launched the AfriStar communications satellite) are said to be supported in part by untraceable funding from parties in the Middle East. Although the company is based in the United States, there is no significant U.S. participation in this multinational venture. WorldSpace plans to launch a CaribStar (or AmeriStar, depending on what source one reads) satellite, the second of three (AfriStar being the first, and AsiaStar being the third) designed to transmit communications signals directly to receivers in developing countries.[52]

Boeing has joined with NPO Yuzhnoye, a Ukrainian firm, the Norwegian firm Kvaarner, and Russian manufacturer RSC Energia to form a joint venture called Sea Launch Company. Sea Launch is making inroads into the burgeoning commercial launch market by launching payloads off a modified Norwegian oil platform after it is towed off the coast of California for polar launches or to the equator near Christmas Island for launches to geosynchronous orbit.[53] Lockheed Martin has joined forces with RSC Energia and Krunichev to create a joint subsidiary called International Launch Systems to market the Atlas and Russian Proton space launch rockets. Starsem is a consortium of four company partners—Aerospatiale, Arianespace, Samara Space Center, and the Russian Space Agency—that will undertake "cluster launches" of several satellites at a time using the Soyuz family of rockets. Late in 1998, the first nongovernmental U.S.-Russian spacecraft blasted out of Cape Canaveral aboard a Delta 2 rocket for geosynchronous orbit. The *Bonum-1* television satellite services, operated by a Russian company, mark the first time the citizens of Russia will be able to see television programming uncensored by the state.[54]

These few examples represent a growing trend in an increasingly trade-friendly world: global competition among global corporations for a share of the global market. To be sure, failure of these multinational corporate ventures will have worldwide repercussions affecting the bottom lines of U.S. companies. In any case, the evolution of public and private partnerships comprising players from several countries to meet global supply and demand is one of the many factors that promises to complicate future U.S. military planning and operations.

How Has the Space Age Affected Diplomacy?

Space activities are a measure of national prestige and an indicator of a country's weight on the scale of global power. Technology, in other words, has politico-strategic manifestations and meaning when artfully employed by statesmen to achieve foreign policy and military objectives.

Reconnaissance satellites are particularly meaningful in this discussion. Indeed, of all space systems, the eyes in orbit have made the most significant impact on the diplomatic scene. Why? Because our intellectual and sensory orientation is naturally directed to what is visible. We relate more easily to what we can see as opposed to what we read or hear (recall the saying "A picture is worth a thousand words"). Missile telemetry, information transmissions that are not translatable into a visual format, and even phone conversations captured by satellites generally do not have the same impact as "spy photos" from space. Revealing images of "denied" areas or objects may be expected to play well in the national and world television media. Moreover, images are useful tools of persuasion and political manipulation.

The use of space reconnaissance satellites for national purposes is increasingly commonplace, and their proliferation has implications for diplomacy. Space imagery means, for instance, that more countries have access to intelligence on other countries. This information transparency may aid arms-control negotiations as more nations become confident in the means to verify international arms agreements independent of the United States or Russia. Or the unprecedented availability of information about other countries may serve to frustrate national goals and strategic objectives. It is increasingly difficult, although by no means impossible, for states to maintain past levels of secrecy about their military programs or activities.

Two areas of military space capability that have captured international attention are the use of satellites for military communications and for the monitoring and verification of arms-control agreements.[55] A case may be made that satellites are directly responsible for the surge in arms-control activities since the late 1960s. It also should be noted that concealment, deception, and denial practices emerged at this time to degrade the effectiveness of the increasing number of high- and low-resolution reconnaissance satellites. Russia, Israel, and South Africa reportedly have done much in the camouflage area and are likely to proliferate their knowledge and expertise.[56] The Russian Strategic Rocket Forces (SRF), for example, have substantial knowledge of U.S. imaging satellite operations and capabilities. The SRF has used this knowledge to conceal the locations of their nuclear-tipped mobile missiles, which, reportedly, are stopped at designated points to evade detection by U.S. overhead reconnaissance satellites.[57]

Not all states have developed similar space capabilities, so that there will be gradations of advantage, with the greatest advantage perhaps going to the diplomats with the greatest number of eyes in the sky. Reconnaissance imagery supplied by U-2 aircraft and Corona satellites showed President Kennedy that U.S. ICBMs far outnumbered those deployed by the Soviet Union (something Kennedy did not know during the Bay of Pigs fiasco), which aided his brinkmanship during the Cuban Missile Crisis. Kennedy also probably knew that Soviet reconnaissance satellites had detected the same imbalance between the two forces, and that Soviet leader Nikita Khrushchev, no doubt, knew how weak his hand really was. This same information about Soviet nuclear forces helped Kennedy successfully negotiate an end to the October 1962 Berlin crisis, when the East German government closed off the city in an effort to force the departure of Western governments. Information supplied by Discoverer spacecraft about Soviet strategic capabilities gave Washington the confidence to stand up to Soviet demands and defeat the blockade.[58]

The Corona program launched the United States into an extended period of nuclear arms control by revealing a wealth of intelligence on Soviet military activities. In 1967, when the United States still maintained that the Soviet Union only sought nuclear parity, satellites revealed that numerous sites for new ICBMs (SS-9s, SS-11s, and SS-13s) were under construction, a strong indication that Soviet leaders were striving for numerical superiority in nuclear forces. Indeed, the Nixon administration eventually took the position that the Soviet Union was developing a first-strike capability.

The intelligence community also could confirm the placement of eight sites around Moscow for the deployment of above-ground antiballistic missile launchers. The ability of reconnaissance satellites to detect Soviet ICBM and ABM sites influenced U.S.-Soviet diplomacy and foreign policy by setting the stage for the Strategic Arms Limitation Talks (SALT) of the late 1960s and early 1970s. The high-powered eyes in the sky also revealed Soviet fighter and bomber inventories, each class of submarine, the presence of Soviet missiles in Egypt, atomic weapons storage sites, Chinese missile complexes, command and control facilities, and activities at missile test ranges.[59] U.S. KH-4 reconnaissance satellites recorded a chain of events in China during 1964 that eventually led to China's first nuclear test explosion on October 16, 1964, at Lop Nor.[60]

Reconnaissance satellites became indispensable to successful arms control, for they provided a capability for scanning extensive territories, 8.6 million square miles in the case of the Soviet Union, and the technology to detect and identify relatively small objects, such as missile silos and ABM launchers. Arms-control agreements expanded the role of National Technical Means for verifying treaties limiting ICBMs, strategic bombers, ballistic missile submarines, and ABM deployments, all of which had support facilities and operational characteristics that analysts could recognize when viewing them from space. Successive U.S. presidents authorized leaks regarding the KH-11s used for this reconnaissance mission in order to reassure a skeptical Congress and public that the United States had the means to verify the rather ambitious arms agreements with a not-so-trustworthy party. A country also may share satellite reconnaissance photos with other countries in order bolster the confidence of allies.[61]

The effectiveness of this means of verification, it goes without saying, depends very much on the monitored country's willingness to be seen and to refrain from employing camouflage, concealment, and deception techniques. In 1990s conflicts against Iraq and Serbia, phoney Scud launchers and rubber tanks were used successfully to throw off U.S. estimates of the enemy order of battle derived from reconnaissance imagery.[62]

The number of monitoring and verification tasks performed by surveillance satellites have multiplied since the end of the cold war. Such missions range far beyond the satellite monitoring requirements of the U.S.-Soviet SALT and START (Strategic Arms Reduction Talks) agreements.[63] Satellites now monitor and verify peacekeeping agreements, treaties restricting the production of chemical weapons, confidence-building activities, and treaties covering the proliferation of nuclear weapons and ballistic missiles. Civil satellites too will assist treaty verification. Landsat and SPOT images helped identify the Soviet Krasnoyarsk phased-array radar as a violation of the 1972 ABM Treaty. The public images eventually persuaded Moscow to acknowledge that the radar had been part of a national antiballistic missile system and eventually dismantle it.

The proliferation of high-resolution commercial satellite imagery could help strengthen the international arms-control process by giving other states a capability to monitor, and thereby verify, arms-control agreements. The general assumption of those who hold this view is that the availability of an independent means of verification would make it more likely that an increasing number of countries would sign and ratify arms-control agreements.[64] Demonstrating this new capability, Space Imaging released in January 2000 a one-meter resolution image of a secretive North Korean missile base, which sparked debate over the extent of the threat posed by the rather primitive looking installations.[65]

In the area of nonproliferation, a major policy objective of many nations today, satellites carry a heavy burden. U.S. national security strategy endeavors to reduce the threats to U.S. and global security posed by weapons of mass destruction and ballistic and cruise missiles, in part by slowing the proliferation of advanced technologies associated with those destructive capabilities. Advanced reconnaissance satellites monitor activities at weapons plants and shipyards, detect preparations for missile launches and nuclear tests, and track the movements of vehicles and ships with an eye to catching illegal transfers. In 1998, images from space captured thousands of North Korean workers excavating an underground construction site that analysts concluded had to be a nuclear facility, a violation of a 1994 agreement to provide North Korea with food aid and oil in return for Pyongyang's agreement to stop the construction of nuclear reactors.[66] Such intelligence influenced the manner in which the United States and its allies (especially Japan) in the ensuing months dealt with the recalcitrant North Korean regime.

Satellites also played a part in the effort to rid Libya of its chemical weapons plants, such as the Rapta facility completed in 1988. The United States verified from space that Libya had attempted to fool the world into thinking the plant had been destroyed. In a separate incident, and in a move uncharacteristic of past U.S. policy, the United States in 1996 used satellite imagery as a card player might use an ace up his sleeve. The United States desired to alert the world to the plans of

Libyan leader Muammar al-Quadaffi to construct a chemical weapons factory inside a mountain at Tarhunah. In order to prepare Libya's neighbor (Egypt) and U.S. allies (especially France) for a possible preemptive military strike by U.S. forces, Secretary of Defense William Perry released to the public a drawing of the plant imaged by U.S. reconnaissance satellites. Quaddafi was able to dispute the U.S. charges when he allowed Egyptian experts to inspect what turned out to be empty tunnels. U.S. policy of not releasing the actual imagery from its advanced spacecraft for fear of divulging the satellite's technical capabilities prevented the United States from adding more realism to the whole campaign.[67]

This experience may have helped modify U.S. satellite photo-dissemination policy. Presentation of satellite imagery became routine during Pentagon press briefings on the 1999 air battle for Kosovo. About 300 "imagery-derived products" were put on display, or, according to one source, "297 more times than ever before." The imagery, with its resolution degraded from roughly five inches to about one meter, depicted with accuracy burning villages and mass graves and is said to have dramatically enhanced the public's understanding of the air campaign in Kosovo.[68]

The United States used its Corona satellites to follow the late 1960s border conflicts between the Soviet Union and China and the war between India and Pakistan.[69] In the mid-1990s, the United States monitored from space troop movements and sanctions in Bosnia and neighboring Serbia. Images of possible mass grave sites in the Kasaba area of Bosnia and in Kosovo were provided to the United Nations (UN) team investigating war crimes in the Balkans.[70] In 1993, the United States reportedly could monitor the delivery of bundles of aid into Bosnia dropped by C-130 aircraft and track shipments of weapons from Malaysia to Bosnian Muslim forces.[71]

The realities of international trade are such that small weapons and inexpensive technologies are readily available to states, terrorists, and crime syndicates willing to buy them. That makes the task of accurately and reliably tracing these illicit transactions from space arduous and sometimes impossible. According to the 1998 Rumsfeld Commission, a bipartisan body of senior defense experts assembled to assess the nature and magnitude of the existing and emerging ballistic missile threats to the United States, the "technical means of collection now employed will not meet emerging requirements, and considerable uncertainty persists whether planned collection and analysis systems will do so." The report concluded that the United States, despite its capability to peer into the back yards of rogue nations, may have little or no warning of threatening ballistic missile deployments.[72]

Failure of U.S. space-based eavesdropping and imagery assets to detect in May 1998 nuclear test preparations in India highlighted a growing concern that the number of intelligence targets was growing to frustrating proportions. This case also underscored the susceptibility of U.S. space assets to deception. Indian officials reportedly knew exactly when U.S. reconnaissance assets would be overhead, allowing them to conceal activities during satellite overflights. U.S. space forces apparently missed the preparations for multiple tests, the setup for which began in 1995. Indian knowledge of the KH-11 and Lacrosse schedules allowed them to time their secret visits to the test site and maximize use of underground storage sites.[73] Verification of arms-control agreements from space still requires a certain level of compliance by the observed party, although it may be possible to verify an attempt to conceal, which in itself may set off alarm bells.

Satellites also may be used to demonstrate national awareness of a particular threat. The orbit of a reconnaissance satellite can speak volumes, especially if an elliptical orbit is used. Elliptical orbits can give a satellite a very low altitude above one side of Earth, and on the opposite side of the planet a very high altitude. Low altitudes permit high-quality and detailed images to be taken of targets under the flight path. High altitudes mean that a satellite will have to dwell over parts of Earth for the longest portion of its orbit. Russia uses the elliptical orbit to give its early warning satellites more time over U.S. ICBM fields. On the flip side, the United States might use an elliptical orbit to get images from the lowest possible points over a targeted area, or to deploy a SIGINT satellite high and long above Russian territory to gather electronic intelligence. A satellite's orbit, in other words, could send signals to a targeted country and indicate which countries the United States considers to be enemies and which to be friends.

Spacecraft, as the ultimate in prepositioned equipment, will be very nearly always first "on the scene" in any future battle. By establishing a virtual presence over a potential battlefield, space forces not only enhance military operations but, for the United States, also contribute to its overall deterrent posture. The very advantages they confer on U.S. troops participating in peacekeeping missions in the Balkans and Middle East will cause any nation considering hostile action against the United States to be concerned about the possible roles satellites will play.[74]

Information on how one side is faring in a war could bolster a nation's confidence, intelligence showing just how it is winning the war or used as a tool to help in negotiations for ending the war. In the 1967 Six-Day War between Israel and its Arab neighbors, a U.S. KH-4 reconnaissance satellite imaged an area from Egypt to Syria over a five-day period. U.S. analysts learned that the war had gone decisively in Israel's favor. A total of 245 Arab aircraft had been destroyed, and many of the surface-to-air missile sites in Egypt were discovered to be unmanned. Satellites also uncovered the fact that a passenger vessel had been scuttled in the Suez Canal as a barrier to unwanted ingress and egress.[75]

Space diplomacy involves signaling American resolve in a crisis or war. During Desert Storm, American DSP missile early warning satellites helped hold the multinational anti-Saddam coalition together. The U.S.-led coalition contained many Arab allies, so there was significant concern in Washington that Israel's participation in coalition military operations would fracture the amalgam of fighting forces enough to hinder the objective—ousting Iraqi forces from Kuwait. The United States sought to prevent this by providing Israel with Patriot antimissile batteries and technical crews to protect Israel from Iraq's short-range ballistic missiles. Because it took the Iraqi Scud missiles less than ten minutes to reach their targets, timely warning of the attacks was required. The only way to achieve this was to provide the Patriot units deployed in Israel information from DSP satellites. These steps, which included the sharing of satellite data, helped defeat the psychological threat of Saddam's missiles and boosted the confidence of the Israelis enough to keep the Israel's armed forces from entering the conflict.[76]

Current satellite-sharing arrangements established by the United States with its allies may not always be suitable in the future, especially if strategic concerns among the parties diverge. The case of Bosnia highlighted the post–cold war dilemma presented to the Europeans. A multinational force arrived in Bosnia in the mid-1990s to stabilize a region in upheaval following the disintegration of Yugoslavia. The United States initially was a part of that large peacekeeping coalition and provided extensive satellite intelligence to support the operation. When the United States significantly reduced its presence in Bosnia, it retasked its reconnaissance satellites, leaving the remaining European forces unable to gather the data required to monitor troop movements and enforce the embargo on Serbia. This not only underscored for the Europeans the fact that reliance on allied space capabilities can be risky business but also reflected how space assets may be employed to influence policies pursued by other friendly nations.

Sharing of satellite data may help prevent a crisis or provide security guarantees to friendly states.[77] There are reports that the United States alerted Soviet leaders Mikhail Gorbachev and Boris Yeltsin of possible coups based on electronic emissions it received from its signals intelligence satellites.[78] The United States anticipated massive failures in foreign military computers on January 1, 2000—the Y2K problem. During 1998 and 1999, Washington negotiated with Moscow, as well as other nations, to share early warning data by the turn of the century to avoid nuclear miscalculations stemming from unreliable command and control computers.[79]

The failure to observe certain events of world significance from space also may be a source of international embarrassment. The United States failed to detect preparations for the May 1998 Indian nuclear tests, the July 1998 test in Iran of a medium-range ballistic missile (although U.S. satellites reportedly were the only intelligence assets that did view the launch), and the August 1998 launch by

North Korea of a three-stage rocket that attempted to place its first satellite in orbit. Such failures have been interpreted by some as a comedy of errors.[80]

Since *Sputnik*, international diplomacy has evolved to incorporate whole new areas of concern. That outer space should be used only for "peaceful purposes" is a politically charged and technically inaccurate concept at the center of practically all discussions dealing with space at the UN. The United States first introduced this concept in January 1957 when its UN Ambassador Henry Cabot Lodge expressed the hope before the General Assembly of the United Nations that "future developments in outer space would be devoted exclusively to peaceful and scientific purposes."[81]

In subsequent months and years, the United States insisted that the question of the peaceful uses of outer space be dealt with separately from space disarmament issues. To accomplish this, the UN proposed the creation of an ad hoc Committee on the Peaceful Uses of Outer Space (COPUOS) to report on the "nature of specific projects of international cooperation in outer space which would be undertaken under United Nations auspices."[82] The earliest discussions using the term "peaceful" found that it was usually used as an antonym for "military" (as opposed to "warlike"). Much of the debate over the use of space, it must be remembered, took place in the context of the cold war competition for political advantage between the United States and the Soviet Union.[83] To this day, the interpretation of the hackneyed phrase "peaceful use of outer space" remains unclarified and a point of controversy.[84]

The emergence of space as an operational environment, moreover, has caused a split in the world, among spacefaring nations and nonspacefaring nations, and has created gradations of influence among space powers. Thus, from time to time in international fora, the arguments familiar to international politics in the 1960s and 1970s concerning space "have" and space "have-not" nations have arisen. The fundamental argument revolves around the issue of whether space belongs to all nations as a common heritage of humanity. Indeed, defenders of the common heritage idea continually pose the question of whether the benefits of space and all space resources should be distributed evenly among all countries, regardless of which country funds which project. At a recent UN International Space (Unispace) conference, leaders from Asia, Africa, and the Americas expressed their fear of the widening space-technology gap, many proposing that the UN set up a special fund to be used for the dissemination of space technology.[85]

The United States formally opposed such an interpretation of its obligations in space when it declined to sign the 1984 Moon Agreement. There are some parallels here in the development of this concept with respect to viewing the sea as the "common heritage of mankind," a concept whereby no state may claim or exercise sovereignty or sovereign rights over any part of the seabed or its resources.[86] Thus far, however, there have been few if any initiatives internationally to evenly distribute the benefits of the sea, leading one to question the Law of the Sea precedent when it comes to "distributing" space resources.

With the establishment of the ad hoc Committee on Disarmament for the "Prevention of an Arms Race in Outer Space," COPUOS no longer debates limits on weapons in space. Today's Unispace conferences focus largely on control and mitigation of space debris, use of geostationary slots, commercial space access, and damage liability.[87] This, however, does not prevent COPUOS delegates from commenting on the subject. When COPUOS opened its thirty-sixth session in 1993, for example, the Chinese representative, Liu Daqun, welcomed the relaxation of tensions between Washington and Moscow, saying it offered "an excellent opportunity to protect the peaceful environment of outer space. . . . We must take effective measures to prevent militarization of outer space and the installation of weapons there."[88]

One need only look at the preamble of any UN resolution on the subject to get a feel for the UN approach to addressing military space issues. Within the preamble, one can be sure to find the statement that the exploration of space and the use of space for peaceful purposes is in the common interest of all mankind. There is usually an expression of grave concern over the extension of the arms race into space accompanied by pleas for nations, particularly those with major space capabilities, to actively pursue the goal of preventing an arms race in space. The UN understanding of space identifies the absence of weapons in space as the essential condition for the promotion of international cooperation in the exploration and use of outer space for peaceful purposes.

Lawyers in the United States and abroad generally argue that UN resolutions and other declarations carry the force of international law on the following grounds: (1) as an interpretation agreed to by all parties, (2) as affirmations of customary law, and (3) as expressions of general principles accepted by states.[89] Yet despite the impressions of many, lawyers do not rule the world. The quest for security and the nature of man will forever work to frustrate the noble but naïve visions of those who believe that when armaments are wrestled out of the hands of national leaders that freedom and safety will be at hand. Evidence of lethal competition in all geographic environments is plentiful in current international security circumstances. One need only look to the continued proliferation of weapons of mass destruction to understand that, no matter how horrible, how destabilizing, or how revolting a weapon is, if it contributes to the military might and political influence of a malicious or opportunistic regime seeking to upset international order or to broaden by force its sphere of influence, it will be sought out with all tenacity and cunning.

What Is Permitted and Prohibited Under International Law?

International relations have never been completely without codes, treaties, and agreements to ensure at least a modicum of order in the world. Recognizable hierarchies (local and state governments) have existed in political bodies since the earliest times, as have codes for exchanging ambassadors, treating prisoners of war, establishing who is a foreigner, and facilitating foreign visits. The development of international law—made up of treaties, conventions, and the customary practices of states—"expresses the broadening of the collective interests of . . . the international system, the increasing need to submit to law the coexistence of human collectivities, politically organized on a territorial basis, on the same planet, upon the same seas and under the same sky."[90] Many treaties tend to assume a conservative quality, as not all parties to it (the losers of wars, for example) are necessarily consenting. Treaties dealing with the space environment, however, tend to reflect a more active assertion among states having similar interests that stability and juridical order on the space frontier are highly desirable goals.

The principal operators in space are governments, international agencies and consortiums, and private enterprises. The transnational character of space activities, the absence of national boundaries in a region common to all mankind, means that these operators are frequently engaged in activities of worldwide interest. A body of international law, consequently, has emerged to regulate the activities of space operators and establish order in this one area of international relations.

Should outer space be considered a sanctuary from Earth's hellish activities? The fact is, it is easy to say space should be a sanctuary because of the logistical challenge of getting there. Would it be considered a sanctuary if space travel were as easy as driving a car, sailing a ship, or flying a plane? Treaties governing the exploitation of Antarctica also benefit significantly, if not solely, from the fact that its climate makes this continent inhospitable and a challenge to reach. Our inability to "get there" inexpensively and routinely has given the space environment a "holy" aura, a sense that it is a special, almost divine preserve. Today, however, we are becoming ever more proficient in our quest to exploit space, and the number of visitors to, and activities in, space will grow with each passing year.

Also wrapped up in this issue of "militarizing" space is the question of what is "peaceful." From a legal point of view, the "peaceful" versus "military" categories for classifying space activities may give rise to at least two very different interpretations. One interpretation views "peaceful" as "nonmilitary." The other interpretation, one more accurate and useful, is closer to meaning "non-aggressive." Experts in space law recognize the folly, however, of trying to make an artificial distinction between peaceful purposes and military or aggressive purposes. "The trouble with these traditional distinctions," according to space law expert Stephen Gorove, "has been that a particular endeavor such as aerial photography may be used for both purposes, i.e., furnishing cloud cover information for meteorological purposes as well as for military reconnaissance."[91]

This is not the place to get into a philosophical discussion about space being a "province of mankind" or "common heritage of all men." Neither is this the place to engage in legal arguments over questions such as where the atmosphere ends and outer space begins or regarding the relevance of national sovereignty in space. Consistent with the defense focus of this book, in these next few pages we will review a sample of some of the more significant legal restrictions placed on military activities in space.

The system of space law regulating military activities is hardly homogeneous, rigorous, or comprehensive. It is made up of fragments of provisions, rules, general principles and norms found in an array of arms-control treaties and conventions, beginning with the Charter of the United Nations. Cold war competition in the 1950s opened up the prospect that the superpowers would turn to space in their search for military advantage. The UN Charter provides modern principles addressing the subjects of international aggression and self-defense.

This charter also lays the international legal groundwork, scant though it is, for military activities in space. The 1967 Outer Space Treaty affirms the applicability of the UN Charter to national space activities. The prohibition in Article 2 of the charter of the threat or use of force for aggressive ends (i.e., not for purposes of individual or collective self-defense) also applies to attacks on space systems. A common interpretation of Article 51 allows for a nation to have a legitimate claim of self-defense; it is a justification for unilateral measures involving force that would be otherwise illegal under Article 2(4). Insofar as these articles are not limited by geography, it would appear as though they also protect the right of nations to engage in activities in outer space involving the use of force for self-defense.

There was little incentive during those early years to use space other than for basic support for traditional national security requirements, namely, intelligence gathering and communications. Yet the growing interest in outer space prompted a series of proposals to govern national appropriation and use of outer space. Out of these proposals, which have been shaped by the political dynamics of the last forty years, there evolved a disjointed body of law comprising a group of treaties and conventions that contains provisions restricting and guiding military activities in space.

Most of the conventions and treaties affecting space do not specifically address the subject of space weapons—and for good reason. Such definitions are snake pits of consternation and controversy. Weapons tend to carry a stigma in the political world and, as they are sometimes objects of scorn, generally are placed at one end of a spectrum with the "tools of peace" (diplomacy and arms-control treaties, for example) at the other end. Yet weapons are not inherently peaceful or hostile. A "weapon" may be offensive, that is, useful for aggression and military initiatives focused on the achievement of positive strategic ends. A weapon also may be defensive to the degree it contributes to more negative or preventative aims, such as a strategy to stem aggressive operations.

Castle moats, armor suits, and space-based ballistic missile interceptors are defensive measures that common understanding recognizes are designed to prevent damage and harm. Yet such weapons, if they are to be called weapons, also may be used as part of an offensive strategy (the moats may be useful for tactical retreats, suits of armor for a brute force frontal ground assault, and ballistic missile defenses as part of an aggressive counteroffensive operation). Obviously, it is futile to assign motives to weapons. Common sense and history tell us that the purposes of man and his governments, not weapons, cause grief in the world. Weapons are inherently neither peaceful nor vicious. With this in mind, we must recognize that weapons (even highly destructive and technologically advanced weapons) can serve the common good, and that it makes sense only for a nation of pacifists to "declare peace" and tear down the city's walls.

Generally, arms-control agreements providing detailed regulation of space activities cover outer space as well as other environments. The Outer Space Treaty of 1967, which addresses such subjects as claims of sovereignty in outer space, on the Moon, or on other celestial bodies, is the only treaty currently in force that speaks wholly, and in a relatively comprehensive fashion, to national activities in outer space. The treaty, though it contains no provision that outer space must be used exclusively for "peaceful purposes," partially addresses military activities. It does not proscribe the use of satellites for communications, navigation, surveillance, reconnaissance, attack, early warning, and mapping, whatever their purpose, nor does it prohibit the use of some weapons in outer space. Weapons without question permitted in space by this treaty include ballistic missiles, nonnuclear antisatellite weapons (ground-, air-, sea-, or space-based), or other space-based weapons, for example, to strike targets on Earth.

"Harmful interference," referred to in Article 9 of the Outer Space Treaty, is not defined, although one can infer reference to any activity causing damage or adverse effects on the legitimate, peaceful space activities of others. Harmful interference, for example, may result from deploying a space object in close proximity to another spacecraft with the intention of adversely affecting its operation. A conflict may be resolved in this instance by establishment of a "keep-out" zone. This, however, raises another contentious issue identified in Article 2: whether a nation can exercise sovereign rights over portions of space.[92]

There are several other treaties having provisions that limit military activities in space. The Limited Test Ban Treaty (1963) between the United States and the

Soviet Union is considered the first significant space arms-control measure, although it does not outlaw orbiting nuclear weapons in outer space. The treaty prohibits nuclear detonations in outer space, in the atmosphere, or in other environments when radioactive debris is caused to be present outside the jurisdiction of the testing state. The 1972 Antiballistic Missile Treaty, also signed by the United States and the Soviet Union, prohibits the testing, development, and deployment of space-based antiballistic missile systems or components and field testing of space-based ballistic missile defense systems. However, fixed ground-based ABM systems, using conventional or nuclear kill mechanisms to destroy space objects and ballistic missiles, are not excluded by existing arms-control treaties. Like other treaties (the Threshold Test Ban Treaty, Peaceful Nuclear Explosions Treaty, the U.S.-Soviet agreement banning intermediate nuclear forces in Europe, and START I), the ABM Treaty also prohibits "harmful interference" with national technical means of verification (the linchpin in modern arms-control agreements). Article 5 of 1991's START I between Moscow and Washington prohibits the deployment of WMD-armed fractional orbital bombardment systems as well as systems to orbit nuclear or mass destruction weapons. Although the present body of space law is spread across a number of international agreements and conventions, it is not as restrictive as one might imagine.

The 1984 Agreement Governing the Activities of States on the Moon and Other Celestial Bodies prohibits the threat of a hostile act or use of force on the Moon, the placement of nuclear or weapons of mass destruction in the Moon's orbit, the establishment of military bases, installations, and fortifications, or the testing of weapons or military maneuvers on the Moon. This agreement broadens provisions already in the Outer Space Treaty regarding military activities on the giant rock orbiting Earth. Unlike the Outer Space Treaty, the Moon Agreement restricts activities involving the use of the Moon's orbit or gravity, specifically for the deployment of weapons of mass destruction. U.S. reluctance to ratify the Moon Treaty revolves around its concerns about the provision declaring the Moon and its natural resources to be the common heritage of all mankind (Article 4).[93]

U.S. military space programs are conducted in strict compliance with U.S. obligations agreed to in many bilateral and multilateral international treaties, agreements, and conventions (an exception here, should one be found, would not undermine the integrity of this assertion). U.S. policy, strictly interpreted, is that any act not specifically prohibited by international law is permitted; there are actually few legal restrictions on the use of space for nonaggressive military purposes. The value of space and the early use of military satellites by the United States and the Soviet Union for force support purposes (e.g., for communications, navigation, surveillance, reconnaissance, early warning) made it certain that space would never be a true "sanctuary" from war or a demilitarized zone.[94]

Are There Bureaucrats in Space Too?

A number of agencies have been established in the United States to develop the rules and regulations that govern U.S. activities in space. This process began with the 1958 NASA Act, which clearly delineated responsibilities for civilian and military authorities. NASA's early duties included supervision and regulation of private space activities. Today, other agencies and departments are involved. The Department of Transportation now watches space launch activities, and the Federal Aviation Administration is responsible for licensing all commercial launches. The 1962 Communications Satellite Act mandated the development of a commercial communications satellite system to serve public needs and national objectives. The United States participates in Intelsat and Inmarsat through COMSAT.

Congress also oversees agencies such as the Federal Communications Commission, an independent regulatory agency set up to control what goes out over the national air waves. Today, the FCC oversees space-based satellite broadcasting and other news services. The commission has been responsible for such regulatory decisions as the "two-degree spacing" rule, a distance requirement applicable to U.S. satellites based in geosynchronous orbit. Other U.S. departments and agencies that have their hand in the space business include numerous DoD agencies (such as the Defense Threat Reduction Agency), the State Department (which issues permits for exporting boosters for launching U.S. satellites outside the United States and holds the U.S. registry under the 1974 Registration Convention), the Securities and Exchange Commission (which must approve space enterprise capitalization ventures), and the International Trade Commission, which protects American positions in space commerce. The Food and Drug Administration, as you might surmise, watches over pharmaceutical manufacturing in space.[95]

Indeed, most countries have national agencies doing similar activities. Any country desiring to play in the space telecommunications field, for example, will have to establish a central office to allocate radio frequencies to individual stations providing telecommunications services. International space regulations tend to suffer from one common defect, however. They are not enforceable and demand no sanctions for violations, leaving nations to exchange bitter letters of protest as their only recourse.

Intelsat and Inmarsat were set up in 1964 and 1979 respectively to manage international telecommunications, with the eighty-four-nation Inmarsat focusing on the improvement of "maritime communications, distress and safety of life at sea communications, efficiency and management of ships, maritime public correspondence services and radio determination capabilities." The participants in these regulatory organizations are telecommunications operators, or government agencies and departments. The purpose of both institutions is to provide technical management and administrative decisions internationally, which includes the equitable distribution of contracts among the participating member governments.[96] The efforts of both these organizations (and of other organizations such as Eutelsat) to stay in step with the times, however, are forcing their respective executives to consider the privatization alternative. Intelsat and Inmarsat currently have plans to spin off private commercial companies.

There also are a number of international agreements that make up a growing body of space regulations applicable during times of peace. Spacecraft are transmitters of information and rely on radio telemetry. The distribution and use of radio frequencies are governed under the customary acceptance of the radio spectrum as a natural resource available to all of mankind. The 1982 International Telecommunications Union (ITU) Convention appointed the International Frequencies Registration Board as the official international registering agent for the radio spectrum, the "custodians of an international trust."[97] Like any bureaucratic agency, ITU also struggles with delays in filing for and coordinating the use of orbital slots.

Most countries are members of the convention and participate in ITU radio conferences to decide upon the allocation of frequency bands for telecommunications services. National radio or communications frequencies are regulated to avoid harmful interference with radio-communication service in other countries. Broadcasting, for example, often results in the "spillover" of radio emissions across national boundaries. The ITU may help to mitigate the effects of spillover by setting distance limits on satellites or by coordinating transmissions. Unwanted radio signals also may interfere with transmissions between ground stations and satellites, in which cases the ITU regulations stipulate that all technical means available shall be used to reduce radiation over countries other than the intended recipient.[98]

The 1972 Liability Convention covers damage caused by spacecraft. Yet this convention is far from perfect, as it provides no recourse against those who damage parts of the environment not under ownership. There are also several difficulties associated with the identification of space objects, particularly how to treat space debris and the problems encountered in attaching fault to an errant hunk of metal.[99] Various incidents involving Soviet and Russian spacecraft carrying nuclear materials sparked intense international incidents. Alarms went off when a Soviet nuclear-powered radar ocean reconnaissance satellite reentered the atmosphere in 1978 over Canadian territory. Similar reentry problems with Russian spacecraft carrying plutonium occurred over the South Atlantic in 1983 and 1996, to no apparent consequence in either case.

The final example of international space regulation is the 1974 Registration Convention, which requires nations to notify the UN about pending space missions. States are requested to furnish the name of the launching state or states, identify the space object, provide the date and location of the launch as well as basic orbital parameters (inclination, nodal period, apogee, and perigee), and reveal the general function of the space object. This convention has served to make the 1972 Liability Convention more effective and limit the number of unidentified objects in orbit. The main defects of this convention are that it leaves to the state to determine the contents and conditions under which registers should be kept, and it does not have any enforcement provisions. It also does not appear to address those objects that make only a brief transit through space, such as ballistic missiles and sounding rockets.[100]

Yes, even in space, the red tape is prolific and unavoidable.

What Is the Bottom Line?

Access to and operations in space have weighty politico-strategic implications. Given its global outlook and plans for global engagement, the status of the United States as the leading spacefaring nation and predominant space power will ensure the continued growth of space activities relevant to international life; although even without the United States, international space relations would continue to evolve, albeit at a slower pace. The United States will continue to influence ideas, concepts, and politics relevant to the exercise of space power well into this new century.

The internationalization of strategic and economic issues, and the increasing number of space powers, will remain major factors for U.S. national security policy. Knowledge of the nature of man (his inclination to be at times maliciously opportunistic), the corrupt and aberrant behavior of some governments, the inherent friction in interstate politics (states sometimes act upon different understandings of what is good and advantageous), and the history of peace and war among nations leads us to presume that strife, alternating with the spirit of cooperation, will characterize international activities in the vast outer realm above us. To be sure, we will continue to seek international laws and other agreed-upon rules of behavior in a ceaseless effort to establish order in space.

Yet we must not overlook the reality, indeed it is practically a truism, that international law is for all peace-loving nations that agree to relinquish a measure of their sovereignty for the common good. If a state believes that its national goals are not served well by the prevalent international juridical order, it may act contrarily and express itself with hostility. In more stressful times, interference by one nation with the satellites of another cannot be restrained by treaties.

War is a permanent feature in international politics. Former British prime minister Winston Churchill, in the pithy style for which he is famous, observed prior to the Second World War that "the history of the human race is a history of war, and the records of thousands of years show only a few uneasy intervals of peace."[101] As yet one more arena within which states will "mix it up," the space environment cannot possibly be excepted from the reality that war forever abides by us. It goes against a long and extensive historical record to regard the spacious area encasing Earth as a sanctuary from man's manifold ambitions and trials. Hence, there is a need to consider next the relation of space to the United States' defense purposes, national security plans, and military capabilities.

3 EVOLUTION OF A SPACE POWER

What Are the Implications of the Space Revolution for U.S. Military Strategy?

This chapter identifies the United States' motivations for exploiting space for offensive and defensive purposes and draws linkages between the multiple uses of space and U.S. military success. In addition, it asks this basic question: What else could space do for American security? Whether space should remain a sanctuary, an isolated environment where governments may find perpetual peace, or whether it should be more fully exploited by the military arm of the country is a fundamental, even philosophical, question germane to sound consideration of the future of American space power. While this chapter examines some options and considers some missions presently not sanctioned by national policy, its primary focus is on official military strategy and present-day capabilities.

The United States' geography, world outlook, national security objectives, and technological inclination dispose her to look to space for solutions to many national exigencies, including security. It is in most every respect a natural space power. The country's manifest devotion to technological excellence assures the means to continue the exploration and exploitation of the outer frontier. We will set aside here an expanded discussion of how technology impacts military strategy, as it is one that has been covered very well by others.[1] Let us simply observe that the United States' world view is a global one, and that with its separation from the rest of the world by distance and water, Americans rely considerably on technological and scientific innovation to maintain international connections and protect vital interests.

Why Is the United States a Natural Space Power?

Americans have always looked outward. Contrary to popular understanding, isolationism never truly characterized U.S. foreign policy or strategic vision. Americans have always seen their fate tied, however loosely, to the progress and fortunes of Europe (the "Old World").[2] This yearning to search beyond national borders for answers to strategic challenges manifested itself as early as the Revolutionary War. The war for independence was a general war, a world war if you will, involving not only Great Britain. France, mortal enemy of England, made common cause with the colonies, having been promised territories in the West Indies. Spain engaged in operations against British seaborne trade from New Orleans, while the Netherlands made Britain its enemy and Russia conspired to assist Americans by disrupting British sea power.

Official America, while at times displaying isolationist tendencies, betrayed its more deeply ingrained internationalism early on. Indeed, President George Washington's famous Farewell Address may be viewed as a check upon American internationalist ambitions. Washington did not so much object to alliances, or to other strategic or political expediencies, as he pleaded with his countrymen to stay out of "foreign entanglements." This advice served to limit American involvement in the wider war raging in Europe at the time, for he feared factions would endanger the young Republic with divisive declarations of allegiance to the belligerents. It is, generally, also prudent policy in any land or age to avoid "meddling" in the affairs of other states and making unwise commitments to foreign sovereigns. Washington sought to steer an independent, not isolationist, course.[3]

The United States first identified strategic interests beyond its territorial borders in 1823, when President James Monroe extended the protection of the United States to the entire Western Hemisphere. The strategic frontier of the United States was extended during the nineteenth century, well beyond the California shoreline north to the Alaskan tundra and west across the vast Pacific to the Philippines. After the Spanish-American War of 1898 the United States became a major Asian sea power. To the east, Washington saw its interests extending from the tip of Greenland to Brazil and Argentina. The United States' defensive limits also were extended north of Canada to Alaska.

This expanded strategic perspective called for new measures to meet security responsibilities in two oceans. America's military vision called for breaking with its continental focus and acknowledging the prospect that the country may have to compete with the sea powers of Europe. In other words, the national fleet had to perform missions beyond ship raiding and coastal defense; it had to mature to control sea lanes and, when necessary, stand up to the strategic ambitions of other naval powers. By the early twentieth century, the United States pursued development programs that enabled it to deploy capable expeditionary forces to far-off regions, though Washington's strategic axis remained within the Western Hemisphere.

While maintaining extensive relationships with foreign nations and expanding its external trade following World War I, the United States chose to ignore many significant events that eventually led to another world war. Indeed, the country turned headlong to naval arms control to limit the very weapons it would need to sustain its commitments in Asia and Europe. A second world war soon revealed that the United States' security interests truly were global in scope. The modernization of its navy, the employment of naval aviation and amphibious capabilities, the development of mechanized ground forces, the emergence of air power, and the development of the atomic bomb came to comprise the backbone of twentieth-century U.S. defense policy.[4]

U.S. national security strategy focused sharply on stability in Europe and the defense of free European states. To meet its escalating military responsibilities using modest capabilities, the United States worked in the face of an anticipated long-term Soviet threat for the establishment of the North Atlantic Treaty Organization in 1949. Within five years after World War II Americans adopted a new strategy to contain and deter a Soviet Union that already had swallowed Eastern Europe and threatened allies in the Asian Pacific. With the emergence of Soviet strategic bombers and ballistic missiles capable of carrying nuclear warheads, the United States lost the accustomed security provided by its two ocean fronts. World War II taught Americans that the United States could no longer afford to wait for events of international and strategic importance to unfold before acting. Rather, Washington saw the need to commit its immense human and industrial resources to the active deterrence of aggression around the world.

In order to prepare the nation's armed forces for the trials of the coming decades, President Harry Truman initiated a fundamental reorganization of the defense establishment (the National Security Act of 1947) and established a new military branch (the U.S. Air Force). The United States' commitment to being the world's leading air power had many positive implications for its ability to maintain a credible nuclear deterrent and its superpower status. The war in Korea (1950–53) further convinced the United States to invest in a sizable rearmament program, continue to take on the obligations of a forward defense in Asia, and work for the development of NATO.[5]

In the postwar world, military forces stood at the ready, at home and overseas, and the political will to use the armed services generally could be found, although the Vietnam War caused much soul searching. In the 1970s the United States' morale, defense budgets, and political will shrank considerably—right along with the power of the presidency. The renewal of U.S. military power and revival of the executive office by President Jimmy Carter (who invested significantly in new weapons systems) and President Ronald Reagan (whose administration led a surging national rearmament, a politically significant U.S. intervention in Grenada, and a commitment to explore the technological feasibility of strategic defenses against Soviet intercontinental ballistic missiles) strengthened the country in the 1980s. The revival breathed new life into Western defenses and resolve. U.S. military operations in Panama (1989) and the Persian Gulf (1991) reinforced the argument that wide-ranging and robust conventional military forces were required in the post–cold war world.

The United States has developed over the past sixty years a strategic perspective that revolves around the idea that its military power and political influence must be transportable to regions around the world. Despite the fact that the decline in the number of American overseas forces and bases has resulted in a diminished U.S. presence abroad, the United States remains the only country in the world capable of conducting effective, large-scale military operations far beyond its borders. By most estimations, it will continue to deploy forces and capabilities required to protect the country against direct threats (wherever they appear).[6] The U.S.-led air campaign against Belgrade in 1999 was only the most recent example of the U.S. reliance on capabilities to project power overseas, capabilities that in this instance involved for the first time flights of B-2 bombers from America's heartland directly to target sites in the Balkans.

Since the United States bears the burden of defending global interests, it must maintain global military capabilities. As an environment for hosting global defense utilities, and possibly one day global forces for active defense, space is a logical, necessary, and unavoidable medium for the United States to exploit. Space would still have an important role for an isolationist or hemispheric-bound United States, but it would not be as important as it is to the United States that represents itself as a world power. The United States, simply put, is a natural space power.

If the trends of recent decades hold, the information-dependent U.S. armed forces will demand in coming years even greater access to more efficient space-based information-acquisition and -handling systems. Although today only a few dare to think along these lines, requirements for increasing the strategic efficiency of U.S. forces and achieving high levels of protection of U.S. and allied troops during military operations may, in the long run, mean putting more missions into space. These missions may include offensive and defensive orbital attack against targets in space or on Earth (reducing, but by no means replacing, U.S. reliance on long- and short-range air- and seaborne weapons platforms).

The medium of space can offer timely access to all points in the world, an irresistible attraction to a great power having vital and major national interests in many regions. Indeed, the growth of space power in the United States may be viewed as one of the most important military developments in this new century.

How Does Space Tie into the U.S. Security Strategy?

"To . . . provide for the common defense, promote the general welfare, and secure the Blessings of Liberty to ourselves and our Posterity": these purposes set out in the Constitution are the basis of the United States' national security strategy. Since the birth of the republic, the federal government has acted to protect the lives and personal safety of Americans living at home or abroad. Serving officials also swear to defend national sovereignty and the integrity of U.S. territories, guard the institutions spelled out in the Constitution, and uphold the standards of political freedom.

The Constitution, of course, does not limit the sphere of defensive activities to particular geographic environments. A constitutional amendment was not required for Congress to provide and maintain an air force when, early in the twentieth century, the air medium took on special military importance.[7] Air power emerged initially to support land and sea forces and eventually evolved to become America's premier weapon in the defense of its economic and strategic interests. There are strong parallels with the evolution of space forces. It follows that the broad sweeping war powers granted Congress in Article 1 allow for consideration of forces to guard and exploit space.

The U.S. national security strategy, together with U.S. commercial interests and foreign policy, provide the architecture within which the major policies of the state are formulated to develop and maintain the necessary powers of action. Elements of the strategy remind the American people and the world that the United States has interests far beyond its shores and that there are many challenges and threats to those interests. Defense of freedom of space ranks next to ensuring freedom of the seas and airways, for the space ways are recognized today as vital lines of communications for our most important national security and economic activities.[8]

Successive U.S. presidents, especially during the twentieth century, have spoken earnestly about the need to retain superior military, diplomatic, industrial, and technological capabilities for coping with a dynamic international environment. Current national security strategy carries forward ideas about the world established more than two hundred years ago, namely, that the United States must have the intellectual fortitude and institutional flexibility to maintain and adapt its armed forces to the strategic demands of the day. Alexander Hamilton established as much in *Federalist* No. 23: "The authorities essential to the common defense are these: to raise armies; to build and equip fleets; to prescribe rules for the government of both; to direct their operations; to provide for their support.

These powers ought to exist without limitation, because it is impossible to foresee or define the extent and variety of national exigencies, or the correspondent extent and variety of the means which may be necessary to satisfy them. The circumstances that endanger the safety of nations are infinite."

Through its strategy, the United States commits politically, militarily, and economically to maintaining its leadership position in space as well as on the seas and ground and in the air. These aspirations, fundamental as they are, may be expected to demand the commitment of national resources to the development of powerful military and diplomatic capabilities to defend U.S. interests in space.[9] A highly fluid, and at times rapidly changing, international security environment will affect U.S. international commitments, political viewpoints, and national security assessments regarding the military utility of space and has direct implications for the way the United States must prepare for, and conduct, joint and multinational military operations.

How Does Space Help Meet Current Military Requirements?

The national *military* strategy and the DoD's space policy stress maintenance of a technological lead in space and the urgent need to protect U.S. interests there.[10] The military arm faces a variety of challenges, some of which are old or new threats, and others of which are operational requirements novel to our time.[11]

The Burden of Power Projection

Power projection is strategically burdensome. Dangers abroad spawn onerous planning, logistical, and operational requirements. For this reason, Washington's ability to project power rapidly and effectively to distant lands, perhaps even to multiple and dispersed locations, is the cornerstone of the country's military preeminence.

On September 19, 1945, when the U.S. Army Air Force executed a nearly seven-thousand-mile flight of three B-29s along a Great Circle route from Tokyo to Chicago, it demonstrated for the first time to the American people and the world (Moscow in particular) that the United States had a long-range military capability. U.S. air power and long-range nuclear bombardment strategy characterized U.S. power projection capabilities throughout the cold war. The enhancement of its capabilities to move and operate conventional forces thousands of miles beyond national borders extended the ability of the United States to reach out to any place on the globe to defend national interests. Today, U.S. forces deploy to regions out of the continental United States and from "forward" inland bases and ports established on U.S. territories and on the territories of allies and friends.

The physical reach of aircraft, ships, and ground forces signals the country's commitment to the security of regional allies and its resolution to protect national interests worldwide.[12]

Along with the decline in the number of U.S. overseas bases after the cold war came a reduction in the ability to apply military power and exercise influence in strife-torn regions. The United States lacks the luxury today of focusing on a single threat, meaning that there is always the possibility that the next enemy will not accommodate U.S. defense planners by choosing the "right site" for a crisis or battle. In so far as we cannot know with certainty where future military operations will take place, the current strategy demands that the nation's armed forces be able to act quickly and effectively even in those instances when it does not have a significant presence in the region or where the transportation infrastructure is poor or non-existent.

The challenge of establishing regional and global communications offers one very illuminating example. U.S. forces deployed overseas rely on an ability to gather, process, and disseminate an increasingly large and uninterrupted flow of information under adverse, fluctuating conditions. Planning assumptions within a power projection strategy cannot be based on guesses as to where a ground communications infrastructure should be developed. The United States certainly can invest in the development of strong fiber-optic cable or microwave "backbones" in the continental United States and Hawaii, for example. The problem is that the nation's forces are not likely to fight on their home turf. Troubles lie abroad. The United States, therefore, must possess the flexibility to jump in and out of regions. Only satellites can provide a reliable real-time and global communication infrastructure.

One of the risks the United States assumes by relying on forward bases and ports is having to fight away from prepositioned U.S. troops and equipment. Allies could deny access to their sovereign territory, including local seaways and airways, for domestic or strategic reasons, including a fear of exposure to devastating retaliatory strikes by an unstable or aggressive neighbor armed with nuclear, biological, or chemical weapons (NBC). As part of a "keep-out" strategy, adversaries of the United States could hit potential landing sites and bases on U.S. allies' territories using ballistic or cruise missiles and weapons of mass destruction.

One may look to the experiences of Desert Shield and Desert Storm in 1990–91. A decision by the United States to roll Saddam Hussein back out of Kuwait required the deployment of a multinational coalition of ground and air forces in Saudi Arabia and the use of in-theater military, transportation, and communication equipment and infrastructures. Without Saudi cooperation, Desert Storm could not have occurred, since any alternative operation to free Kuwait would have had to rely heavily on much more risky amphibious and airborne operations.

Political restrictions apply in every geographic environment—on the land, on the seas, and in the air. That is, they apply to all environments *except* space, where missions may be accomplished well above spheres of sovereignty. There are no recognized political boundaries in space; indeed, the establishment of such boundaries would amount to sovereign claims, which are prohibited by the Outer Space Treaty. Satellites also allow the United States to monitor events from space and "bring into" another country systems and processes for navigation, position location, and communications to support military operations.

Although U.S. military leaders consider overseas presence to be *visible* presence, it is possible to establish a virtual presence using satellites. Space assets are usually not visible to the naked eye, yet friends and foes alike are aware of their presence. If they are appropriately used, satellites can provide constant surveillance of a region, leading some U.S. defense planners to conclude that "any nation contemplating an action inimical to U.S. national security interests must be concerned about U.S. space capabilities."[13] There is also a measure of safety in distance. Naturally, satellites are, at present, largely above the fray and do not impinge upon the territorial sovereignty of any nation. Satellites also allow reconnaissance and communications without placing troops in harm's way. Space, in other words, permits us to do things differently.

Great Britain's insularity granted it immunity from direct attack as long as the Royal Navy controlled the seas. London's mastery of important sea lanes of communication multiplied its military options. Access to space similarly is fast becoming a fundamental prerequisite for the projection of U.S. troops, weapons platforms, and support equipment to far-off and dangerous regions. Some have postulated that, should our space assets ever be threatened, the United States could turn to advanced Earth-bound systems to perform the communications, surveillance, and navigation missions. Fiber optics, microwave, and switched network communications, it is argued, would provide the operational connectivity. Reconnaissance packages may be placed on manned and unmanned aircraft. Inertial navigation systems are said to be improving in accuracy and portability.[14]

Yet while terrestrial alternatives certainly exist, they literally are not in a position to replace satellites. As the United States' former national space architect, Maj. Gen. Robert Dickman, notes, "Could the armed forces conceive of a way to get out of space? Probably yes. But would you want to? No. Could I do away with air as a part of the ground battle? Yes, but why would I, when it makes my operations so much more effective and efficient? Why would I do missile warning or reconnaissance in denied areas in any where but space?"[15]

The advantages U.S. armed forces receive from space are so extraordinary and the national investment in space is so great that the option of retreating back to Earth is no option at all. Indeed, such a move would leave space to hostile forces and relinquish significant efficiencies in military operations. Through space, the

United States has constant and, for the moment, unhindered sensory and functional access (for communications, observation, navigation, etc.) to all strategic regions on Earth at all times. Without world-circling assets, power projection becomes a much more arduous enterprise.

Strategic Agility

The United States must respond to military crises with what the Joint Chiefs of Staff call "strategic agility." Current planning guidance states that not only must the country move hard-hitting forces and equipment quickly (ideally within twenty-four hours)[16] and steadily into distant regions of the world, but the country's very commitment to the defense of global interests means it must be prepared to confront a range of invitations to battle.

Since the early years of the Clinton administration, the United States has planned, according to the national military strategy, "to deter and defeat nearly simultaneous, large-scale, cross-border aggression in two distant theaters in overlapping time frames, preferably in concert with regional allies." For most of the 1990s, the United States maintained a readiness to conduct two simultaneous major regional contingencies, one in East Asia (possibly against North Korea) and the other in the Middle East (possibly against Iraq or Iran).[17]

U.S. forces also must have on hand a wide range of capabilities to engage in smaller-scale contingency operations. World events in the 1990s demonstrated a clear need for the United States, for at least the next ten to fifteen years, to engage in military operations falling below the threshold of a major war. Indeed, that decade was one of the busiest on record for the U.S. armed services (short of all-out war), with smaller-scale contingencies being the major reason for heightened activity.[18] The physical presence of soldiers, sailors, and airmen executing a mission or providing security and order in a country may be expected to contribute more to "victory" in smaller-scale conflicts than high-tech precision weapons, tanks, destroyers, and attack aircraft.[19]

Space-based intelligence systems will be required in major conflicts to locate in a timely manner threats to U.S. and allied operations and possible targets, maintain general situational awareness of combat and associated support operations in all geographic environments, track troop movements, and undertake periodic battle damage assessments. Surveillance and reconnaissance satellites will be used to monitor infiltration routes as well as assess terrain and escape routes for rescue operations.[20] Satellites also will monitor weather conditions critical to the progress and pace of terrestrial military operations.

Communications and navigational hookups to space are vital parts of a continuous logistics train that reduces the burden on sea, land, and air transportation systems, requires fewer resources to defend, and is more difficult for an enemy to detect and target. Available in jungles and cities, in mountain areas and deserts, satellite communication links may be maintained in all environments and military scenarios. Satellites enable secure two-way communications to and among all fighting forces, provide critical information to the National Command Authorities, and support U.S. diplomatic operations.

Living with Weapons of Terror

Nuclear, biological, and chemical weapons (NBC) have the potential for significantly disrupting U.S. operations and logistics.[21] The distinguishing characteristic of NBC weapons is their indiscriminate, and often extreme, effects. Properly employed, they are fairly reliable producers of death, disability, and destruction. The physical and psychological damage caused by weapons of mass destruction is unfocused and random, which lends to their capacity to terrify.

Weapons of indiscriminate destruction offer hope to weaker states and nonstate actors that the conventionally superior United States may be defeated on the battlefield or intimidated to the point where it avoids engagement or capitulates.[22] U.S. defense planners are then challenged to develop new capabilities and engage in new thinking in order to deter or mitigate the consequences of the use of such weapons. Military responses to the use of WMD will require not only a range of nuclear and conventional response capabilities, in addition to passive and active defenses, but also supporting command, control, communications, and intelligence assets.[23]

Improved satellite intelligence collection can assist U.S. efforts to prevent or slow WMD proliferation and to protect U.S. forces from attacks. The last decade saw (in many cases through satellites) a veritable mosaic of WMD and missile production and dissemination activities in Northeast Asia (North Korea and China), South Asia (Pakistan and India), the Middle East and North Africa (Iran, Iraq, Libya, Syria, and Egypt), and Eurasia (Russia, Ukraine, Kazakhstan, and Belarus).[24] Surveillance and reconnaissance satellites perform unique (though by no means foolproof) verification functions for arms-control regimes and export controls, which must cover impossibly vast areas. Illicit weapons transfers also may be detected and monitored from orbit by tracking pending or in-progress shipments of critical technologies and materials.[25] Space systems also may detect an impending attack and communicate this information to commanders. In addition, they may help deal with the dreadful consequences of their use.

U.S. space assets comprise a central element in plans to develop and deploy ballistic missile defenses (BMD) to protect the American homeland from long-range missiles and troops who may find themselves engaged with one or more of a growing number of countries possessing short-range (thirty to one thousand kilometers) and medium-range (one thousand to twenty-seven hundred kilometers) ballistic missiles. What began in 1983 as the Strategic Defense Initiative (SDI) to exploit space to defend against a major Soviet ICBM attack evolved in 1993 to a collection of less-ambitious programs featuring the deployment of ground-based interceptors to defend U.S. troops against limited, shorter-range missile attacks. Policy constraints limited defense planners to developing architectures that kept interceptors of short- and medium-range missiles based on the ground, at sea, and in the air.

The possibilities of accidental or unauthorized launches, a limited ballistic missile attack from China or Russia, and the spread of long-range missile technologies to countries such as North Korea, India, Pakistan, and Iran have helped make the case for deploying a national missile defense (NMD) system. In January 1999, on the heels of North Korea's 1998 Taepo Dong 1 launch, when Pyongyang demonstrated to the world that it was well on the way to developing an ICBM, the United States moved a step closer to acknowledging that "a new strategic threat" had emerged and, consequently, closer to deployment of a limited NMD system.[26]

The United States is able to transport its theater missile defense units into a region of conflict, knowing full well that it can plug into a robust space-based warning, intelligence, battle-management, and command, control, and communications architecture. Observing from the high ground of space, it is possible to recognize when a missile has been fired against U.S. troops or the homeland, and from which locations. Today, surveillance assets using infrared sensors systematically observe space and Earth to detect (providing tactical warning and attack assessments) and track boosting ballistic missiles.

Critical elements of the BMD battle-management, command, control, and communication architecture also rely on satellites, especially for long-haul communications. All of these functions are needed to intercept and destroy a hostile missile or warhead in the atmosphere or in space.[27] But the possible uses of space to defend against ballistic missiles does not stop there. The U.S. Army, for example, has been experimenting with commercial satellite paging systems as a means of quickly warning troops of incoming missile attacks.[28]

Given the growth in missile threats, today's leaders in the BMD community anticipate that, one day, the United States will move into space to provide ballistic missile defense, that this move is "inevitable."[29] The ballistic missile threat of tomorrow could come from many directions and carry a variety of payloads, ranging from conventional to nuclear weapons. A capability to subdivide NBC payloads ejected from ballistic missiles using submunitions and multiple independently targetable reentry vehicles (MIRVs) and the increased use of countermeasures will complicate missile defense and strengthen the case for fielding space-based capabilities to intercept ballistic missiles in their boost and ascent phases.[30] Few envision that the United States will ever deploy nuclear-armed, Nike-equivalent NMD sites all around the coasts in response to this growing threat (although this perspective could change if the United States were attacked by ship-launched ballistic missiles tomorrow). The best answer, once again, may lie in space, where given the appropriate technologies and intercept systems, the mission could be accomplished more effectively.

Information Operations

It should be clear to the reader by this time that the United States has established as one criterion for military success the ability to integrate quickly, reliably, and accurately information from multiple military and commercial sources. Naturally, there is a counterpart requirement: to deny this same ability to the enemy. Information, properly processed and employed, contributes to overall battlefield situational awareness, an advantage upon which the U.S. armed forces now depend.[31] Satellites play a key role in information operations, channeling large amounts of data relatively inexpensively, with dedicated military satellites enabling protected and undisrupted transmissions.

There are parallels here with the discussion of space control below, although major differences exist between information operations and space operations. Like the nation's economic and commercial activities, military data-processing operations rely heavily on the national information infrastructure. Information operations amount to a *process* of exploiting systems of sensors, networks, and weapons. This process represents a condition of modern warfare that presents both opportunities and vulnerabilities for U.S. armed forces.

Space operations, on the other hand, incorporate viable military concepts such as space control and force projection. They concern themselves more specifically, but not exclusively, with the flow of information through conduits in space. The space-control mission one day may evolve to ensure access to space by protecting space assets through active measures. The deployment of platforms carrying projectiles for striking targets in space or on Earth, moreover, may characterize future space operations. Space vehicles may also perform a logistics function. Space power concepts, in other words, extend well beyond the gathering and handling of data for the purposes of improving intelligence, early warning, navigation, and communication in support of nonspace forces.

The United States' dominance in high-speed data processors, communications and surveillance technologies (which, again, rely heavily on space elements) confers an unmatched capability to integrate complex information systems. Information may be translated into relatively low-cost alternatives for deterring potential aggressors, strengthening diplomacy and foreign policy and enhancing the nation's military power. The United States may share its military information and enhanced awareness with war-fighting partners or allies to whom it would like to provide security assurance, thus ensuring its position as a natural coalition leader.[32]

Modern warfare revolves around information. It is the basis for seamless and coordinated operations among the United States' services, friends, and allies.[33] Information operations allow enhanced coordination and reliable communications among different forces in separate locations, affecting response times and general operational efficiency. Artificial intelligence combined with databases fuse, process, and tailor information to the needs of the airman in the cockpit, the soldier in the tank, and the shipboard seaman. One challenge will be to use new satellite-based data-dissemination networks, such as the global broadcast system and global command and control system,[34] to deliver only information that the war fighters need to do their job—and to do it with speed and accuracy and in highly readable and intelligible formats. High-level officials have speculated that, in the future, information could be on demand from anywhere in the world, and DoD should provide it, totally protected and assured, "in two seconds."[35]

Air- and space-based electro-optical, infrared, and radar sensors are deployed on platforms such as the high-flying U-2 or SR-71, AWACS (airborne warning and control system) and JSTARS (joint surveillance target attack radar system) aircraft, unmanned aerial vehicles (such as the Predator)[36] as well as satellites. These assets allow friendly forces to "see over the next hill," to view the entire battlefield in its many dimensions.[37]

Navigation aids provide data on the positions of friendly and hostile forces. They enhance the probability that weapons can be placed on targets across great distances. Digital three-dimensional maps provide a representation of the battlefield that may be updated routinely and disseminated to the war fighters to assist in the targeting mission or synchronize actions among all operational units. The May 1999 destruction of the Chinese embassy in Belgrade by GPS-guided bombs, which reportedly used coordinates based on outdated maps, demonstrates that this process has its limits.

New threats may require operational concepts relying heavily on the United States' information superiority. For example, since a potential enemy is likely to have many more missiles than missile launchers, it makes sense to set gunsights on the enemy's mobile transporter-erector launchers (TELs), even after they have launched their missiles. There will be a window of opportunity of only three to five minutes for detecting and then successfully attacking them. The ability to engage these targets obviously requires highly efficient information systems.[38]

Desert Storm also underlined the need for interoperability both among the U.S. services and among U.S. allies. The ability of fighting partners to "talk" to one another during battle will be critical.[39] Integration of space, air, sea, and ground sensors among the services is an expected challenge, and there is a growing need to share surveillance and reconnaissance imagery with coalition partners.[40] The United States spends much more on information technology than its allies, leaving open the possibility that a more technologically advanced United States will some day leave its allies behind. Perhaps even more important than "the technology gap," however, is the gap that may result in the areas of doctrine and operational concepts, resources for modernization, strategy, and threat perception. Although technological equality is not a necessity, an unkind combination of these areas could make a true partnership in war all but impossible.[41]

The vulnerability of the United States' information operations is a source of growing concern. The level of sophistication of information warfare techniques is rising steadily, while defenses against cyber-threats are still in their infancy. The tools of defense are still under development, and rules of cyberwar are still being written, but they will include techniques to ensure data encryption, authentication, data signature, and data integrity. As in other aspects of warfare, successful computer warfare also will require offensive operations to disable an enemy air defense network, shut down city power and phone lines, and feed false information to enemy troops. Cyber-warriors are still struggling to understand the implications of information warfare, including just what the cascading effects on decision making (at home and abroad) will be.[42]

Strategic information warfare puts yet one more twist on America's quest for information assurance. U.S. military and national infrastructures rely heavily on the evolution of cyberspace. The U.S. defense information infrastructure is part of a larger national infrastructure comprising complex management systems and other infrastructures, which are highly dependent on information resources for electric power, finance, oil and gas, telecommunications, emergency services, and air traffic control.[43] The objective of an attack on these infrastructures would be to cause catastrophic system failures in the United States and in allied countries.

It may be expected that the vulnerability of the United States over the coming decades will be greater than that of the potential enemy, making the need to prepare for and counter the threat of strategic information warfare fundamental to American security. Together with other departments and agencies, the Defense Department will be compelled to pursue strategies and programs that ensure the flow of information.[44] The use of space will be an essential part of any risk assessment to determine the vulnerability of the national and defense information infrastructures to information warfare.[45]

Controlling Space

Space is a medium common to all nations, unbounded by political markers or geographic features. As strategic and tactical operations increasingly look to incorporate space functions, the stage is set for what may be regarded as the next natural step in the evolution of military science. In the future, as a prelude to any engagement, there may be contests to control orbits. It is difficult, if not impossible, therefore, to talk about the varied uses of space and the growing importance of space power without at some point addressing the more fundamental requirement of space control.[46]

The well-established operations the Department of Defense must be ready to perform are space support, force enhancement, space control, and force application. Current policy specifically states that the United States will develop, operate, and maintain space-control capabilities to ensure freedom of action in space and to deny such freedom to a potential enemy.[47] Generally, space control addresses who operates in space and in what manner. Without reliable control over that environment, any military campaign that derives its advantage from space will be at greater risk of failure.[48]

The United States considers the space systems of any state to be national property with the right of peaceful passage through space without interference.[49] While freedom of space is the general goal, there are several compelling reasons for developing capabilities to *control,* if not dominate or claim ownership over, space orbits. According to the national military strategy, "It is becoming increasingly important to guarantee access to and use of space as part of joint operations and to protect U.S. interests." With growing national dependence on space systems comes a prerequisite to maintain access to space and to deploy, maintain, and rapidly augment or reconstitute adequate constellations, in part by ensuring that the United States improves its ability to conduct reliable launches.

Space control also requires ground- and space-based capabilities to monitor routinely and accurately satellites and their flight paths. U.S. and allied satellites, or more accurately, their satellite *missions* must be survivable in major conflicts. Satellites may be disabled or destroyed, but as long as the essential jobs get done, what else matters? Given the growing military utility of space for all nations, it is hardly a strain on the imagination to project that future enemies will seek to degrade U.S. satellite system functions. Our enemies know that the United States relies on space for information collection and distribution as well as navigation.

Nor can the country afford to ignore the importance of "denial operations," which may require capabilities to eliminate selected space assets (possibly commercial satellites) used by an enemy. Unimpeded and protected access to space is not the same as being able to prevent its use by an enemy. Although the satellite capabilities of the United States' enemies are at present inferior, as we have seen, a state need not own formidable space forces in order to gain military advantages through the space medium.

Gradually, the idea that space power is as important as sea power and air power is working its way into the minds of those who are accountable for the nation's military successes and failures.[50] The emerging importance of space for military activities and military space threats have highlighted the need to develop capabilities to engage the enemy in space, a current deficiency in the U.S. national security space program.[51] Gen. Howell Estes (ret.), USAF, the former triple-hatted commander of the U.S. Space Command, the U.S. Air Force Space Command, and the North American Aerospace Defense Command, has stated that the United States should be thinking more about the concepts and doctrine needed to conduct operations in space. Space, he indicated in the U.S. Space Command's landmark 1998 publication *Long Range Plan,* must be added to a list of "vital national interests," alongside regions such as Europe and the Persian Gulf, and space should become its own "area of responsibility."[52]

The commercialization of space has complicated the space-control mission.[53] Commercial satellites may be used by an enemy to secure increased operational efficiencies and, indeed, may become fair game in satellite warfare. Privately run space systems (especially telecommunications systems) pose two fundamental problems. First, the task of identifying the right space assets and linking the use of those assets to a hostile party could be daunting. It is expected, in any case, that future adversaries will attempt to cover their tracks by hiding in the commercial constellations. But even if a commercial satellite or ground space component is located, there is still a second problem: Can it be targeted? If the United States is serious about developing capabilities to selectively deny the enemy access to satellites by degrading or destroying them, it may well have to take aim at satellites owned by U.S. or foreign companies or by an international commercial venture group.

This military technology revolution led by space systems affects all nations. If we are undergoing a true transformation in warfare, defense thinkers in other nations are likely to be alive to the growing relevance of space to military victory and to understand that, when the terms of combat shift from a less effective to a more effective approach, those who are slow to adapt will be all the more vulnerable to defeat. They also may come to understand that the supreme advantage of the United States, with some ingenuity and determination, may be turned into a supremely exploitable vulnerability.

How Can the United States Satisfy Future Military Requirements?

Strategic and technological developments, and accompanying changes in doctrine, will offer new, more efficient ways to defend the country. Many of these changes cannot be foreseen. In the year 2010, what positions will Iraq and Iran have in the world? Will Russia have recovered economically, and how will that regime have postured itself vis-à-vis the United States and NATO? What of China, which is expected to have a more modernized military force by 2010? Will North Korea exist? Will the U.S. armed forces be preparing for major encounters in Middle Eastern or North African deserts or in the jungles of South America? Will the United States and NATO continue in a strong alliance that maintains European stability, or will pivotal events in Sir Halford Mackinder's "Heartland" of the world, Eurasia, pose new sea, land, air, and space threats to the United States and its allies across the North Atlantic?

Many changes at the political-strategic level may be expected to occur over the next ten years, influencing the way the United States plans for and conducts wars. The geography that concerns us most may change as new threats emerge in, or disappear from, the land, sea, air, and space environments. Will we find our most worrisome defense challenges in the air, at sea, or in space?

Amid all of this uncertainty, we can look at existing trends, and be sufficiently confident in bureaucratic inertia, to know that many of the weapons we use and the fighting doctrines we rely on are not likely to change significantly over the next ten years. We may find further evidence of future military requirements in the vision statements issued by the individual armed services and by the Joint Chiefs of Staff. One also may look to the *Long Range Plan* and the *Vision for 2020* published by the U.S. Space Command in 1998 to understand at least how space commanders and operators see the role of space in the decades ahead.

Joint Vision 2010 (1996), along with its associated documentation and the recently published *Joint Vision 2020* (2000), are perhaps the most authoritative statements available on future U.S. military requirements. These documents assume that the United States must continue to maintain a superior strategic and operational deterrent capability to counter threats. In the event of war, U.S. forces would strive to achieve an early foothold in a region of strife and be dominant in ensuing battles. "Power projection and overseas presence," the report notes, "will likely remain the fundamental strategic concepts for our future force."[54]

Visionaries on the U.S. Defense Science Board, who completed a study looking at the integration of capabilities to implement *Joint Vision 2010,* see future military requirements revolving around abilities to project military power *within hours* anywhere in the world, to follow this early show of force with more substantial operations within twenty-four hours, and to undertake sustaining operations even in regions where there is limited local infrastructure. This is no mean undertaking!

In the estimation of many on the Defense Science Board, offensive space forces could play a prominent role alongside some terrestrial systems, such as hypersonic cruise missiles, in a U.S. military strategy that calls for a capability to strike enemy targets within an hour. Potential far-term capabilities included the use of a space laser beam in addition to orbital and suborbital munitions. Space assets supporting the early target mission might include an all-weather, all-time, and continuous global surveillance capability. A suborbital space operations vehicle might be used for targeting or reconnaissance.[55]

A number of assumptions pervade current defense thinking concerning the potential contributions of space. Among these are ideas that space can and will play a major role in placing doubt, hesitation, and confusion in the minds of future enemies. The more complex the United States can make warfare in the future, the more it employs superior technologies, the more effective may be its conventional deterrent. The enemy may come to see that he has no hope of overcoming all of these complexities. Space brings to the table greater complications for Iraq, North Korea, and China; in the coming decades, they also will have relatively few resources to engage in the space arena.[56]

While there is much to be said for this approach to deterrence, it does assume that the United States will be facing rational opponents. It further assumes that the enemy will choose to fight on U.S. terms. Neither of these assumptions, however, is a foregone conclusion. History is full of examples of the weak successfully taking on the strong. Chapter 6 provides a fuller discussion of this subject.

By 2010, the United States may be expected to become even more information-dependent for its military successes. Indeed, *Joint Vision 2020* reaffirms the central importance of the information revolution to future military operations. The Joint Staff's vision acknowledges the unqualified importance of assured information or knowledge superiority, of keeping a step or more ahead of potential enemies in the technology race.[57] Defense officials desire to compress the now-lengthy planning and execution cycles into a more continuous and dynamic process, a transformation that will rely on improved information flows to the commander, who must have up-to-date knowledge of the battle situation. New and old space technologies will provide the required enhancements that will permit satellites to play a unique role in each of these operational concepts.

Capabilities for dominant maneuver would allow forces to gain a positional advantage over the enemy by bringing to bear an array of air, land, sea, and space capabilities from widely dispersed locations at a specified time, thereby creating the effects of mass. Precision engagement will require aircraft, ships, and ground platforms using low observable technologies (or stealth) that are responsive, agile, and capable of striking with a high degree of discrimination from increasingly longer ranges. New generations of "fire and forget" weapons will exploit GPS positioning and navigational signals and satellite communications will aid responsive command and control.[58]

Military planners today see a growing need to provide full-dimensional protection to U.S. forces and facilities from adversary attacks while ensuring freedom of action during deployment, maneuver, and engagement. Overhead surveillance and missile-warning space assets will help assure battlespace awareness and alert commanders to imminent attacks and provide information useful for distinguishing friendly and neutral forces from enemy forces. The full range of space assets may be used to provide the battlespace awareness required to deliver timely and tailored logistics packages where they are needed.

Service visions also are indicators of future military requirements. The air force is the Pentagon's executive agent for most space programs. The chief air arm of the U.S. armed services leads the integration of space capabilities into joint military operations. As a leader in this area, the air force has articulated doctrinal guidelines for maintaining U.S. space dominance.[59] The air force vision for the future recognizes that the United States depends upon the ability to both ensure and deny superior intelligence, advanced technology, and precision firepower.

The U.S. Army brought the United States into the space age with its 1958 Jupiter launch, and it continues to this day to demonstrate an avid interest in satellites. Indeed, this oldest U.S. service established its First Space Battalion in Colorado Springs in December 1999. According to Lt. Gen. Edward Anderson III, USA, former commander at the army's Space and Missile Defense Command (SMDC), there is joint interest in designating space an "area of responsibility" and working out the growing interoperability related to space among the services: "As the military uses of space expand in scope and importance, the likelihood of such a struggle can be expected to increase. Already we can clearly foresee improvements in the redundancy and responsiveness of space platforms, the clarity and resolution of sensor information, and the volume and speed of data transmission. The next 20 years may also witness practical means of applying force directly from space, increased use of space for rapid global transport, and the shift of significant portions of battle command and control to space-based platforms."[60]

The "Army after Next" games have established for the army the importance of space for war fighting beyond 2010. The insights gained from these games included: (1) space capabilities were high-value targets and vulnerable to disruption, denial, degradation, deception, and destruction; (2) space and information operations were linked; (3) space provided precision to maneuver, firepower, and logistics; and (4) assured access to space was a priority.[61] With programs such as that developing a kinetic energy antisatellite (ASAT) capability, the army underscores its belief that future success will depend mightily on space control, especially control of foreign imagery satellites. By implication, such army programs also highlight a rather decided unwillingness to rely too heavily on the air force to develop, deploy, and operate weapons required to deny enemy access to space.

Finally, the Department of the Navy is slowly coming around to acknowledging the role space must play in the areas of communications, navigation, and surveillance. The U.S. Navy has always had a healthy regard for satellite reconnaissance and the impact such systems could have on war at sea. The navy's 1998 program guide, *Vision... Presence... Power,* nevertheless gives slim recognition to requirements to defeat adversarial intelligence, surveillance, reconnaissance, command and control, and strike capabilities. Whereas the 1992, 1994, and 1997 vision statements practically ignore the space medium, navy officials acknowledge the important strides made in *Joint Vision 2010.* In *Forward . . . From the Sea,* the navy recognizes the importance of acting jointly with space forces as a requirement for achieving decisive military power.[62]

Over the years, the navy has not exhibited a solid understanding of the strategic meaning of space—incredible, when one considers that the U.S. naval fleet could not function properly without the assistance it currently receives from space. Nor is there sufficient acknowledgment that space is but another war-fighting medium.[63] Concerned about the future of the navy, Vice Adm. Jerry O'Tuttle (ret.) underscored in 1994 that "space is the fourth environment of warfare . . . after the land, sea, and air. As we enter the 21st century, we in the Navy need to focus on charting a course for warfare in this fourth exciting environment."[64] The navy, though readily acknowledging the role space must play, still has significant work to do in the areas of doctrine and strategy.

What Is in the Space Arsenal?

This section will focus on the orbiting segment of the present-day U.S. space arsenal (a term used here to describe all U.S. military space assets), although the reader should be aware that the largest part of a functional space system actually resides on Earth.[65] The ground segment comprises user equipment, including items such as fixed and portable terminals, long-haul fiber optics, command and control stations, data-handling software and operations, support activities, management agencies and organizations, exploitation programs, sensor systems and platforms, and processing nodes and communications links. It should simply be recognized that the ground segment is an extensive and costly part of the space system. A comprehensive review of ground elements in these pages, however, would be an exhausting, unproductive exercise for our purposes and would not contribute materially to the reader's understanding of space power.

This section also will review briefly some of the modernization efforts currently planned. Most of the defense space business revolves around the maintenance and improvement of information-handling systems. This leads one to ask whether we are going to reach, at some point in the next ten years, diminishing marginal returns in the improvements of the military uses of space *as we know them today,* although it may be expected that new uses of old technologies may be found. Barring a revolutionary technological development, such as the invention of a hyperspectral space sensor that can see through jungle foliage or a networked system of optical sensors providing a distributed surveillance capability (which would allow operators to monitor events all over the world on a real-time basis),[66] we ought not to have dramatic expectations for space—at least for information-handling missions.

We are likely to experience incremental improvements in space force enhancement systems, as, for example, new uses for GPS are discovered, communications are improved, and satellite services become more personalized. Stewards of U.S. resources might question whether we ought to invest great sums of money to get that little bit more of a return from existing force enhancement satellites (maintaining that what we have is good enough), to achieve that extra half-meter in resolution in order to find that *nth* Scud launcher. There may not be, in other words, a continuous, sensational growth in space capabilities over the next few decades. Two things, however, may turn this around: a breakthrough in technology or a policy decision to put weapons in space.[67]

No state will match U.S. space dominance over the next ten to twenty years. Foreign capabilities are simply not there in quantity or quality. In the area of information gathering and dissemination from space, the United States may be considered well, perhaps even redundantly, equipped. Notable deficiencies in the arsenal, however, do exist. The following four military space mission areas were established during the 1980s and still provide the country's defense planners a useful topology for classifying significant military space activities.

Space Support Capabilities

Space support activities make it possible to do the three other space missions. Launch is a critical support function. National launches (commercial and national security) are expected to increase dramatically over the next decade. Basic U.S. launch technology, however, is many decades old. While some believe that there are too many different types of boosters and too little standardization, a situation that reduces the degree of flexibility in U.S. launch systems, an argument may be made that diversity prevents reliance on a single system—a potential source of vulnerability. Current launch system capabilities are plagued by delays, beset by

Table 2. Space Support Capabilities

	Current Systems	*Planned Systems*
Launch complexes	• Western Range (Vandenberg) • Eastern Range (Cape Canaveral) • launch support and facilities • range standardization and automation	
Expendable launch vehicles	• Titan 4, 2 • Atlas 3, 2 • Delta 3, 2 • Titan 2 • Taurus • Pegasus • inertial upper stage • Centaur	• evolved expendable launch vehicles
Reusable launch vehicles	• space shuttle	
Satellite control	• U.S. Air Force Satellite Control Network • U.S. Navy Satellite Operations Center	

soaring costs, able to provide only moderate reliability, and unresponsive to changing requirements. Present-day and future military requirements, including the requirement of space control, will increasingly demand prompt and reliable access to space and an ability to reconstitute satellite constellations.

The United States still relies on two old and vulnerable space ports, both of which operate a costly launch infrastructure dedicated to the different boosters. Overall, the armed forces require a launch system that provides accurate, sufficient, predictable, and repeatable performance in operation.[68] The development of the evolved expendable launch vehicle system is intended to meet some of these requirements. Today, NASA and several private companies are investing in the development of reusable launch vehicle technology, technology that could solve many of the problems interfering with current launch operations, although we are unlikely to see a viable RLV for at least a decade. There has not yet been an engineering feasibility demonstration of the cost effectiveness of this technology. The space shuttle, for example, requires 500,000 man hours of processing to prepare for the next launch.[69] We may find in the end that expendables are the most economical (if not operationally desirable) systems for most missions. A flexible, survivable launch infrastructure will be required to maintain and reconstitute satellite constellations in a timely manner.[70]

Force Enhancement Capabilities

At present, U.S. space power is limited to providing data that improves the performance of nonspace forces. There is every indication that the United States will undertake the necessary modernization efforts to continue to expand and refine its ability to gather and transmit information fundamental to future military success.

According to Keith Hall, director of the National Reconnaissance Office, the United States' emphasis on information superiority ultimately requires moving from a reconnaissance architecture to a surveillance-like architecture. This, in turn, will require a greater number of more capable satellites. Currently, NRO is planning for such a surveillance satellite architecture, the Future Imagery Architecture. With such a system, stated Hall, "the number of satellites available for tasking goes way up, and the contention for each asset goes down, allowing significantly greater assuredness that if you task the system you're going to get a response."[71] The U.S. Air Force also is sponsoring the development of new sensor technologies to be the "eyes and ears" in the future environment, including space-based radars.[72]

In the communications area, modernization plans focus on providing advanced extremely high frequency (AEHF) systems for secure and reliable global command and control links using a spacecraft that is 40 percent lighter than *Milstar 2* satellites and having at least five times the capacity. Current Defense Satellite Communication System satellites will receive service life enhancements, while commercial gap-filler satellites are planned to bridge the time between now and the deployment of the advanced wideband system. Requirements for mobile UHF communications and a global broadcast capability will depend on UHF follow-on and commercial satellites. Also expect greater use of commercial mobile communications satellite systems.[73] The GPS constellation and ground systems already are scheduled to receive upgrades to provide increased capability and more enduring versions with the objective being to fend off possible competition from the European Galileo and Russian GLONASS systems and keep GPS the global standard for satellite navigation. U.S. early warning and weather monitoring satellites also eventually will be modernized.[74]

Table 3: Force Enhancement Capabilities

	Current Systems	*Planned Systems*
Reconnaissance	• IMINT (KH and Lacrosse series) • SIGINT (Vortex 2) • ELINT (Magnum, Orion, Trumpet, Mentor series) • MASINT (Nuclear Detonation Detection System)	• Future Imagery Architecture (NRO)
Surveillance, warning, and tracking	• Defense Support Program • Space Surveillance Network	• SBIRS high • SBIRS low
Nuclear detonation detection	• Integrated Operational Nuclear Detection System (GPS satellites)	
Weather	• Defense Meteorological Support Program	• National Polar-Orbiting Operational Environment Satellite System • GEOSAT • GEOSAT follow-on
Communications	• defense satellite communications system • Milstar • UHF follow-on system • gap-filler • leased services	• global broadcast system • advanced EHF • advanced wideband system • commercial satellites
Navigation	• global positioning satellites	• upgraded follow-on systems

Force Application Capabilities

The emergence of new military technologies in foreign lands (e.g., in precision attack systems) would compel greater U.S. consideration of improved surveillance capabilities, more robust active missile defenses, deception operations, and the employment of dispersed long-range forces (with perhaps the greatest degree of dispersion and possibly even protection being gained from forces based in space). Given constant national security objectives, a radical elimination of major parts of the U.S. overseas basing structure (e.g., from Europe, Japan, and Korea) would increase dramatically the value Americans place on intercontinental conventional land and air forces, mobile basing at sea, and the uses made of space.[75]

Current military and political requirements, budgetary and programmatic decisions, and the United States' continued reliance on basing options have led defense planners increasingly to place emphasis on (1) long-range strike options (ICBMs and strategic bombers), (2) stand-off attack (to improve the protection of forces), (3) rapid execution of the mission, and (4) precision targeting. Space weapons might help meet each of these requirements—indeed, they may possess operational advantages allowing them to outperform most conceivable terrestrial-based options. According to Ivan Bekey, who drafted an engaging paper on technologies for force projection from space, "Satellites present a presence over battle areas that is difficult to deny, and do so repeatedly and frequently enough from LEO, or continuously from GEO, so that force application using them could have a marked strategic as well as tactical effectiveness on the conduct and outcome of conflicts. This force can be applied anywhere rapidly, with minimal risk to U.S. forces, and at all levels of conflict. It is equivalent to artillery and strike support with infinite range and moving at 25,000 mph., with the added advantage of enjoying complete surprise."[76]

It is conceivable, in other words, that the space-based force application option may be even more advantageous than the conventionally armed ICBMs and intercontinental hypersonic aerodynamic options, including long-range cruise missiles, currently being explored by the air force, some defense agencies, and NASA.[77] In the eyes of some analysts, these terrestrial weapons, flying at supersonic (Mach 1 to Mach 5) or hypersonic speeds (Mach 5 to Mach 25), would "make the entire world vulnerable to attack from any point and completely reshape the geography of surface warfare." Some Pentagon officials are speculating that the next U.S. strategic bomber, the follow-on to the B-2, may need to be totally different, suggesting that they may be hypersonic or suborbital vehicles.[78]

But all the attention going to land-based options for destroying targets from a long distance with conventional explosives or kinetic energy impact begs the question: If we are willing to invest great sums in the development of structures, materials, propulsion, guidance and controls for hypersonic cruise missiles, why not explore the feasibility of doing the same mission from space, an environment that might allow dramatic improvements in overall military and strategic efficiency?

Hermann Oberth, the German rocket theorist, fantasized during the 1920s about using huge orbital mirrors in space to focus the Sun's rays into a lethal beam of energy to destroy Earth targets—one of the earliest concept designs for a space weapon. Since that time, history records occasional references to other conceptual space-based weapon platforms, including German scientist Eugene Sänger's "antipodal bomber," RAND's 1946 concept for a "satellite missile," the U.S. Air Force's designs from the 1950s and 1960s for a nuclear-armed bombardment satellite, the bomber-missile, the DYNASOAR bomber, and a hypersonic aircraft that would take off from the ground and rocket into orbit. During the 1960s, the Soviets developed a so-called fractional orbital bombardment system (FOBS), a nuclear warhead placed briefly in orbit and brought down from a low altitude upon an enemy target before the completion of one full revolution around the earth. Besides developing and testing a co-orbital antisatellite weapon, in 1974, Moscow reportedly mounted a "machine gun" or cannon on a military space station, which might have been used against an Apollo capsule or space shuttle (then under development) in retaliation for on-orbit spying.[79]

Technical hurdles (especially in the directed energy programs) and long-running associated political problems are major reasons the United States shies away from considering the tactical and strategic utility of space weapons. But these things are also true: politics change, new threats will emerge, and technological advances will be made. The ability to reach targets quickly and overcome "the tyranny of distance" would enable near-real-time command, control, and strike capabilities by shortening the time it would take to observe a target, orient attack forces to the target, decide to attack, and finally attack the target. Such capabilities also could improve U.S. nonnuclear deterrence.[80]

Over the years, the air force has considered a number of space plane designs. More recently, the United States' "aerospace" service has invested in the development of technologies to build a suborbital space operations vehicle (SOV), or a vehicle that could extend its range by exploiting the effects of lift in the atmosphere as well as venture into the exoatmosphere, and a more genuine spaceship design, the space maneuver vehicle (SMV). Using what is called a common aerovehicle to deliver the lethal and nonlethal power, an SOV would be capable of striking targets anywhere on Earth in less than sixty minutes. A space maneuver vehicle may get its ride into space aboard the SOV, where it would conduct space asset protection, satellite resupply, ballistic missile defense, or strike missions in space or from space into the air, land, or sea environments. The SMV also could be orbited over any point on Earth in less than sixty minutes and provide near-real-time reconnaissance capability.[81]

One should also recognize that space weapons do not need to be deployed in space, on call at all times. Assuming flexible, reliable, and rapid launch capabilities, force-application (and force enhancing) assets may be launched during times of crisis to demonstrate resolve and contribute to overall military preparedness should an engagement prove necessary.[82]

Space weapons might facilitate lightening strikes against WMD storage and production facilities and associated launch platforms and wreak havoc against the bases of terrorism. Indeed, most military targets—fixed or mobile, land-based, sea-based, or even airborne and orbiting—would be held at risk by space weapons. Space-based weapons may be one means available to future defense planners to defeat "hard and deeply buried targets," a mission directed in the 1999 Defense Planning Guidance.[83] In light of these advantages, the Defense Science Board has recommended initiation of a demonstration program to show the feasibility of highly precise, hypervelocity reentry of long slender and short rods made of heavy material into the atmosphere from space.[84] These guided or unguided weapons may be launched from space planes or satellites.[85] To be sure, questions surrounding the utility and advantages of space strike weapons to U.S. national security deserve extensive analysis to determine the full range of political, diplomatic, military, and economic implications associated with their deployment.

Speed-of-light, or directed energy, weapons also hold out intriguing possibilities. Much of this work has been conducted in national laboratories under the auspices of Strategic Defense Initiative (1985–93), Ballistic Missile Defense Organization (BMDO), and U.S. Air Force programs. Today, speed-of-light weapons are concepts that remain on laboratory drawing boards, in part owing to the high costs and technical challenges associated with placing in Earth orbit very heavy systems that, depending on the target, would quickly use up allotted energy in just a few shots. The problem of refueling a laser weapon in space is a unique one, although autonomous space rendezvous operations for refueling are well within the state-of-the-art. Long-duration nuclear reactors offer another possible source of fuel. As the costs of reaching space decrease, however, and the United States refines its ability to jump in and out of orbit routinely, lasers in space may be more practical.

Space-based interceptors could be a critical part of effective defenses against future tactical and long-range ballistic missile threats and serve as the backbone of a future ASAT capability. Such interceptors may use kinetic energy (such as the "Brilliant Pebbles" interceptors once researched as part of SDI) or directed energy system to destroy in-flight theater ballistic missiles soon after launch, like the space-based laser (SBL) concept currently under exploration. Today, there are no plans to go beyond treaty-compliant research into advanced technologies for space-based BMD systems.

Depending upon the availability of a power source, a number of directed-energy weapons already explored may one day have military utility.[86] Lasers may sufficiently heat an ICBM or tactical ballistic missile booster to cause the missile to lose its structural integrity in flight. Particle beam or high-power electromagnetic (HPM) weapons might be directed against other space or terrestrial objects to disrupt or destroy the target internally. HPM weapons could blow out, jam, spoof, and disrupt electronic equipment on Earth or in space as well as disseminate disinformation.[87] Lasers also may be used to blind the electro-optical sensors of intelligence-gathering platforms in all geographic environments.[88]

Today the United States does not deploy space weapons for power projection or missile defense. A few visionaries in the defense community in and out of government argue that the energy and mass that may be delivered from space could yield great tactical and strategic benefits on Earth. For political and technical reasons, and secondarily because of the high cost associated with some of the force applications systems mentioned above (especially high-energy lasers), one ought not to expect that such capabilities will be a part of the U.S. arsenal anytime in the near future.

Space-Control Capabilities

The practical implications of space control, as a critical mission of U.S. armed forces, are several. A wide range of approaches to achieving space control, political and military, is desirable and possible. Indeed, in practically any endeavor, it is preferable to have more than one approach.

A detailed intelligence picture of space activities is a fundamental prerequisite of space control. By definition, a satellite threat is a global threat. Space surveillance assets, therefore, must attempt to cover a vast area, much of it dark. In order for the space-control mission to succeed, system operators must be informed of threats and other activities in space that may impact space operations. They must be able to differentiate between a returning space object and a reentering ballistic missile warhead.

The growing number of identifiable objects in space severely complicates the task of detecting, identifying, and tracking orbiting objects. More than nine thousand detectable manmade items are in Earth orbit today, less than 7 percent of them active satellites. The remaining objects are space debris, much of which will burn up upon reentry into Earth's atmosphere.[89] The existing debris includes exploded rocket boosters, pieces of junk formed by the ASAT activities of the 1970s and 1980s, and disintegrated satellites, among other things.[90] Bits of material from discarded boosters and inactive or broken up satellites may strike satellites or space shuttles and cause severe, even life-threatening damage to a spacecraft. Every now and then, there will be an incident involving space junk that brings the issue to the fore. Debris apparently wrecked a Soviet satellite in 1978 and again in 1981. More recently, in a most dramatic and unprecedented incident, a French military satellite was permanently damaged when it struck a ten-year-old, burned-out French Ariane rocket stage!

For these reasons, surveillance of space is a daunting task. At present, the United States operates the Space Surveillance Network (SSN), a collection of dedicated and secondary optical and radar (deep-space and near-Earth) sensors that can track manmade objects up to geosynchronous orbit. The SSN does not monitor objects at all times, but rather "spot checks" them by using a predictive technique. The U.S. Space Command's Space Control Center, located deep inside Cheyenne Mountain in Colorado Springs, Colorado, receives all SSN information, including orbital data on geostationary communications satellites from the ground-based electro-optical deep space surveillance (GEODSS) sites, located at Socorro, New Mexico; Maui, Hawaii; and Diego Garcia, in the British Indian Ocean Territories. These telescopes operate only at night and reportedly see objects ten thousand times dimmer than the human eye can detect and as small as a basketball twenty thousand miles high.[91]

Conventional radars use immobile detection and tracking antennas to transmit radar energy to space and capture the reflected energy. A narrow beam is then sent up to the space object, the beam then tracks it to provide orbital data. Phased array radars located in Massachusetts, Greenland (Thule), United Kingdom (Fylingdales), California, Alaska, and Texas detect and provide early warning of ballistic missiles in flight. Yet the ability of these radars to scan the horizon reaches well into space, making them a secondary source of intelligence. Major U.S. Space Command detection and tracking radars are located in Alaska, Turkey, the Ascension Islands, and Antigua. There are also electro-optical and radar sensors and laser trackers (such as the Starfire Optical Range located on Kirtland Air Force Base in New Mexico) not owned by Space Command but operated by services, agencies, or laboratories that contribute to the space surveillance and cataloguing mission. The combined powers of all SSN sensors permit more than eighty thousand satellite observations to be made each day.[92] A future space surveillance architecture may include space-based optical or radar sensors that are part a system capable of monitoring more objects for longer periods of time, allowing timely intelligence of objects of high interest.

Space control also involves protection of the satellite system, measures that must be taken in order to ensure the survival of critical satellite missions. This subject will be addressed more fully in the next chapter. For now, the reader need only be aware that growing U.S. dependence on space for military and economic activities will create a center of gravity in space—and centers of gravity are attractive targets of extraordinary opportunity for military adversaries.

The United States relies on passive measures for defense of its satellites (e.g., hardening and shielding against radiation, maneuvering capabilities to avoid attacks, stealth technologies to avoid detection) and of the satellite links (for example, data encryption, antijam technologies). Many of the commercial satellites the United States will rely on in the future are not likely to employ such passive defense technologies. Active measures may be employed to defend ground sites and facilities. Technology development efforts currently are underway to develop satellite sensors for detecting, identifying, characterizing, and reporting radio frequency and laser interference with spacecraft operations.[93] Detection and reporting on threats to satellites is one area in which improvements are required, for current systems lack necessary sensors that would help characterize anomalies to isolate manmade threats.[94] Protection of satellite missions also will demand attention to capabilities to reconstitute and repair space assets in a timely manner.

Negation is another key space-control operation. While the United States has developed a number of technologies critical to the development of an antisatellite system, it currently does not deploy systems to do physical harm to an enemy's satellites. Pentagon officials have underscored that physical destruction of satellites is "not the preferred approach" because they could undercut U.S. commercial space interests, which depend on global cooperation. For these reasons, "terrestrial negation" may be "more consistent with long-term American interests." Hence, the Clinton Pentagon aimed only to achieve what it called "tactical denial" of an enemy's space capabilities, that is, denial through jamming and other short-lived disruptions.[95]

Countersatellite capabilities might include nuclear, directed energy, or kinetic energy ASATs. In 1959 the United States conducted the world's first ASAT test, called Project Bold Orion, when it launched a missile from a B-47 to intercept the *Explorer 6* satellite (the missile came within four miles of the satellite). Interest in air-launched ASATs quickly turned to ground-launched possibilities. The United States deployed for a short time in the 1960s and 1970s nuclear-tipped Thor missiles on Johnston Atoll in the Pacific as part of a ground-based ASAT system intended to destroy Soviet satellites.

Concerns about maintaining the integrity of the Partial Test Ban Treaty (which bans atmospheric nuclear explosions) and the 1967 Outer Space Treaty (which bans deployment of nuclear weapons in space), along with a growing understanding that nuclear detonations in space would hit only small numbers of targets and would not discriminate between Soviet and U.S. satellites, convinced planners by 1970 that such ASAT tactics should be abandoned (although "Program 437" remained in place until 1975). Nuclear ASAT weapons were very unlikely to be used short of dire emergencies, and legitimate doubts arose about their overall utility to the nation.[96]

The United States also briefly experimented with an air-launched, nonnuclear, kinetic energy ASAT. In the late 1970s, the Pentagon sponsored a program to use miniature homing vehicle technology to guide a hit-to-kill ASAT weapon launched from an airborne F-15 platform. Threats posed by Soviet reconnaissance satellites to the U.S. Navy prompted the decision to invest more than $1.5 billion in this program. The Pentagon canceled it in 1988, however, due to diminished funding and a congressional ban on testing ASATs in space.

Today, the U.S. Army has a small program to develop a ground-launched kinetic energy antisatellite (or KEASAT) weapon, which would destroy a satellite by impact. The KEASAT successfully completed a hover test in October 1997, when its sensor acquired, locked on, and remained locked on to a simulated moving target. Funding for this program was halted in 1998 by President Clinton's controversial use of the line-item veto, an executive decision later overturned by the Supreme Court as unconstitutional. Funds to KEASAT have been reinstated, and low-level research and development by the Boeing Company continues.[97] However, the precarious political support evidenced today for the ASAT mission on Capitol Hill, the lack of support in the administration, and the meager funding for this program create considerable doubt as to whether KEASAT, or a similar system, will be fielded anytime soon.

There are latent ASAT capabilities in a few other weapon systems deployed or currently under development in the United States. In October 1997 the U.S. Army tested the ground-based midwave infrared chemical laser (MIRACL) against an air force Miniature Sensor Technology Integration-3 (MSTI-3) satellite at the White Sands high energy laser static testing facility in New Mexico. The army and TRW developed MIRACL in the 1980s as part of the research for the Strategic Defense Initiative. The test involved firing the laser to illuminate, or "lase," the dying satellite. According to the Department of Defense, the primary purpose of this test was not to destroy the satellite, but to compare the data collected with computer models that are used to develop methods and technologies for protecting U.S. satellites from such damaging outside interference. Since lasers are available worldwide, defense planners strongly argued that it behooved the United States to identify critical vulnerabilities in U.S. satellite sensors and then develop appropriate passive defenses and operational countermeasures.[98] Although it was never advertised, MIRACL is said to have had since the mid-1980s a contingency mission to negate hostile satellites.[99]

The theater missile defense systems capable of exoatmospheric interceptions currently under development, such as the army's Theater High Altitude and Area

Defense (THAAD), the Navy Theater Wide, and the National Missile Defense system, also would have latent capabilities to hit satellites in low Earth orbit. Most satellites follow a predictable orbital path, making their interception less of a challenge. While still an experimental program, the air force is developing technologies for an airborne laser (ABL), which could be used to strike enemy ballistic missiles just after launch. While not its primary or even secondary mission, ABL may also be tasked to patrol the skies to find and kill hostile satellites in low Earth orbit during times of crisis.[100] Some technologies that go into the ABL also may benefit the development of a space-based laser, which might also have an ASAT mission.[101] I am unaware, however, that any concept of operations or doctrine for employing currently planned BMD systems has been written to fortify the space-control mission.[102]

There are also political avenues that may be taken separately or in conjunction with military efforts to achieve space control. The United States could use export controls and international cooperation to discourage a state's or a foreign company's commitment to, and investment in, space, thereby disrupting or undermining a nascent or maturing space capability in a rival state. Bilateral or multilateral arms-control agreements also might succeed in reducing, prohibiting, or geographically restricting space denial weapons. An emphasis on a political remedy would *require* a consensus among key national leaders upon long- and short-term space-control strategies. There must be unity of purpose in defense and foreign policy planning circles for denying potential enemies certain space capabilities. Because these political avenues necessarily require years to execute before there is a tangible result, the executors of policy will have to have the perseverance to see their political space-control strategy through over the long term.

Technology control may be effective in the short run. Restrictions on high-resolution space-based imagery are gone forever. The proverbial genie is out of his bottle. Nevertheless, there are technology restrictions that may help to preserve the U.S. technological advantage over the next few years (it is folly to assume that the spread of technology can be stopped over the long term). Equipment and software useful to satellite imagery interpreters, for example, is one class of technology that may be restricted, although even this equipment is increasingly available commercially.

Negotiations with potential competitors in space is another political avenue. Reaching agreement with commercial imagery firms or foreign governments on digital degrading, delay in product delivery, or greater government control over what gets "imaged" are within range of the possible. The United States may even take steps unilaterally to deny imagery to an enemy. For instance, by saturating imagery companies with well-timed orders for its product, the United States may block other countries out of the market and effectively seal off their access to some space products (although one may expect that even here the market will respond accordingly).

The political avenues suggested above have a rather checkered performance history and ought not be relied upon solely. Neither arms control nor export controls nor international cooperation should be expected to yield more than limited, short-term benefits to the United States. By delaying or preventing the spread of technologies, these arrangements may be marginally useful as part of a larger space-control strategy that includes active and passive military measures. Programs involving the development of space combat weapons, however, have never been fully supported in the United States and remain politically controversial.

What Obstacles Lie before U.S. Space Power?

The United States has made great strides in space over the past forty years. In the 1950s, the country jumped into space. In the 1960s, U.S. satellites revolutionized telecommunications. The 1970s witnessed remarkable improvements in space imagery, and the 1980s brought about a revolution in navigation and geographic positioning. The 1990s, shaped from the outset by the world's first space conflict (the 1991 Persian Gulf War), may become known as the decade when interoperability and user-friendly space systems became the watchword.

Significant progress has been made since 1991 in smoothing out the process of getting information collected from space down to the tactical level, although much work still remains.[103] The defense leadership also has made remarkable strides in educating the services about what space has to offer, to integrate space into training, war gaming, and tactical operations, and to convince space operators that their "wings" and "squadrons" are real war-fighting units. What additional challenges now lie before the country?

National Policy

National policy is the subject of part 3 of this book, where it will be addressed comprehensively. Let us simply state here that there are outstanding fundamental questions surrounding the development of U.S. space strategy and doctrine and investments in a force structure that would allow engagement with the enemy in and from space. Senator Robert Smith of New Hampshire acknowledged in November 1998 some of the implications of a refusal to recognize space as a combat medium: "If we limit our approach to space to just information superiority, we will not have fully utilized spacepower."[104]

Indeed, other than the characteristics of the medium itself, there is no difference, certainly no difference that would impact military strategy, between warfare in space and warfare in the other geographic environments. There is a tendency in the policy world, however, to think that space is different. A working assumption of Senator Smith was that space power may provide faster, better, and cheaper offense and defense. This would be truly revolutionary. If he is right, it may be argued that neglecting the development of the policies, doctrines, and enabling tools of space power would amount to an act of high irresponsibility.

No Space Advocate

Perhaps one of the thorniest problems the country faces in the next decade is the decision on the establishment of a space force, or even a Department of Space. The U.S. Air Force became the Defense Department's executive agent for space in 1961, yet one does not need to search far and wide to find someone, in or out of government, who has a grievance against the U.S. Air Force for its lack to leadership in the space arena. Many are quick to observe, including the former commander in chief of U.S. Space Command, General Estes, that the air force has been slow to integrate the air and space missions, largely because of the service's traditional image of itself, but also because of the high costs associated with space operations. Indeed, constricted budgets, immature space weapon technologies, and an unsupportive national policy (current written policies are severely undermined by vigorous political opposition to the development of space-control and space-weapon technologies). According to Estes, in the far term, a new space service may be appropriate and desirable.[105]

The air force is still struggling to figure out how space fits into the service mission, as the recent reversion back to the use of the term "aerospace" force might suggest. "Aerospace" is a curious term, invented in the late 1950s by air force chief of staff Gen. Thomas D. White, that lumps together two vastly different environments. This idea of an indivisible medium is not one, however, that works well in existing air force doctrines, which clearly separate operational air from space forces. While the current air force leadership is trying to prevent a stark split in its own organization between "air" and "space" forces, admittedly a bureaucratic (and budgetary) nightmare, it is clear that the dire need for a national space power advocate to fight the resource battles and ensure America's preeminence in this area may well force this issue. When asked whether the United States needs a space force, retired USAF general Bernard Schriever paused thoughtfully for a few seconds, then responded, "In the past, I have said no. But I'm tempted to change my mind on that. I don't think we have a service today that is really fighting the issues. We have to make an adjustment from once having the oceans to protect us to a situation today where they don't mean anything as far as a major war is concerned. We [all three services] continue to spend most of our money in traditional areas."[106]

Clearly, the United States is not exploiting space nearly as well as it could. With evidence piling up that the air force and the other services cannot fight without space, the air force views space primarily performing a support function. Yet none of the services has made the philosophical jump to ask how they would fight a war in a fundamentally different way. This is a doctrinal question that goes back to asking the fundamental questions: How do you think about air power? How do you think about sea power? Indeed, how do you think about space power? There are many questions that will arise as the country struggles to implement *Joint Vision 2010* and the more recent *Joint Vision 2020,* documents that reinforce the need to work jointly and to achieve information superiority, but nonetheless documents that some might argue are still too traditional in scope and approach to warfare.[107] Questions about organization, roles, missions, and, basically, "who's in charge" inevitably will arise. Until these issues are sorted out, the United States cannot hope to institute the changes required to mature as a space power.

Strategic Thinking

Being rather young in the age of space, it goes without saying that much more needs to be done concerning how we think about the strategic uses of space and the strategic implications of fighting in space. Modern-day war gaming strives to achieve such an understanding, but there is even here a temptation to look primarily at operations, doctrine, and employment of futuristic systems rather than examine the larger strategic and political picture. Some have noted a tendency among wargamers and analysts to consider using space weapons like current weapon systems without thinking about new operational concepts or doctrines.[108] How might a system such as SBL, for example, be used in a joint operational doctrine, not just for the purposes of "strategic" strike, but to leverage the total offensive force (not only against enemy missiles, but also satellites and aircraft)[109] and work with terrestrial systems to increase overall efficiency?

There is a self-fulfilling prophesy at work here. At present, consideration of these matters at the military level is rather constrained. By and large, the services tend only to think as far as the information support they receive, for example, from the Milstar satellites, GPS, and space-based intelligence assets. The commanders and operators understand that they do not have a supportive policy or the capabilities to undertake operations in and from space. So why think about it? Point well taken. If space is simply going to be free terrain, but nontraversable and unavailable for combat missions, then one does run into an *intellectual limit* on doctrinal and strategic thinking. Not too much thought has been given to these questions in the military outside the space-interested organizations or in the higher counsels of government, and in part this is due to the fundamental immaturity of the country's defense space policy.

Resources

A frustrating factor for many who are interested in expanding American investment in space is resources. There is a limited pie, and these days every department and agency is being asked to do more with less. There are problems associated with allocating existing money not only to the services in general but also within each service. Generally speaking, the dollars still go for the traditional items (tanks, destroyers, aircraft). Fundamentally, according to the air force chief scientist, the air force science and technology budget has not been oriented toward space, and whatever is spent is done so without the benefit of strategic (and, I will hasten to add, policy) guidance.[110]

Until we change that paradigm, the budget will remain a source of frustration. The leadership in the air force and the country at large must be convinced of the critical importance of space, that space capabilities are as important as maintaining the nation's tactical fighter capability, if a paradigm shift is to be achieved. Ultimately, the country requires resources to deal with the issues of paramount concern, such as an adversary's ability to counter U.S. space capabilities in the future, or the U.S. ability to maintain its leadership in the space environment. Clearly, an investment plan is needed to transition some military capabilities to space, while, in other areas it may make sense to look to the commercial sector. A politically blessed and holistic understanding of U.S. space power would be an instrumental resource allocation guide.

What Is the Bottom Line?

Discussion of future national security requirements gets off on the wrong foot unless it begins with a consideration of space power. In chapter 1 we established that space power may be defined as the competitive use of space for national purposes. "Competitive" is a key word here because it assumes a struggle for advantage, that there are other "players" over which to exert power. There is an analogy here with wrestling—a wrestler is out of his element and pointless if he stands alone. The concept of power implies an opponent.

As an emerging military concept, space power promises to alter the terms upon which future wars will be waged. The much publicized evolution of information warfare, a method of warfare that depends critically upon space for efficiency and effectiveness, has dominated the thinking of analysts and represents but one expression of space power. As a part of the Revolution in Military Affairs, information operations are often characterized by the enhanced quantity and quality of information that may be brought to National Command Authorities and military forces at all levels. Much of the analysis and discussion of information warfare, however, scrutinizes primarily the technology components and emerging techniques, downplaying or failing to consider the policy, strategic and war-fighting issues at hand.

Whether Desert Storm represented the first true space war in history is a frequently asked question that misses a larger lesson. There are good arguments on both sides of this debate about that 1991 conflict. On the one hand there was no enemy to contest the coalition's use of space. On the other hand, space greatly enhanced coalition success and, after all, the United States did engage in active space-control operations to deny Saddam access to satellite telecommunications. What is most important to recognize is that we are in the earliest stages in the now-brief history of space power, when space assets are serving an increasingly important supporting role in the way modern armed forces conduct their business.

Space power is today in a similar position to air power in 1914, when aircraft were undertaking only reconnaissance missions and gradually becoming an indispensable adjunct to regular forces.[111] What we are seeing today is the progressive exertion of influence of the satellite on the terms of deterrence and combat, to the point where space has become indispensable to the timely and reliable delivery of information of military value.

Although we cannot know now the totality of the strategic value space will offer the armed forces of the future, the experience of Desert Storm tells us that we would be remiss if we did not see that space power has the potential, when compared to other types of military power, to be the biggest military contributor to victory in the years to come. While startling new developments in information-handling systems are likely to be few over the next decade, satellites nevertheless are likely to offer the most significant return on defense investments, for the United States *and* its enemies.

PART 2 ▸ IN THE ARENA

Part 2 of this book is a "reality check," a no-holds-barred examination of the external limitations and constraints that may be placed on U.S. space activities described in part 1. In defense language, this portion of the book addresses the broad and sometimes ambiguous subject of space threats. Objectivity in this analysis is crucial, and I have made every effort to deal evenhandedly with the facts, all of which have surfaced in the public domain. Although the conclusions offered in this part will suffer somewhat *in the details* from not having had a window into the classified world of the national intelligence estimates, available public information is good enough to help us arrive at useful findings with respect to the national space vision and defense policy.

The new international security environment demands that policy makers and defense planners come to grips with a basic question: What is a space threat? United States' leaders will engage the public on security and space subjects through an understanding of the threat. Yet there are several reasons why getting our arms around this subject might not seem as easy as it looks, four of which the reader should consider.

The space threat is complex. In trying to understand present and future dangers involving the space environment, it is important to recognize that a foe of the United States could use space in two very different ways. An enemy could exploit, first, vulnerabilities in U.S. satellite systems. There are proliferating high- and low-technology antisatellite capabilities, from information warfare to direct-ascent ASAT weapons to nuclear-generated electromagnetic pulse (EMP). We can thus, according to the U.S. Joint Chiefs of Staff, "expect some adversaries in 2010 to have the ability to attack low-earth-orbiting satellites."[1]

Second, a foe could use space to further its strategic and military objectives by exploiting information derived from satellites. The number of dedicated military satellites is growing. Commercial satellite services and markets are expanding. Some satellite services (such as the signals from the global positioning system satellites, some communications satellites, and the Internet) are universally available and relatively inexpensive or even free.

Today's space threats are immature and sparse. The profile of threats to U.S. space systems and threats from foreign space systems is relatively low—so low, in fact, that there is no consensus regarding the gravity of current and projected dangers. As a result, little attention has been paid to implications of hostilities involving space operations in times of peace and war. Concepts for space warfare also are in the early phases of development in the United States and abroad. The conclusions presented in part 2 will underscore that as more satellites are placed in orbit, as more military missions move into space, and as the United States and its allies grow increasingly dependent on satellites to perform functions vital to national existence and the creation of wealth, the frequency and intensity of specific space threats will change over time. For these reasons, it may be argued that now is a time when heightened awareness, rather than complacency, is called for.

The space threat is evasive. Space threats lack a strong definition in the minds of policy makers. Aside from the explosion of nuclear weapons in space, there are no "scary" weapons to command the attention of U.S. leaders. Multi-use and commercially available technologies make most space activities possible. There are, however, no Scud-like weapons to provide a dramatic visual effect or to drive home the true menace behind threats to the United States' ability to exploit space or the enemy's ability to use space against the United States. Moreover, many countries have access to space technologies, meaning that there is no firm basis for defining a "rogue nation."

Space threats are viewed generally to be nonlethal. Satellites are generally seen as conduits for information, not vessels for people. The odd exception is the use of the space shuttle and, possibly one day, the *International Space Station.* Since modern-day satellite operations deliver information, not firepower, the fact that satellites contribute significantly to terrestrial warfare may be lost on most people. Unless one is careful to draw this linkage, one may come to the erroneous conclusion that the loss of satellites will not result in the loss of blood, land, or treasure.

The chapters that follow will explore the most significant aspects of this subject matter. Chapter 4 looks intently at the possibility that U.S. access to space could be denied by a resourceful enemy. There are some steps, or countermeasures, the United States can and should take to better prepare for a more perilous future in space.

Chapter 5 examines the implications of space having become an open arena, a medium exploited by an increasing number of countries. The proliferation of space technologies offers foreign governments and nonstate entities unparalleled opportunities to enhance their diplomatic and military influence over the United States and to strike with strategic effect. Foreign and commercial space capabilities present unique problems to defense planners. It is in the interest of the policymaker, the defense planner, and the commander to ensure that an adequate number of reasonably good defense options exists to cope in a politically dynamic and militarily diverse security environment.

And finally, chapter 6 explores the importance of understanding the political and strategic dimensions of the space threat. Future enemies might not fight agreeably on terms dictated to them by the United States, striving instead to achieve a military victory through indirect or unconventional means. Military science may be expected to highlight to foreign defense leaderships that space, as a new strategic frontier, could offer a rather promising path to victory over superior U.S. armed forces. U.S. officials responsible for planning and executing tomorrow's wars must recognize that there are new opportunities inherent in space and counterspace operations for future adversaries to undermine or at least frustrate U.S. strategic and military objectives.

4

Survival in the Twenty-first Century

Is There a Credible Threat to U.S. Space Systems?

Simple mathematics can demonstrate what otherwise might be clouded by more sophisticated defense analysis. Although not nearly the final word on this subject, numbers do talk. The space budgets of NASA and the Department of Defense have gone from 1959 levels of $1.266 billion and $2.377 billion, respectively, to 1998 levels of $12.321 billion and $12.359 billion (fiscal year 1998 constant dollars).[1] In July 2000, the United States had a total of 741 operational and inactive satellites in orbit.[2] New remote-sensing, scientific, meteorological, and dedicated military satellites also will swarm the skies. Orbital traffic, in other words, is expected to become heavy.

Ten years hence and beyond, when the United States will have larger and more intricate space architectures in low, mid-, and geosynchronous Earth orbits, reliance on satellites may be expected to increase by one or two orders of magnitude. The steady rise in the number of space systems necessarily increases opportunities for accidental failures or intentional disruptions, and it raises fundamental issues pertaining to satellite system frailty and vulnerability. The law of averages will ensure adversity a place in space.

The United States' expanding, boundless trust in space-based assets to perform a full spectrum of military, civil, scientific, and commercial activities parallels its growing inability to act on Earth without them. This dependency on the orbital engines of the modern information revolution may even rival the country's twentieth-century industrial reliance on electricity and oil. Although hailed today as the backbone of national economic security and crux of its military power, the elevated exposure of U.S. satellite constellations to the malignant workings of misfortune or the malice of future enemies is a condition that should light a fire of concern in all Americans.

Chance and animosity are continually at work, so that even under the best conditions, the United States cannot always be guaranteed access to space. Indeed, in the past fifteen years, some eye-opening experiences have shown how tenuous its hold on space actually can be. Incidents dating back to the mid-1980s, moreover, are most instructive to the defense planner, who, in the absence of any guarantee that the country's space systems will remain unmolested indefinitely, may best be counseled to take into account the consequences of failure or interruption.

An unfortunate string of launch failures in 1985 and 1986 highlighted the folly of (1) taking anything having to do with space for granted and (2) limiting one's options unnecessarily. For much of 1986 the United States lost a major space capability—its ability to place heavy objects, major defense space systems, into orbit. Most will remember the January 1986 catastrophic and tragic loss of the space shuttle *Challenger.* This was a painful national experience in more than one way. Until the loss of *STS-51L,* the national space transportation strategy featured a steadily increasing reliance on the semireusable space shuttle at the expense of expendable boosters. Indeed, based on the shuttle's success record, the wheels were set in motion to stop production and even dismantle existing expendable launch vehicles. The country was in the process of consolidating all of its eggs into one basket—and then the basket fell . . . and the country lost one-fourth of its shuttle orbiter fleet.

After the *Challenger* disaster, a statistically predicted event, the future of the space shuttle program was uncertain. The now-three-orbiter fleet was deactivated for an indefinite period of time. Both the Pentagon and NASA turned to McDonnell Douglas to restart the Delta expendable launch vehicle production. But it takes time to reorganize specialized production teams comprised of highly experienced professionals. McDonnell Douglas indicated that it would be unable to deliver new launchers for at least eighteen months, while administration officials at the time estimated that the waiting period would be two to three years.[3] The air force also considered increasing Titan 34D production, which, from beginning to end, is a two-year process.[4]

Not only did the country have a frightening booster shortage and serious payload backup problem developing from the radically diminished number of rides into orbit, but before 1986 was over, a series of expendable launch vehicle failures lifted the veil on a nagging secret the United States had been keeping from itself—the country did not have reliable access to space. In August 1985, a Titan 34D booster carrying a KH-11 reconnaissance satellite failed to reach orbit. In April 1986, another Titan 34D exploded 8.5 seconds after launch—it too was carrying an intelligence satellite. When in May of that same year, a Delta 2 rocket exploded in flight, the failure rate, three out five launch attempts, began to take on dire significance for national security.

To make matters worse, France's Ariane booster, which U.S. officials considered using on an emergency basis to launch certain critical payloads into orbit, also suffered failures in September 1985 and May 1986.[5] Indeed, the incidents of Western launcher failure may be extended in time if one were to include cases involving the United States' remaining heavy-lift vehicle, the Atlas/Centaur. Its upper stage failed to boost a communications satellite into geosynchronous orbit in June 1984. And in March 1987, another Atlas/Centaur launch vehicle, which was carrying a fleet satellite communications (FLTSATCOM) spacecraft, had to be destroyed by range safety personnel. These were not good years for rocketry.

At least for a short while back in the mid-1980s, launch accidents significantly impeded the United States' ability to access space. Donald J. Kutyna, then a major general in the air force, confessed the precariousness of the U.S. launch situation when, following the May 1986 Titan failure, he noted that "we never want to be as vulnerable as we are today again."[6] For some time after the Titan and shuttle fleets were grounded, some aging U.S. military spacecraft, including U.S. missile warning and military communications capabilities, were forced to operate on their final backup systems. The country's top space commander in 1987, Gen. John Piotrowski, USAF, referred to the fragility of the U.S. position in space when he admitted that "we have been fortunate that industry has provided us reliable satellite systems with redundancy, but every day that goes by my concern goes up that we may have a failure."[7] The concerns expressed above amount to sound advice, to be sure, against leaving major security matters to the whims of fortune.

A series of launch failures in 1998 and 1999 were poignant reminders that the United States still does not have the capabilities to assure, with a very high degree of confidence, its access to space. The explosion of a Titan 4A shortly after liftoff on August 12, 1998, destroyed what was believed to be a National Reconnaissance Office strategic reconnaissance satellite; on August 26, 1998, a Delta 3 rocket suffered a similar fate, resulting in the loss of a PanAmSat communications satellite; a Titan 4B placed a Defense Support Program satellite in the wrong orbit on April 9, 1999; the failure of a protective fairing to separate from the satellite vehicle on April 27, 1999, spelled doom for what was to be the first commercial satellite capable of delivering one-meter resolution pictures from space (*Ikonos 1*); on April 30, a Centaur upper stage placed the highly protected and jam-resistant $800 million Milstar 2 satellite in a useless orbit; and, finally, on May 4, 1999, a Boeing Delta 3 placed the Loral Orion 3 communications satellite into the wrong orbit. This series of launch problems was so disconcerting that several high-level investigation teams and committees were established to assess weaknesses in U.S. launch capabilities.[8]

The history of GOES satellites, which have provided the United States with continuous weather monitoring since the mid-1970s, provides another lesson in how frail the space business can be. On the heels of the 1986 *Challenger* disaster, in May of that same year, a GOES satellite was destroyed when its Delta 2 booster blew up during launch. Another GOES satellite failed in orbit in 1989, and for the first time weather watchers seriously contemplated having to cope over an indefinite period of time without a home-grown continuous weather monitoring capability.

Delays and technical problems in the GOES-NEXT program, which hampered the development of the next generation of advanced weather satellites, did nothing to soothe mounting anxiety. Until the successful launch of *GOES 8* in April 1994, the United States was dependent on a single geostationary satellite, *GOES 7*, which limped along well past its design life and perilously close to being out of fuel. Until the *GOES 8* finally ended the disconcerting lapse, Washington had to request access to observation data on the Atlantic and East Coast areas from Europe's *Meteosat*, while *GOES 7* monitored western areas. It was observed prior to the launch of *GOES 8* that a launch failure would mean the United States could have found itself without a working weather satellite in geosynchronous orbit.[9] *GOES 8* did make it to orbit, as did subsequent platforms, although the on-orbit failure of *GOES 9* reminded us once again that we should not take our presence in space for granted.

The agony of having to cope with vulnerability in space resulting from misfortune or human oversight offers us some glimpse of the kind of pain an adversary might want to cause the United States. I offer a note of caution, however, before we proceed. The reader will not find in these pages revelations of imminent threats to U.S. national space assets; I am in no position to uncover such looming or portentous dangers. There are neither references to secret National Intelligence Estimates warning of impending plans by China, Russia, or Iraq to take our eyes out in space nor sensational insights into an impending failure of our national space system (something akin to the Y2K panic of the late 1990s). This chapter is not a prophecy of doom, nor is it intended to stoke the ardor in alarmists who desire to revive a cold war call to space arms.

Essentially, this chapter sharpens general awareness of a serious and expanding defense concern. The basis for such a discussion is elementary. If reasonable people can agree that, given the right strategic context, an attack against a U.S. satellite or a ground station can provide an attacker with a military or political payoff, that technological capabilities do exist to attack all three segments of a space system, and that such counterspace operations are generally within the financial and technical means of foreign entities, then it makes perfect sense not only to study this problem but also to prompt the United States to undertake requisite countermeasure programs.

A proper consideration of these propositions concerning foreign means, motives, and capabilities must be part of an assessment of threats to U.S. space systems. We can be certain that the United States' reliance on space has not been ignored by other countries, terrorists, or criminal organizations. We can assert with confidence that those who possess ill will for the United States today and tomorrow will endeavor to come up with ways to knock the country off of its orbital balance. This chapter provides a basis for thinking objectively about how U.S. access to space might be threatened or undermined, today and in the future, and for considering the consequences of collective and protracted indifference in the United States to survivability in space.

The 1991 Persian Gulf War demonstrated in no uncertain terms the importance of viable, space-based systems for command, control, communications, and intelligence. Never before in the history of warfare has the quality and quantity of information been such a decisive determinant of military advantage. Never before has one side in an armed conflict structured its campaign around complete reliance on the free flow of information at the tactical, operational, and strategic levels of war. In the past, unreliable communications, while a hindrance, usually did not mean that operations would be halted or the outcomes of battles materially altered. In fact, historically, communications have been slow and disruption expected. The extensive space assets deployed by the United States for the war in the Middle Eastern desert, by contrast, assured a high volume of information, routine collection of intelligence on ground, sea, and air activities, and impeccable navigation in otherwise featureless deserts.[10]

The quest for information-based military forces is simply a chase after greater efficiency and effectiveness. Over the years, the defense leadership of the United States has sought to realize some rather traditional military strategies and goals through space-based information architectures that allow significant economies of scale to be achieved, more efficient and cost-effective operations, and the fielding of smaller forces. Although the bridges back to more traditional ways of engaging in war have not all been burned, the country's dependence on space has advanced considerably. For all practical purposes, there is no turning back. Consequently, there will be tradeoffs from more traditional approaches to warfare, and with these tradeoffs come new vulnerabilities that the country must learn either to reduce or abide.

How Shall We Understand Threats to U.S. Space Systems?

Attacks on space systems are only meaningful to the defense planner if they undermine the satellite mission. Pinpricks by the enemy, the loss of a satellite here and there, may be inconsequential to the course of a military campaign if the overall function of the satellite system remains intact. These same attacks, however, despite being operationally negligible, could have a profound effect at the levels of policy and strategy by negatively impacting public perceptions in the United States or diverting the attention of commanders. The loss of a few satellites during hostilities will not mean that the war is lost, although history does show that little events have a way of producing big surprises. The loss of a single spacecraft providing timely photographs or positional data during a critical maneuver could profoundly affect operational and tactical developments on land.

The politico-strategic context (who's fighting, for what, when, how, where, and why) ultimately gives meaning to questions surrounding threats to space systems. If the enemy can achieve his war objectives on land quickly, for example, then whatever pressure he might seek to exert by attacking U.S. space systems becomes irrelevant. Why? Because the seat of decision and power is on the land. If the objective of the conflict has been achieved, then what happens in space is of no direct consequence.[11]

All three elements of a satellite system—orbital, communications links, and ground—should be considered candidates for attack or disruption. Indeed, there is a spectrum of specific potential threats against each segment, as well as an array of possible technical, diplomatic, and military ground countermeasures the United States may implement to resist these threats. We must also consider the sources of threats. In some cases the threats to satellites flow from natural phenomena, while in others they are a result of hostile intentions. It is also important to recognize that a state need not own space infrastructure in order to develop space weapons.

Rapid commercialization has put space-based technologies within the grasp of many more countries as well as subnational organizations. As technological advances in high-power microwaves, lasers, sensors, and tracking systems are matched by improved launch capabilities, U.S. space systems, including ground-based elements and the interconnecting electronic links, are facing increased levels of risk. All public evidence points to a conclusion that today's antisatellite threat to U.S. space platforms is marginal. This is not to deny, however, that the proliferation and maturation of space technologies and the enduring realities of strategic and political competition on Earth eventually will propel countries to consider ways to degrade the U.S. space advantage. Future enemies of the United States will attempt in time of war to destroy the most valuable and militarily significant national assets, whether they are confined to Earth or circling overhead.

Natural Menaces to Orbital Systems

While the use of the term "threat" in this discussion generally will be reserved for those actions in which an intent to commit damage or injury is expressed, there are some natural hazards to orbital systems to consider. Although it may seem prosaic, the most likely and possibly most disabling threats to orbital systems in the short term are natural.

Orbital debris is comprised of freely moving manmade objects and particles that, though not strictly natural and not placed there intentionally to do injury, nevertheless can wreak havoc upon all but the most hardened systems. A large chunk of space debris could destroy any of the multi-million- (or billion-) dollar space systems fielded in orbit today. When one considers all of the hardware associated with rocket launches—fragmented stages, explosive bolts, and the tiny bits resulting from a series of mechanical operations—it is easy to understand how a single spacecraft launch can cause a multitude of hazardous objects to be released in space.

Orbital debris disperse randomly. Today there are approximately 150,000 debris objects in orbit ranging in size from one to ten centimeters. The vast majority of these fragments is not trackable by the country's Space Surveillance Network, which can only account for roughly nine thousand orbiting objects. Though comparatively tiny, these objects represent the greatest threat of damage to spacecraft resulting from hypervelocity impact.

Most of the scattered fragments in space are so long-lived that even if near-perfect mitigation techniques were implemented, the existing debris might not be reduced significantly for decades, although studies currently underway may help us to discover new techniques to depopulate Earth's orbits.[12] The problem is so much a part of space operations, it may not be feasible or cost-effective to protect large structures such as solar panels and large antennas against impacts because of the high-impact velocities. Debris in geosynchronous orbit, where we field large communications spacecraft, are less of a problem, although the menace to all space operations will grow with the increased levels of space traffic.[13]

As space operations mature and the servicing and upgrading of reusable space systems become routine, there will be a need to control approach and departure corridors, at least around large space facilities and in the more heavily populated orbits. Indeed, a space traffic control system, one that goes beyond just monitoring debris and incorporates international involvement, similar to that used to control aircraft flights, may well arise early in this new century. Even today, shuttle launches are delayed in order to avoid potential conflict with debris or other spacecraft in orbit.[14]

More purely natural hazards to satellite activity include such things as solar flares and meteoroids. Spewing high-energy electrons and protons, solar flares are potentially lethal to humans and can damage the microelectronic circuitry of satellites, causing, for example, temporary malfunctions in computer memory cells, control systems to be switched into unrecoverable modes, and electronic circuitry to be burnt out. During the least stormy times in the solar cycle, or the solar minimum, million-mile-per-hour solar "gales" of electrons and protons last up to two weeks. Space storms can cause communications disruptions, power losses, and disorientation in precision-guided munitions due to degraded GPS performance. Between 1983 and 1999, thirteen satellites failed due to space weather.[15]

Meteoroids, a regularly occurring hazard, are naturally formed, high-velocity space particles and debris that emanate from the wake of comets passing through our solar system and other natural sources. For example, the so-called Leonids, a "dust" trail created by the comet Tempel-Tuttle, comprise a sixteen-million-kilometer-wide swath of particles that traveled toward Earth at a relative velocity of 158,000 miles per hour. These fine grains could strike satellites with enough density and force to penetrate aluminum skins, pit solar arrays, or introduce powerful electrostatic charges, causing irreparable damage to spacecraft. The loss of European Space Agency's *Olympus* communications satellite in 1993 has been attributed to the Perseid meteor shower. These regularly occurring events serve to highlight to satellite operators how vulnerable their orbital assets are to space debris and other limitations imposed from without.[16]

We Cannot Hide . . . At Least Not for Long

While it is an exaggeration to claim that the skies above us are entirely transparent, it is even more unrealistic to believe that distance and darkness are an adequate cloak for satellites. Stealthy technologies, skillful maneuvering among and within the different orbits, and other deceptive practices may allow satellites to operate periodically "in the black." Yet U.S. satellites are hardly out of sight.

When considering vulnerability, it is important to know how easily satellites may be identified and tracked by other nations and hostile entities. Using stealth technologies together with other deception practices, it may be possible for a satellite to remain undetected for a long period. Nevertheless, detection techniques are improving. The detection and identification of space objects does not require the very expensive and extensive network of radars deployed by the United States. Techniques for locating and tracking satellites are improving with the aid of increasingly sophisticated computer processing software available on the open market, which are accessible to countries and subnational groups having very limited resources.

Amateur satellite surveillance clubs maintain very accurate catalogues and ephemeris data of practically all satellites, many of which are posted on Internet bulletin boards.[17] Together with information on orbital elements, such as inclination, altitude, and timing, provided by online services, it is possible for an amateur to use off-the-shelf telescopes, sensors, and software to track and sometimes photograph nearly any satellite from any point on the globe using the stars as reference points. The rather large U.S. KH-11 and Lacrosse intelligence satellites have been spotted and photographed using this minimal technology approach. According to one amateur, "At any one time, there's probably about 800 different satellites above us, and probably 80 of them are in low Earth orbit. By hitting a key, typing in the name of the satellite, the telescope tracking can quickly switch from satellite to satellite . . . those that are visible from our location at that very time."[18]

The United States' sophisticated and costly intelligence satellites usually receive special attention among satellite watchers, if only because the secrecy surrounding their operations makes these objects, which are capable of being maneuvered in orbit, more challenging to identify and monitor. It is a game among amateurs around the world, from hobbyists to foreign intelligence officers, to locate U.S. spy satellites armed only with personal computer software, binoculars, celestial charts, and stopwatches. Before dawn and after dusk, a satellite, as it makes its journey from one horizon to another, may catch the Sun's rays just right and reflect them back to Earth, making it visible against the sky's dark background. A satellite several hundred miles high can be seen with the naked eye from the ground below, especially if the spacecraft is very large. The large, radar-imaging Lacrosse satellites are said to "glow" as they streaks across the black sky.

Reportedly, when the Pentagon released imagery taken from space of destroyed Iraqi targets following the December 1998 Desert Fox air campaigns, officials provided enough information for one analyst outside the government to produce a schedule of classified satellite passes over Baghdad in the days to follow. It is also believed that Indian scientists avoided detection by the United States of its 1998 nuclear tests by calculating when U.S. intelligence "birds" were scheduled to be overhead and timing their ground preparations for the tests accordingly.[19] With amateurs and semiprofessionals demonstrating how it can be done, it is clear that virtually any country can organize a reasonably effective search for objects in LEO, especially if radar detection techniques, a 1950s technology, are also used in conjunction with optical observations.

GEO is harder to monitor than LEO, although the technology that will make this task much easier is evolving rapidly. Charge-coupled devices enhance the optics of ground-based cameras that may be used to detect larger satellites in GEO, even during daylight hours (the traditional watch period of satellite watchers being near twilight). Advances in this technology may be expected in time to make this capability accessible to virtually all countries. In the future, satellites in any orbit will not remain undetected for long.[20]

Antisatellite Weapons

Japan, Western European countries, and, ironically, the United States, are the biggest threats to proliferate technologies that could one day hurt the United States in space. The technical threat to satellites is made possible by the purveyors of dual-use technologies, technologies having civilian and military applications, found in international space businesses at home and abroad. Today, countries having well-developed technology and manufacturing bases have optics, telecommunications, propulsion, fusing and guidance industries, and these are hooked into global markets. These technologies are critical to the development of space surveillance, target acquisition, and target tracking capabilities.[21]

Hitting Moving Targets

With the fall of the Soviet Union, the United States' concerns about experiencing sophisticated attacks on its space assets also fell. The Soviet Union developed a co-orbital antisatellite system, a satellite interceptor programmed to destroy a target in space with shrapnel as they approached one another, was once considered to be the primary threat to U.S. systems. The co-orbital ASAT, which became operational in 1971, was last tested in 1982 and remained operationally ready at least until the regime change of 1991. Although it is believed that the Russians no longer maintain this ASAT system in a state of combat readiness, it is possible to keep parts of the system operationally ready without resumption of intercept trials by testing the S-11 booster and associated ground components.[22] Technology cannot be uninvented—coorbital ASAT, or future versions of this weapon, may make another appearance in the future.

In the near- to midterm future, however, it is difficult to imagine a lesser developed country being able to afford development of such a system. So-called direct-ascent systems are far more likely acquisition candidates. The rockets required for a direct-ascent ASAT mission are comparatively simple and cheap, easy to procure, and can be produced and employed in large numbers.

States with access to ballistic missile technologies have a potential ASAT capability. While foreign assistance is a major factor in nearly every case, countries may be grouped into four tiers: (1) those having an indigenous ballistic missile fabrication, assembly, and deployment capability (the United States, Russia, China, the United Kingdom, France, India, Israel, and North Korea); (2) those needing foreign assistance to develop ballistic missiles (Argentina, Brazil, Egypt, Indonesia, Iran, Iraq, South Korea, South Africa, and Taiwan); (3) those that, having no development capability, are entirely dependent on foreign-supplied ballistic missiles (Afghanistan, Algeria, Chile, Cuba, Libya, Pakistan, Saudi Arabia, Syria, and Yemen); and (4) those that may acquire a capability in the future following a political decision to do so (Australia, Belgium, Japan, Canada, Germany, Italy, the Netherlands, Poland, Romania, Spain, Sweden, Switzerland, and Ukraine).[23] Several countries are likely to be able to manufacture missiles in relatively large numbers, and some are willing to export them.[24] Conceivably, transfer of these technologies and systems could be achieved without detection by U.S. intelligence, although some might argue that complete surprise is not possible.[25]

Even staunch advocates of arms control in space acknowledge that relatively unsophisticated "weapons capable of threatening a handful of low-altitude reconnaissance satellites might eventually prove to be within reach of any space-faring nation."[26] In relative terms, LEO is so close to the earth that it offers very little security to spacecraft and space facilities. The farther out one can place a satellite in high Earth orbit, which extends from LEO to GEO at thirty-six thousand kilometers, the harder it will be to reach a moving target from Earth. Earth's midlatitudes, or locations not too far distant from the equator, provide excellent opportunities for engagement of LEO payloads. Analyses have shown that a launch site near Tehran would have had launch opportunities within a one-week period in 1992 against four of the satellites identified and tracked by amateurs—two KH-11 and two Lacrosse intelligence satellites.[27]

A ground-launched ASAT need not be complex to reach targets in low Earth orbit.[28] Simple, kinetic-kill space weapons can be very effective against space targets. Kinetic-kill weapons, as opposed to beam weapons, use some solid mass to strike a target. Explosive warheads are not necessary. A rocket could be launched from anywhere on the globe off of any platform, including the deck of a ship directly under the orbital path of the target, making these ASAT systems much more operationally flexible and potentially more survivable than co-orbital interceptors.

Assuming an adversary could locate and track satellites and other systems, and could fire a significant payload into their vicinity, it is then left to the attacker to maneuver the payload close enough to the target to damage it. The crucial part of direct-ascent ASAT systems is the terminal tracking, guidance, engagement, and fusing mechanism. Given the technical challenges this poses, multiple launches of inexpensive, direct-ascent weapons may be expected against high-value targets.[29]

The main use of explosives, particularly nonnuclear explosives, is to disperse shrapnel. A two-hundred-pound payload of common nails or "B-Bs" on top of a sounding rocket could be dispersed by a small explosive charge at the apogee of a sounding rocket, or the highest point it reaches above Earth, at a known time after launch. Certainly, solar panels, heat radiators, and antenna would not survive. A strategically placed blast would have destroyed *Skylab* and no doubt would tear apart structures belonging to the future *International Space Station.*

Fragmentation warheads permit greater miss distances but require fusing accuracies that vary with the ASAT's closing speed. The Soviet co-orbital ASAT was a fragmentation weapon, whereas the F-15 miniature homing vehicle pursued by the United States in the 1980s relied on direct impact, or kinetic energy, to destroy its target. Direct-ascent ASATs can reach their targets faster than orbital ASATs, the latter possibly requiring ninety to two hundred minutes to intercept a target in LEO and five hours or more to reach GEO. A co-orbital interceptor matches orbits with the target. A "co-planar" approach shares the target's orbital plane, but not its orbit. A nominal co-orbital ASAT might be a "space mine" designed to orbit near its potential target, awaiting a signal to close and explode. Space mines, satellites with high-yield explosives placed into orbits close to the desired target, are an effective means of simultaneously putting out of action several assets.[30]

The ASAT could be an attractive weapon, not least because the problems posed by a hostile satellite may be most effectively banished by attacking a single target in space rather than a series of potentially dangerous assaults on numerous and dispersed Earth surface targets.[31]

Nukes in Space. A nuclear weapon carried by a ballistic missile could be used deliberately to destroy space assets or disrupt satellite operations, although an attacker could lose many spacecraft as well (if he has any). The United States learned during the summer of 1962 during the "Starfish Prime" experiment that inadvertent collateral damage to satellites could result from nuclear bursts, when electrons become trapped in the Van Allen magnetic belts and the propagation of electromagnetic pulse can disrupt or damage electronic systems on the ground, in the air, and in space.[32] Starfish Prime involved the detonation of a 1.4 megaton nuclear weapon 248 statute miles (400 kilometers) above Johnston Island in the Pacific. This burst of radiation permanently damaged the solar cells of three U.S. and British satellites in 600-mile orbits.

Today, many analysts view the threat of a nuclear explosion in space as one of the more likely ways an enemy might wage asymmetric warfare against the United States. LEO satellites appear to be most vulnerable to the effects of tactical nuclear weapons in space. There seem to be few architectural, operational, or procedural means to lessen this threat. The only remedy is to harden them sufficiently to protect the satellite circuitry. Yet even hardening may only delay the onset of degradation.[33] Today, replacement satellites are generally unavailable, yet even space assets flown up immediately after the explosion would suffer the same degradation of capability for months owing to lingering trapped electrons. Military and commercial architectures are likely robust enough to ensure survival of some satellites for intelligence, warning, communications, navigation, meteorological, Earth observation, and communications purposes following a high-altitude nuclear burst. But which satellites survive and for what periods of time may be the critical questions on a commander's mind.

Despite international agreements banning the use, production, and transfer of nuclear weapon systems and technologies, nuclear capabilities have spread. In

1998, India and Pakistan undertook a series of underground nuclear explosion tests. While not a declared nuclear power, Israel is believed to have had a nuclear bomb for more than twenty-five years. U.S. officials are concerned about the theft or clandestine transfer of nuclear weapons and nuclear material out of the former Soviet Union, which lacks rigorous export controls to prevent their illicit sale and transport. Iraq, Iran, North Korea, and Libya have had active programs to acquire nuclear weapon technologies and materials. All these states also have active ballistic missile programs, meaning that they could have the expertise and launch capabilities to carry nuclear material beyond the upper atmosphere.

The proliferation of nonnuclear weapon systems and technologies for generating an EMP also will be a source of concern. Russia is thought to have tested a high-altitude weapon that discharges electromagnetic pulse as recently as April 1999.[34] Analysts also believe China could develop its first high-power microwave weapon by 2015, which, like EMP, would give it a capability to disable electronics on Earth and in space through a single burst of microwaves.[35]

Beam Weapons. Lasers are coherent beams of electromagnetic radiation. A laser can damage a satellite by overheating its surface, by "blinding" key on-board sensors or puncturing the outer surface of the target. Ground-based laser weapons, using ground-based energy sources, could pose a threat to LEO spacecraft, although they are subject to atmospheric deflection, dispersion, and absorption. Significant cloud cover, for example, would greatly reduce the military effectiveness of a laser system.[36] Particle beam weapons consist of large accelerators that propel charged or neutral particles at great speeds toward their target. Unlike a laser, a particle beam could immediately penetrate the surface of a satellite and disable its internal components through heat and radiation damage.

Ground-based laser threats to U.S. satellites were most prominent in the 1980s. At Semipalatinsk, the Soviet Union operated an explosively driven pulsed iodine laser. The French SPOT satellites produced imagery that showed elements of Soviet strategic lasers at Sary-Shagan and Nurek. Some Soviet lasers were known to possess a lethal range of 287 miles and damage capability to 460 miles.[37] Because LEO systems are so close to the surface of the earth, usually less than 690 miles high, and can be accurately tracked, they are vulnerable to large pulses of energy. Certainly for optical systems this is a concern because to function they must concentrate incoming light on small, sensitive detectors.

More recently, lasers have been cited as an emerging threat. The press picked up on a 1998 Pentagon report to Congress that the Chinese People's Liberation Army (PLA) is building lasers to destroy satellites and already has lasers capable of damaging sensors on space-based reconnaissance and intelligence systems. A report released by the House National Security Committee (since renamed the House

Armed Services Committee) said the PLA has acquired a variety of technologies that could be used to develop an ASAT weapon. A committee investigating Chinese theft of U.S. technology chaired by Representative Christopher Cox asserted that China may be developing space-based and ground-based antisatellite laser weapons.

According to then–Lt. Gen. Lester L. Lyles, USAF, former director of the Ballistic Missile Defense Organization, China's apparent laser capability surprised the analysts. "When you think about how third world countries can learn from Russia and China how to blind, inhibit, disrupt, and deny space capabilities," according to Lyles, "one can assume that they will desire to acquire this capability very quickly."[38] Even in the next couple of years, we may see a tremendous growth in this area. The development and deployment of these systems could be hard to detect. With present technology, these circumstances make it nearly impossible to prevent or preempt enemy laser attacks.

Ground-Station Attacks

Missile-, air-, or sea-launched weapons or special forces may be used to disrupt space capabilities or access by attacking ground sites, the locations and functions of which are widely known, effectively preventing the replenishment or augmentation of space capabilities. Coastal facilities, such as the Vandenberg and Cape Canaveral sites, are very accessible to attackers from the sea.[39] Targeting unmanned, or lightly manned command and control sites, processing, and support facilities would neutralize a large number of on-orbit space assets, forcing satellite operators to bypass those stations and undertake repairs, tasks that complicate military operations.[40] Although, for the most part, there is no single point failure on the ground, meaning there would likely be graceful degradation in space capability. But degradation is degradation, and at some point, if it lasts long enough, the system will not be useful when it is needed.

Most U.S. satellites are operated from a single ground station. Spacecraft have varying needs for updates from the ground. Newer designs such as NAVSTAR GPS systems may have more on-orbit station-keeping autonomy, but for many applications, such as reconnaissance, tasking must change frequently and requires instructions from the ground.[41]

During Operation Desert Storm, U.S. forces set up a telecommunications logistics network that depended heavily on two nodes—a single switching station at Thumraite in Oman, south of Saudi Arabia, and a satellite relay between Thumraite and the United States. Saddam Hussein was unable to attack Thumraite, limited by the range of his ballistic missiles and his conventional inability to mount an attack over ground, through the air, or from the Arabian Sea. The fact is, however, that destruction of the centralized telecommunications node at Thumraite, apart from the survivability of the relay satellite, could have ground to a halt U.S. operations to distribute war materiel essential to high-tempo military operations.[42]

Disturbing and Manipulating the Communications Link

Although many countries may want a "hard-kill" ASAT weapon, the "soft-kill" alternative, attacking the electronic ground-space links, is more readily available because of the simpler technology and industrial base it requires. Unless they have been fitted with specified countermeasures, so-called electronic warfare (EW) is perhaps the easiest means of neutralizing satellite operations. Indeed, one can counter U.S. GPS navigation capabilities with a simple one-watt jammer.[43] This type of space warfare already has occurred, as there have been various instances when Indonesia, Turkey, and Iran have jammed satellites using primitive radio technologies.[44]

A data stream is assumed to be true and legitimate, but in fact there are means of intruding or infiltrating it.[45] Modern-day communications links achieve robustness through redundancy, so that the whole network cannot be easily disrupted. Yet it would be possible to get in and disrupt some links. A technologically less advanced country would have to decide where to attack, because it could only get into a part of the network. An advanced country, on the other hand, could put a "telebomb" in a major computer system that could go off twenty-five years from now.[46]

There are three possible points of attack on communications systems. First, there will be ground-based gateways that allow interaction with the public network and the Pentagon. The hand-held communicator will interact with the satellite; the satellite will interact with the gateway; the gateway will interact with someone on the ground. Any of those links can be attacked. An adversary could take out the satellite, intercept the signal to the satellite or from the satellite, or the gateway could be attacked. Although some commercial systems are able to shift frequencies and store information should they encounter signals interference, they are not designed specifically to resist deliberate jamming.[47]

Terrorism

State-sponsored terrorism by Iran, Iraq, Sudan, Libya, North Korea, Syria, and perhaps China cannot be discounted.[48] Many terrorist organizations have the ability, with support from various states, to attack U.S. space systems through intrusion in communications links, cyberwar, or physical attacks against less rigorously defended parts of the United States' space infrastructure (including Earth orbits). It is not too outlandish to believe a terrorist could purchase a space launch capability in order to fire a missile having a homing capability with some kind of explosive, possibly a mine, up into space near U.S. satellites or worse, near the *International Space Station*.[49] A terrorist's motives are usually political or ideological, not military, so a single, well-aimed shot against a satellite or critical ground station might yield a tremendous payoff.

Diplomacy

Foreign states also may restrict U.S. space power using diplomacy and arms-control initiatives to persuade Washington to restrict of its own volition a capability. In this manner, progress in the development of weapons can be stalled or a whole class of weapon eliminated from the U.S. arsenal. Some analysts believe, for example, that President Clinton's 1997 vetoes of space weapon programs may have been bolstered by an arms-control proposal from Russia's then-president Boris Yeltsin, which arrived just days before his decision to ban antisatellite weapons.[50] A political decision not to pursue a particular weapon system may take years or even decades to reverse—hence, this approach will be very attractive to potential adversaries of the United States.

How Can the United States Assure Its Space Operations?

Space is a common area. Space vessels increasingly are in need of protection, protection against organized enemies and modern-day "pirates." As then–air force colonel "Pete" Worden reminded me, the reason sea pirates are less of a threat today throughout much of the world is not because we live in the twentieth century, but because there are a few highly competent navies to oppose them. This may be an instructive analogy as we begin to consider the approaches for maintaining freedom of space and enforcing the right of peaceful transit through that environment.

The space elements of the U.S. C^3I network must be capable of absorbing losses and at the same time retaining functionality to perform vital missions. Space control, or satellite protection, does not mean the ability to operate without disturbance or losses, but the ability to use the space environment without crippling losses to key systems. The correct defense planning objective, therefore, is to enhance the survival of key assets long enough to contribute to mission success.

Although it is true that satellites are inherently fragile and there is no place to hide permanently, there is a straightforward, logical suite of technical and nontechnical countermeasures to threats to space systems. Resource constraints and the general perception that threats to U.S. space systems today are minimal have left us in a situation where passive and active defenses are not being worked adequately. As we become more dependent on commercial assets, the time may come when they too may need protection, lest the risks of disruption and security compromises become too high—even for businessmen with five-star insurance plans.[51]

Designing, developing, and integrating countermeasures is something the young commercial industry would rather not deal with at this point in time. Corporations must struggle mightily to keep costs down, and the costs associated with countermeasures would take away significantly from the bottom line. As the industry develops, however, countermeasures may make sense as the surest form of protection available in a hostile environment. When the *Galaxy 4* telecommunications spacecraft went down in 1998, the loss of this single satellite caused widespread communication and credit card problems. Operations and system analysis will be needed to determine the tradeoffs for building in protection versus reliability through architectures to give assured access and survivability, which are important objectives, especially if you're a large financial organization where dependability is critical and just as valued as it is in the military.[52]

Survival measures are generally synergistic. Indeed, it is pointless to protect one part of a system and leave the other parts vulnerable. By increasing ground terminal redundancy and satellite autonomy, for example, the weak links in the space system may be minimized. Survivability measures should be implemented in proportion to the value of the satellite's mission.[53] Ensuring survivability naturally will become harder and costlier as antisatellite weapons become more numerous and sophisticated. Properly categorizing the type of attack, timing evasive maneuver and switching efficiently to other assets, all require adequate training.

Passive Satellite Protection

Although the techniques discussed below will offer good synergism when employed together, some will be more costly than others. The costs associated with the proliferation of satellites or having at the ready capabilities for rapid, multiple launches are obviously high. There are considerable costs associated with maneuvering, including the opportunity cost, or the temporary loss of mission caused when a satellite is moved out of its more desirable planned orbit. Others, such as the incorporation of antijamming and hardening measures, are more affordable and, consequently, more readily adopted.

Using Shields. Hardening a satellite involves providing it physical resistance to various attack effects, including nuclear, laser, electromagnetic energy, and kinetic intercept. Electronics hardened against nuclear effects are being incorporated in systems such as Milstar, the Defense Department's supercommunications satellites. Hardening against lasers and electronic warfare is desirable if it does not unreasonably compromise system design and cost. Physical shielding may involve the use of ablative or reflective outer materials, shutters and filters for sensitive sensors, and the use of nuclear power supply rather than solar arrays. Satellites cannot be protected against a deliberate nuclear attack, but protection measures could increase the number of weapons required for a successful attack. Hardening and mitigation technologies incorporated in the earliest phases of satellite development will help moderate the cost of passive defense measures. Expect to pay handsomely, however, to retrofit a satellite after its manufacture with the same defenses.[54]

Data-Stream Battles. The main challenge in electronic warfare is the control and use of the data stream, the ability to maintain the integrity of command, control, communication and intelligence systems. Spoofing, or giving a satellite false commands, is also a concern. Worries about electronic warfare have prompted moves to higher frequencies, electronic antijam techniques, and antispoofing techniques such as encryption and command authentication. Satellite crosslinking will reduce susceptibility to jamming and dependence on overseas ground stations, a data repeat capability will help overcome interruptions, and more onboard autonomy will reduce dependence on ground control.

Imagery satellites can be protected from electronic warfare and laser damage by use of shutters that protect the photo sensors and electronic circuitry. A good suite of measures can be found in the Pentagon's DSCS-III satellites, which provide wideband communication service. The superhardened Milstar system employs encrypted data and command links, advanced jamming protection, cross-links, multiple satellites in different inclination, GEO orbits, and hardened, dispersed ground stations. The United States also is investigating ways to prevent a future enemy from jamming the global positioning system.

Although many in the satellite business downplay the threats to commercial space systems, concerns about satellite survivability are legitimate, particularly threats posed by EMP. The U.S. armed forces have turned to commercial communications satellites out of their own desperate need for more bandwidth. But the business community approaches the problem of satellite survivability differently from the defense community. If you are going to rely on a system, from a military planner's point of view, you better make sure it works. But, as of today, there is no business case for integrating costly satellite protection measures into the satellite design beyond what may be required to protect against natural phenomena. Countermeasures significantly raise the costs to the commercial operator—and they may slow him down. A delay of a year or two in getting a constellation deployed could put a business at a significant competitive disadvantage, even were these improvements subsidized by the government.

Tougher encryption technology and more advanced security features are increasingly a part of the satellite telecommunications industry, so much so that interference with communications is not an issue that appeals to private companies.[55] The use of space and the interference posed by electronic warfare will for some time be a major issue in future government/industry relations. The space commanders of the future may well perform a "911" service for the commercial sector, which may have to look more and more to the government for protection of the space-based "bit stream" upon which our economy depends so heavily.[56]

Sending in the Reserves and Using "Stealth" Orbits. The U.S. government and the U.S. space industry traditionally build satellites for operation in a peacetime environment, expecting only random failures and an occasional need to expend funds on replacement vehicles. Based on the presumption of a benign environment, satellites generally are expensive and designed to last seven years or longer, so they do not require frequent replacement. Wartime changes this calculus, possibly with dramatic effect. In times of hostilities, when space objects become fair game in a life-and-death struggle, satellites will likely fail by more than chance.

In hostile situations, the United States could protect itself if it had a capability to proliferate its space assets. Reconstitution of lost assets requires available ground spares and quick, reliable launch capabilities, which currently do not exist in the United States. U.S. satellites tend to be large and heavy and do not lend themselves to rapid replacement. Multiple satellites and spares can minimize the effects of disruption and degradation and allow for operational replacement or reconstitution of combat losses.

Orbital parameters also can be used to achieve an advantage since assets may be widely dispersed. Both GPS and the U.S. highly jam-resistant Milstar, for example, use constellations of satellites at altitudes beyond the distance from Earth at which the Russian coorbital ASAT has been tested. Satellite spares also can be stored in a dormant mode in orbits above a few hundred miles, and may be moved into operational position when needed. Satellites also may achieve a degree of "stealth" through concealment in very high orbits, where they would be more difficult to detect and track.

The United States is limited today to two sites for major satellite launches, both of which lie exposed on the coasts—easy objects of sabotage. A shot from a hidden position against a firing booster would not only destroy the launch vehicle and payload, but the resulting inferno could envelop and destroy much of the launch pad itself. Because it must orbit in the middle latitudes, the space shuttle is restricted to launch out of the site in Florida. Disable its two launch pads, and the United States' only manned space vehicle is grounded.[57] U.S. launch ranges, moreover, are not designed to be secure and accommodate the requirement for high numbers of rapid launches. Part of U.S. space strategy must be to upgrade existing launch ranges, which currently rely on the same communications, command and control, tracking, and surveillance equipment used to place the first satellites into orbit.[58]

The more satellites deployed by the United States, the more challenging it would be for an attacker. Space architects could allocate satellite capability to a distributed network comprising many low-cost satellites rather than a few high-value satellites, reducing reliance on any single platform. Yet although a distributed architecture (similar to an Internet configuration) may be a smarter way to operate, there are certain operations (nuclear command and control) that still will require unique protection.

Deceive the Enemy. As long as there has been warfare, men have sought to avoid attack by frustrating the surveillance and reconnaissance of the enemy. Reducing the radar and optical signatures of the spacecraft by incorporating low-observable technologies would make it more stealthy. Such technologies may include optically transparent compounds, smooth contours (to reduce radar cross section), absorptive and nonmetallic materials, heat shields, emission controls, and other countermeasures that alter or minimize enemy sensor returns.[59] Decoys having signatures simulating those of high-value spacecraft may also be used. The enemy's best offensive weapons are worthless if he cannot find the target.

Satellite Servicing and Autonomy. On-orbit repair and refueling capabilities would help extend the life of a satellite.[60] It could allow more survivability features to be designed into some satellites by virtue of the potential savings realized by extended lifetimes. Greater autonomy means fewer connections to Earth, and fewer links to Earth reduces susceptibility to intentional data disruption or manipulation.

Protection Begins at Home. Ground stations and terminals may be hardened against collateral nuclear effects and protected against sabotage in part by establishing independent power supplies and security against unauthorized access. Mobile ground-, air- and sea-based terminals would minimize risks of catastrophic failure if a space system's fixed ground-based nerve centers were destroyed. Other steps include interactive computer systems with data bases to assess spacecraft "anomalies" and better communications with satellite operators and military exercises. Hardening measures for ground facilities also may include hookups to multiple, diversely routed land-line fiber-optic cables.

Staying on the Move. On command from the ground, from another satellite, or by command of their own threat sensors, satellites could maneuver to evade some ASAT threats, or debris, if properly designed. Reaction times for LEO satellites will be extremely short, given that an ASAT launch first must be detected, identified, tracked, and possible aim points selected before the commands can be sent to the target to get out of the way. Limited fuel supplies mean that each satellite has a restricted number of maneuvers it can make. Each time it maneuvers to avoid a threat, it gives away some portion of its life. Space systems based at high altitudes are not only farther away from Earth-based threats but also less affected by the gravitational pull of Earth, meaning that maneuvering capabilities in these orbits are more easily accomplished and at less cost to the life of the satellite.

Advanced U.S. intelligence satellites reportedly have some capability to maneuver. Moving the satellite, however, must be balanced against the need to replan later usage, the prospect of temporarily disrupting the mission, and the expected decrease in satellite lifetime caused by the expenditure of limited fuel. For example, optical reconnaissance satellites cannot go very high without losing needed resolution. Indeed, even the Hubble Space Telescope, placed in a 380-mile low Earth orbit for scientists to view celestial phenomena billions of miles away, would only have a resolution of eight meters when used to view Earth objects from GEO.[61]

Active Countermeasures

Fundamentally, defenses grant a country and its defense planners options. Options are good. It is distressing to a commander to enter a crisis period knowing full well that no active defense or counteroffensive option exists. By deploying an effective national missile defense system, for example, the country's leadership would have options beyond high-risk attacks against an enemy's ballistic missiles prior to launch in order to erase the threat. In any case, to expect that the preemption mission could be performed to perfection each time is ludicrous. Manned U.S. aircraft or cruise missiles could miss their targets, or arrive late or not at all, after which the only recourse available to the American people would be to take shelter and pray.

The same points can be made about defense of U.S. space systems. The ability to destroy enemy ASATs in space or satellites may circumvent a need to hit targets on enemy territory, a potentially dangerous and politically hazardous mission. Active countermeasures necessarily would increase defense options by complementing the passive techniques discussed above.

Sounding the Alarm. Effective satellite defense begins with timely recognition of an attack. Monitoring events in deep, black space *requires* a sensor network to detect, identify, and track space objects. A commander needs "situational awareness" before he can act.

The United States' current warning and surveillance sensors, radar and optical, are largely confined to Earth. The most critical location to place attack detectors is on the spacecraft itself. U.S. military *and* commercial satellites must have sensors integrated into their design that will enable detection of an attack while in orbit. Without such a detection capability, one cannot know whether interference with satellite operations resulted from natural phenomena or the satellite itself was hit by a laser, or exposed to a nuclear or conventionally generated electromagnetic pulse.[62] Similarly, satellite controllers have no way to know whether a spacecraft was spoofed or whether an attempt had been made to bypass link encryption. In large measure, we remain today literally, and figuratively, in the dark.

Satellite Self-Defense. Self-defense concepts include defensive satellites (or DSATs), although none is under development or planned now. The concept is simple. Satellite constellations might include weapons platforms to attack enemy ASAT systems, or other enemy satellites in nearby orbits. Defensive satellites could be positioned near high-value systems and use lasers, missiles, or electronic counter-counter measures to eliminate attacking ASAT weapons. DSATs do not seem very promising if the defended satellite is subjected to multiple attacks by low-cost weapons that disperse small, inert projectiles.

Space-based missile defenses could be used to attack threat missiles carrying nuclear warheads for detonation in the space environment for the purpose of propagating X-rays, gamma-rays, neutrons, and electrons. By destroying a warhead in its boost phase, or early ascent phase, its interception would take place before it reached the desired altitude. There are other methods to thwart an attack. A satellite may be turned off to block access temporarily. Short-life decoys having a similar but stronger signature than the target satellite are effective if the satellite is being attacked by a radar or optical self-guidance system. Intelligent antennas may, when detecting jamming, locate it in their reception pattern and modify the gain distribution to null, or "not to listen to," the jamming.

Nontechnical Countermeasures

A country's policies and diplomacy can provide yet another dimension of synergism and another layer of protection options. Deterrence policies and strategies, sovereign declarations, and diplomatic initiatives make up at least part of a national space-control package. Indeed, without such top-level positioning and support, defensive strategies lack coherence and focus.

Deterrence. A fallen guard encourages aggressors. Retaliation for attacks against ground elements, data linkages, and orbital systems, whether it be against the enemy's spacecraft or Earth-bound targets, always obtains. There is a significant protection role for a comprehensive deterrence strategy. Deterrence is inherently dependent upon the minds of the leadership being deterred. It assumes leaders will act rationally to protect what they value. Deterrence should work if the deteree values those assets that are projected by the United States for destruction. In a space deterrence strategy, the most valuable assets may not be hostile space systems.

By an equal measure, the feasibility of deterrence of hostile acts against U.S. satellites hinges upon the value Washington places upon its space assets. An adversary may believe that attacks against systems deployed in distant space ("out of sight, out of mind") will not be responded to very vigorously by the National Command Authorities because of the absence of direct and lethal damage on Earth. If deterrence is to work, U.S. policy makers will have to be unambiguous in their declaration to potential adversaries, and resolute in any retaliatory response, to communicate that an attack against U.S. satellites is an attack against its vital interests.

Not all states, however, are subject to deterrence. It is debatable, for example, whether the leaderships in Germany and Japan ever could have been dissuaded from pursuing their aggressive designs prior to the outbreak of World War II. This reality should only serve to reinforce the following observation: If ever a country's leadership judged that the elimination of the United States' advantage in space were *critical* to the achievement of its strategic objectives, it too may be beyond deterrence.

The deterrence business is a highly uncertain and complex one, given the range of political, psychological, military, social, cultural, and economic variables at play. Moreover, no one can know if deterrence ever succeeds (although we can make educated judgments about this). We can only know when deterrence fails. It is difficult to foresee how the space environment will be treated in the context of future international rivalry and how deterrence will play there. Yet we may draw some lessons from experiences in the other geographic environments. It is hard to imagine that Washington would simply tolerate aggression against its vessels on the high seas or attempts by another power to strike its aircraft with surface-to-air missiles at will without a vigorous counterresponse.

We can conclude that U.S. officials increasingly rely on satellites to undertake many menial and critical tasks. It is logical for any prospective enemy to view them as vital national assets of the United States. It is reasonable to believe that adversaries of the United States could be deterred from attacking them if they understand the stakes and inherent risks of space warfare. Policy and strategy must support a no-holds-barred deterrence posture, within which such things as "keep-out" zones around critical space assets may play a key role.

Keep-Out Zones. Often in international negotiations, a favorite means of precluding potential conflicts is to establish so called "rules of the road." One approach for providing a measure of protection for space assets might be to establish a "rule" and create, either by negotiation or unilaterally, "exclusion," "self-defense," or "keep-out" zones around designated "must have" satellites. Such zones, which may involve establishing a circumferential distance around all sides of a satellite, may be more easily monitored in GEO, where the space around a satellite can be more readily defined in relation to Earth.

The purpose of declaring such a zone would be to guard against the surprise intrusions of space mines or ASATs. One way a potential attacker could destroy a large number of satellites, thus causing a sudden collapse of the satellite mission and making military adjustments on Earth by the United States more difficult, would be to activate numerous space mines emplaced near the satellites and to employ space-based ASAT weapons. Thus, any satellite stationed in, or for that matter passing through, another zone would be subject to inspection, disruption, or counterattack. Although the monitoring challenge is a monumental one, a plan for the enforcement of this zone would have to be incorporated into a rather ambitious deterrence strategy.[63]

Arms Control (Maybe). International arms control, a popular pastime of twentieth-century policy makers, also has been touted as a credible approach to reducing the likelihood of war in space (thereby protecting U.S. space assets) and to minimizing the political and economic costs of preparing for it.

There are a number of well-known political and technical challenges facing the implementation of any antisatellite arms-control plan. First, the vastness of space imposes enormous verification challenges. Detecting tests of space weapons would require unprecedented intrusiveness. Such intrusiveness would require negotiating with secretive societies, thus further complicating this task. Second, there is no way to separate offense from defense in space any more than it can be done on Earth. The difficulty of defining what constitutes a space weapon means we cannot know with confidence whether a satellite launched by a potential enemy is on a peaceful mission. If one cannot define it, how can governments control it?

Third, an arms-control plan covering space systems does not provide protection to the communications links, ground stations, and launch facilities. Fourth, any arms-control proposal is unlikely to address the threats posed by foreign nonweapon systems, such as imagery satellites, which, left free to operate in times of war, would enhance the effectiveness of the enemy's armed forces. Fifth, effective arms control and cooperative activities, including confidence-building measures, would necessitate the release of sensitive information on highly classified satellite programs. Finally, inadequate verification and monitoring capabilities mean an adversary may continue to research and develop weapons over a period of years, possibly putting itself in a position to "break out" of the treaty.[64]

Potential benefits include limiting specialized threats to satellites, reducing the political and economic costs of arms races, raising the political threshold for attacks against satellites by restricting threatening activity (thus increasing stability), and meeting international concerns regarding the weaponization of space, especially the concerns of alliance partners.[65] The touted benefits of arms control must be evaluated in light of particular circumstances and, of course, the details of the subject international agreement.

Why Attack the United States in Space?

What are the motivations for threatening or attacking U.S. space systems? What do countries or subnational organizations hope to gain by developing or showing off capabilities to assail the United States in space? Would they not have more to lose by offending a superior power? Why would an adversary want to open up the gates to U.S. retaliation and provide the United States with a reason for weaponizing space?

All of these are good questions, but they are questions that fundamentally apply in all situations involving the use of force against another sovereign state. So we must ask an even more basic question. What are the motives for waging war in the first place? The reasons states attack other states on the ground, at sea, in the air, or even in space are basic and timeless. For centuries polities have clashed over access to and control over natural resources, territorial dominance, or the size of populations, which contribute ultimately to a state's economic strength and its power on the battlefield. States go to war to garner prestige and glory or to establish regional hegemony and strategic position. Men also take up arms to spread religious faith and ideology. I see no obvious reason to assume that life in the twenty-first century will be any different or that the place called "space" must be excluded from military calculations conceived to secure these objectives.

Yet the specific reasons an Iran, China, or North Korea would commit aggression are strictly circumstantial. If the United States is in a state of war, jeopardy will have a hold on all national assets. Short of war, other calculations may drive an enemy to take the offensive in space. The very novelty of space warfare in our age may permit a hostile state to test the U.S. response or to make a bold statement against Washington's regional policies by striking out at sovereign assets that otherwise result in no direct U.S. casualties (political and situational ambiguities may in fact soften forceful language in current policy regarding interference with U.S. satellites). We should expect our adversaries, stated the chairman of the Joint Chiefs of Staff in early 2000, to "attempt to create conditions that effectively delay, deter, or counter the applications of U.S. military capabilities."[66]

Why might U.S. satellites or the link and ground segments come under attack? States may engage in a modern form of piracy or blackmail, by stealing information passing overhead or using U.S. satellites for illicit purposes. Simple escalation in local hostilities may involve strikes against targets in space or on Earth (some of which may belong to the United States), a collateral result of clashes, say between China and Taiwan. Sporadic disruption of telecommunications satellites could precipitate a global economic crisis—what better way for a terrorist to get his message across. A state may desire to establish some control over the conduits of commerce or prepare for war against the United States or a neighbor. In the end, nothing definitive can be said about motivations unless one establishes a politico-strategic context. The reader need only accept at this stage that if something can be done, it may be done.

Is There a Credible Threat to U.S. Space Systems Today?

Those skeptical that there is a requirement to prepare for military engagements in space might ask a number of pertinent questions. What would be the point in taking out the commercial satellite infrastructure by using, for example, a nuclear weapon? The military benefits are not so self-evident, and the attacker would risk disruption of his own systems. Moreover, there are regional and global risks in setting off nuclear weapons in space. After all, a launch is a launch, and its motivation may not be discernible. How much would an attack using a direct-ascent rocket accomplish, given that only LEO constellations would be affected? Certainly, it is impossible to take out everything. Indeed, is it possible for a country to do more than simply "poke holes" in our constellations?

Yet the circumstances that define a given situation make it more or less fragile. Much depends on the criticality of the satellite's mission over a particular period of time and its relevance to certain events on Earth. For an attacker, success probably will not be so demanding that it requires "everything" to be taken out. Even partial success might affect the course of activities on Earth. A B-2 bomber on a bombing run might not receive the latest status report on a newly located surface-to-air missile site; botched GPS navigation signals could make it impossible for that same B-2 to use its GPS-aided JDAM (joint direct attack munition) bombs or cause separated U.S. forces to lose synchronization and tempo; the lack of timely pictures from space may cause tank crews to hesitate before climbing over that next hill. Even partially stressed information architectures, in other words, might increase the fog of war just enough to raise significantly the psychological and physical costs for U.S. forces on the battlefield.

We can state with certainty that we will be challenged in space, in some form or manner, simply because it makes military sense to do so. Space operations, much like air operations in the beginning half of the twentieth century, have become an enabling force and a critical part of modern military strategy. Desert Storm demonstrated the contribution of information channeled through space and intelligence gathered from space to the efficient prosecution of war. Much has been published in the United States and abroad recognizing this linkage, and foreign space experience grows day by day, so we can be assured that these findings will not be lost on defense leaderships in other countries.

Whether U.S. space systems can be assaulted, degraded, disrupted, or destroyed is not in dispute. They can be. The question that remains then is this: Is there a credible threat to U.S. space systems today? Addressing the issue of *credibility* is really a matter of speaking to two constituent factors: *probability* of attack and *capability* to attack. Understanding probabilities requires understanding of political and strategic factors affecting its relationships with potential adversaries. This will always be a variable, one which must be accounted for in the threat assessments made in high-level defense documents, like the *National Security Strategy, Defense Planning Guidance,* and National Intelligence Estimates. What remains to be answered is what offensive capabilities are available to potential adversaries. An assessment of capabilities should include a look at the types of weapon technologies available and the number of weapons deployed—the number and types of weapons will tell us to what extent foes can damage U.S. national assets. An assessment of tactics will capture operations that range from low-technology terrorism to more sophisticated ASAT operations.

Countries today may marshal some capabilities to threaten or degrade U.S. systems, the most onerous potential menace being a nuclear detonation in space. This could be the most usable threat tool for a China, India, or Pakistan, for example. Using a surface-to-air missile (such as the Russian SA-12) or other radar-/infrared-guided technologies available on the commercial market, together with the right software, a determined foe would have a kinetic energy ASAT for use against LEO targets. Ground-based directed-energy weapon systems also may be considered a near-term threat. And many threats, of course, may not be directed against space targets. Cyber-attack represents one example.

This question of how likely it is that a threat will materialize in the near and far terms may be counterbalanced and tempered by another: does the Sword of Damocles have to be hanging over one's head before space threats can be taken seriously? If one waits too long to address an emerging threat with the development of appropriate technical countermeasures and doctrines, which may require several years in lead time, one may be faced with a significant intervening period of time when the country is vulnerable. Judgments about how long the United States may bide its time regarding some of the most significant concerns will have to be made by political leaders and defense planners.

According to the former space architect, Maj. Gen. Robert Dickman, USAF, the regret for being wrong on the question of U.S. vulnerability in space today is not necessarily all that high. "Wrong" for the leading military power in the world, which has other reconnaissance capabilities, means losing one or two or three low-altitude reconnaissance satellites, in General Dickman's estimation. As capabilities proliferate, however, and space is featured ever more prominently in foreign offensive and defensive doctrines, this presumption is not likely to hold true, especially if the United States must fight a protracted war or two major wars simultaneously.[67] Much, of course, depends on the types of threats confronting U.S. forces on the land, in the air, and at sea at the time.

What Is the Bottom Line?

The inexorable quality of technological growth and our maturing dependence on space to perform missions essential to national security and general economic prosperity provide us with the rough-hewn evidence we need to conclude that threats to U.S. space systems are evolutionary in nature and, to some degree, unpredictable. William Graham, former science advisor to President Reagan, illustrated for me one day how the expansion of applied science is itself a fascinating subject, for technology has a way of relentlessly pushing us forward and expanding our vision of the possible. Technology not only allows us to do what we used to do better, it also extends our reach a little bit further with each new advance.

Consider that the invention of the car not only displaced the horse and buggy, it also allowed man to travel across the land much faster than he had ever imagined. The multiplication of fast, overland transport vehicles led to opportunities to travel much greater distances over shorter periods of time, and with this came a new desire to branch out to different parts of the country, which in turn led to the development of the interstate highway network. What Americans took for granted beginning in the 1950s, rapid and easy transit to most points in the United States, forefathers just two generations earlier could not even imagine.

We may expect that new space technologies and novel space applications will have similar, forward-pushing, unimaginable social, economic, and military effects. Already we are beneficiaries of unforeseen progress owing to recent developments in space-based navigation and communications. The question is, how will technology's invisible hand work for or against U.S. national security tomorrow and in the decades ahead.

Many centuries ago, the first seagoing vessels sailed and rowed unopposed, until such time, of course, that it became expedient to deny the seafaring advantage to one's enemies. Similarly, early-twentieth-century aircraft skirted the blue skies unmolested, until such time they became connected with military purposes, after which other countries thought it advantageous to invest in their own aircraft and antiaircraft capabilities. According to one account of air warfare during World War I, "The relationship of opposing pilots of observation aircraft was quite friendly at first. As they crossed the front lines each morning, German and French pilots usually waved and returned friendly smiles. This, however, did not last long. As the fighting dragged on . . . more and more aerial incidents took place as bricks were thrown and a few shots from hand guns exchanged. Hand dropped bombs were developed, making the airplane a real offensive weapon."[68]

As the exploitation of the seas and air became increasingly advantageous, it dawned on an increasing number of nations that ships and aircraft may have significant influence on the strategic-diplomatic course of events on the land. As the numbers of air- and seagoing vessels increased, much as one would expect, control over the sea lanes and command of the air became in many instances important military requirements, and ships and aircraft were built to attack the platforms of the enemy (the first aircraft was shot down in October 1914). Thus, the rise of navies and air forces added new dimensions to warfare.

Much as the use of observation balloons in late-eighteenth-century and nineteenth-century conflicts, and of reconnaissance aircraft during World War I, eventually prompted initiatives by the enemy to counter them, so we can expect that the operation of space platforms to intrude upon the enemy will inspire forceful counteractions. Today, much like the early air pilots of World War I, spacecraft from many nations ply the ocean above day after day, hour after hour, in a friendly and disinterested manner, with space operators giving little consideration as to who is flying nearby and for what purposes. But as military applications of spacecraft evolve and the number of such spacecraft expands, so too will occasions for contention over the uses of space arise more frequently. According to the Pentagon's long-range thinker, Andrew Marshall, the head of the respected Office of Net Assessment, "The long-term trend is that nations are seeking new forms of strategic attack." Attacks against space assets, stated Marshall, are "inevitable."[69]

The United States' exploitation of a global environment means that its satellites must operate in close proximity to potential threats. Indeed, the mission requirements of many satellites demand that they be flown over and near the lands of possible adversaries. Ships and planes are free to navigate in their respective environments, and in most cases are driven into enemy seas or air space at the pleasure of their respective commanders and pilots. Satellites are bound by orbital chains; the laws of gravity command them to circle the globe in a direction and at a velocity that precisely project their orbital paths to all who are watching on Earth. Unlike air- and seagoing vessels, the travel of spacecraft is usually very predictable. Even satellites with a maneuver capability have limitations imposed on their movements; they still must make complete revolutions around Earth, and, until such time as a refueling capability is developed, they are constrained in their already limited freedom of movement by the amount of on-board fuel.[70]

Inevitably, we come around again to the issue of space control. The cold war has skewed many perspectives about what space control entails. Indeed, it is more than simply destroying satellites. A critical part of the space-control mission adopted by the United States is to maintain access to and functional use of space. Space control means the ability to prevail over the enemy's hostile use of spacecraft when and where it matters, to be able to use space at will for such purposes as the establishment of an efficient command, control, communication, and intelligence network. The function of that network today assumes that friendly space systems will be on call when needed.[71]

The focus on threats to U.S. space systems will evolve over the next decades from ones primarily centered on nonhostile, natural occurrences, such as orbital debris or solar activity, to those intentionally posed by nations and individuals who wish to harm the United States and its allies. In 1999 "there is relative harmony in space," according to General Richard Myers, USAF, "but there have been instances of jamming." We also know, the general stated, that people are working on the technology that may be used against U.S. satellites.[72]

There will be many policy and strategic challenges ahead to coping with the United States' dependence on space and remedying its space vulnerabilities. The answers may require bold steps to maintain connections in space, including the use of ground-attack and space-combat options. Again, defenses are good precisely because they provide options. By doing what is required to maintain security in space, by continually improving defenses, by avoiding situations where the U.S. leadership concedes the use of space to the enemy, the United States will have the tools and options it needs to act with strength and conviction on an increasingly dynamic and multifarious international stage.

5
The Shattered Sanctum

How Might Space Be Used against the United States?

The twentieth century is notable in history for the United States' rise to world leadership, economic dominance, and military preeminence. Paradoxically, technological developments in that same century also tempered U.S. strategic leverage by creating new vulnerabilities, for no longer could Americans rest blissfully behind shelters propped up by distance and supported by the surrounding seas. Alexander Hamilton was one of the first national leaders to recognize that, although abundantly blessed by geography, new technologies and the dynamics of international politics one day would impose new national security challenges. In *The Federalist* he noted, "Though a wide ocean separates the United States from Europe, . . . various considerations . . . warn us against an excess of confidence or security. . . . The improvements in the art of navigation have, as to the facility of communication, rendered distant nations, in a great measure neighbors. Britain and Spain are among the principal maritime powers of Europe. A future concert of views between these nations ought not to be regarded as improbable. . . . These circumstances combined admonish us not to be too sanguine in considering ourselves as entirely out of the reach of danger" (No. 24). Hamilton's warning, published more than 210 years ago, foreshadowed a time when future inventions would strip away the natural ramparts that Americans took for granted.

New technologies and weapon systems established conditions in the world that compelled U.S. military forces, despite possessing capabilities to establish ground and sea control and command the air, to plan for sudden assaults emanating from distant regions (especially by ballistic missiles) and enemies having enhanced space-battle knowledge, especially in overseas theaters. As a medium providing global access, outer space is today a "flank" of concern for the armed forces. Space is a big common area, or a "new" ocean, as John F. Kennedy referred to it in 1962.[1] Many foreign countries and companies (domestic and foreign) are determined to chart it, ply it, and learn how it may be turned into national advantage and profit. This is the reality for which twenty-first-century defense planners must account and plan.

As foreign national and commercial capabilities to reach and exploit circumterrestrial space spread, we must try to understand what others can do in space to harm the United States. Iraq, North Korea, Pakistan, India, Brazil, Russia, or China fires an object into space—what is it and who owns it? Is it an advanced satellite for gathering intelligence or for communications? Or does it carry a payload of deadly agents? Are our satellites under attack? Can we, should we retaliate? Can we, should we shoot it down? Can we, should we stop other nations from using space in hostile ways?

In the not-too-distant future all enemies will have access to mobile communications with encryption technology, one-meter resolution imagery from commercial providers, GPS and maybe even Russian GLONASS or European Galileo satellite navigation systems, as well as other space technologies developed in the 1960s, 1970s, and 1980s (with a smattering from the 1990s). Space launch services will be available and foreign rockets will proliferate. Common sense tells us this trend marking the growth in space expertise and capabilities will carry onward, that the one-time superpower space monopoly now belongs very much to the past, and that proficiency in space operations is widening steadily to the point where it now reaches every corner of the earth.[2]

U.S. commanders and stewards of national security should anticipate that, in some circumstances, adversaries will be watching the United States at ports unloading materiel, counting F-15s deployed at local air bases and tracking B-2s as they arrive in theater for their bombing runs, observing troop movements, and communicating with their own forces in the field with a level of efficiency once achievable only by a handful of advanced countries. The one-time sanctum, where only a few countries blissfully operated just a short time ago, has been shattered once and for all.

What Foreign National Space Capabilities Exist?

With computing power doubling every eighteen months, countries today have much more processing capability available to them than did the United States or the Soviet Union during the cold war. Steady growth in the space commercial sector and increased investments in the cultivation of space expertise means that large space infrastructures are not required.[3] Advances in highly sophisticated processors almost certainly will usher in an era when smaller spacecraft are commonplace, including in the area of Earth observation. Future satellites will have tremendous computing power with the ability to interlink constellations to cover the globe for a variety of telecommunications, scientific or intelligence missions. Naturally, smaller satellites would greatly lower per satellite launch costs, allowing several satellites to hitch a ride on one large launcher.

Since technology and knowledge do not recognize earthly boundaries, no country can have the slightest confidence that it can monopolize the fruits of technological progress. The United States is still a world leader in satellite technology, but the Europeans are trying hard to catch up, and Japan is successfully doing basic research in satellite communications. Significant advances are being made in the areas of antennas, power systems, chemical propulsion, on-board processing, laser communications, and robotics. Computers and the steady increases in processing power will shape satellite capabilities and size in the future. Satellites also are being configured so customers only pay when they use them. And it is anticipated that the Internet will bring a greatly increased global demand for mobile communications.

Technologically, there should be little dispute over whether man will become an ever more proficient exploiter of space. He shall. Moreover, proliferation of space technology and knowledge cannot be stopped. There is no way to prevent enemies or friends from exploiting space, in effect joining the United States, today the country most committed to make use of that environment.

A Look, Country by Country

The purpose of this section is to familiarize the reader with space developments in foreign countries. Although no other country can match the levels of space expertise, investment in space systems, and space infrastructure of the United States, it is important to note that the rest of the world is also very active in space and acquiring day-by-day impressive space proficiency. We will look here at a broad range of countries that have exhibited a notable commitment to space.

Chapters 1 and 2 emphasized that a country need not own space systems to derive benefits from space, that participation in international ventures and the rise in the number of space vendors will provide more than adequate telecommunications, remote-sensing, and weather-monitoring services worldwide. This section will focus mainly on communications, remote sensing, and launching capabilities, paying special attention to military uses of space where appropriate. The many national and multinational space science and technology development activities are not recounted here.

Argentina. Argentina, the second largest country in South America, has taken several steps toward the development of indigenous space-based telecommunication and remote-sensing capabilities. Together with Brazil, Argentina is in the forefront of otherwise very modest South American space programs. The Nahuelsat program, legally authorized by Argentina to operate a private, commercial system of satellites, provides direct-to-home television signals and other telecommunications services to the southern half and central region of South America. Argentina's Earth observation capabilities rest on the development of its SAC, or scientific applications satellites. Plans exist to build ground stations to access commercial remote-sensing data. Buenos Aires has given some consideration to reactivating the Condor 2 missile program, which would be the technological basis for a national space launch vehicle.

Australia. Australia's Joint Project 2008 will develop by 2005 a national satellite communications infrastructure useful for strategic and tactical communications. Australia currently operates the Optus series of communications spacecraft, one of which currently carries a transponder to facilitate mobile communications by the Australian Defence Force (ADF), including naval ships and aircraft. Optus satellites also provide services to countries in the Southwest Pacific. Currently, the ADF makes extensive use of U.S. X-band military satellite communications services but also has plans to acquire its own military communications capability.[4]

According to reports that surfaced in 1995, the Australian Department of Defence may be examining space-based surveillance options under Joint Project 2044 in order to monitor the outback and vast surrounding oceans more closely.[5] The government is keen on developing a remote-sensing data network and services to exploit data from Landsat, SPOT, ERS, and NOAA satellites.

Australia is striving to stay on the leading edge of the space launch business as a provider of commercial services. Parliament passed laws to create a licensing and regulatory regime to accommodate foreign launches. Among the first to jump at the possibility of using Australia as a launch base was Kistler Aerospace, the reusable launch vehicle pioneer. The company plans to operate its new K-1 two-stage RLV to boost payloads into low Earth orbit from a range in Woomera.

The United States has a strong space partnership with Australia, where space tracking activities are undertaken at Canberra, the capital, and early warning satellites are monitored at Pine Gap near Alice Springs. Other activities at Pine Gap include the operation of a signals intelligence station.[6]

Brazil. Brazil's extensive geography, comprising large remote areas (Brazil borders ten countries and has a coastline forty-six hundred miles long), is a powerful driver behind its quest for access to space. As a partner in the *International Space*

Station, Brazil has demonstrated a serious political commitment to establishing a space presence. With its spending for space on the rise, analysts anticipate Brazil will continue to concentrate on remote sensing. Launched in October 1998 aboard a Pegasus rocket from Cape Canaveral Air Station, its *Satellite de Coleta de Dados-2 (SCD-2)* follows the successful operation of *SCD-1* (the first "made in Brazil" satellite, deployed in 1993) to gather data from orbit on forests, national resources, oceans, and climate. Brazil also has plans to launch the first satellites in the world to circle above the equator at altitudes below GEO in order to monitor the agricultural and mineral resources of the country. Brazil's satellite receiving station handles data delivered by *Landsat 7, Spot 4,* Canada's *Radarsat 1,* and Europe's *ERS-2.*

Brazil is in league with China to develop capable remote-sensing satellites within the China-Brazil Earth Resources Satellite (CBERS) program, which will give both countries a capability to view with militarily significant accuracy objects on Earth that are twenty meters across. Launched from Taiyuan Space Center, China, CBERS imagery will provide environmental data, allowing officials to monitor more closely the Amazon. Successors to CBERS reportedly on the planning board would carry multispectral cameras having a resolution of three meters and improve vastly Brazil's and China's reconnaissance capabilities.[7]

Brazil took another important step in its development as a spacefaring nation when it launched *Brasilsat B1, B2,* and *B3* communications satellites, which are operated by the state-owned telecommunications company, Embratel. Brazil may invest in a constellation of eight satellites, an ambitious and financially stressing project called ECO-8, an eight-satellite system to provide mobile voice and data communications services for the Brazilian government. Currently languishing for lack of funding, ECO-8 would provide commercial services to the rest of the world's equatorial regions.

Blessed with an equatorial location that provides the country with an ideal coastal launch site, objects may be inserted into low latitude orbits from Alcantara using less power and at less cost than is possible from more northerly or southerly latitudes. Alcantara is operated by the Brazilian air force, but it soon may be open to rocket launches by foreign clients. Indeed, as many as nine independent launch sites are under consideration.[8] The launch of Brazil's indigenous booster, the veiculo lancador de satelites (VLS), has been delayed now for a number of years as engineers at the Brazilian Space Agency have struggled to overcome technical problems. The follow-on VLS 2 is a commercial small booster intended to put two-hundred-kilogram payloads into LEO.

Canada. Canada became a spacefaring nation when in 1962 it sent aloft *Alouette 1* to undertake ionospheric studies. The Anik series of satellites, first launched in 1972, gave the country a domestic communications capability and will continue to service Canada well into this decade. Although they are not likely to buy their own satellites, the Canadian armed forces are looking to acquire a dedicated military satellite communications capability, probably by acquiring terminals to U.S. military communications satellites.[9] Canada is also considering development of a fleet satellite communications capability to facilitate communications among Canadian and allied ships.

Canada's space program "jewel" is the *Radarsat 1* imaging satellite, launched in 1995 to map ice conditions, monitor crops and forests, and chart oceanic and geographic formations. The advantage of *Radarsat 1* is that it can produce imagery despite cloud cover, smog, or darkness. It performs very well in the remote-sensing market, a fact that no doubt convinced the Canadian Space Agency to push ahead plans for development of *Radarsat 2* and *Radarsat 3.* It is expected that *Radarsat 2,* now planned for launch in 2002, will be able to image objects on Earth at three meters' resolution, meaning that it will have a capability to detect and track surface vessels, vehicles, and even slow moving air targets.

Radarsat 1's many imagery distributors are worldwide and include such companies as China National Remote Sensing Center, ODC Pakistan, and Soluciones Integrales GIS CA (Venezuela). *Radarsat 1* is so successful that, by 1999, it is said to have imaged Earth three times.[10] Plans to make commercially available radar imagery worry U.S. defense officials, who believe (correctly) that the radar's ability to operate under all weather conditions and to detect objects buried under sand will harm U.S. national security should they fall into the wrong hands at the wrong time. U.S. policy at present restricts domestic operation of radar imagery satellites having a resolution better than five meters. Canadian officials share some of the United States' concerns, and as a result have considered legislation to give Ottawa "shutter control" over Canada's remote-sensing satellites.[11] The Canadian military also is interested in strengthening U.S.-Canadian defense cooperation in the North American Aerospace Defense Command by supplementing U.S. space surveillance capabilities with Canadian space-based sensors.[12]

China. Space technology has become a high national priority, President Jiang Zemin stated in 1996 in the Great Hall of the People. Chinese premier Li Peng declared space to be a critical component of national defense and economic growth.[13] According to Zhu Yilin of the Chinese Academy of Space Technology, three space development challenges lay before China. First, China must meet the increasing domestic demand for telecommunications services across its vast regions. Remote-sensing satellites, autonomous navigation capabilities, and space-based assets to enhance national defense are urgent national priorities. Second, China wishes to attain standing in the international community by participating as a competitor in the burgeoning space marketplace. China desires to join the elite company in this field of the United States, Russia, the European Space Agency, and Japan. Third, Zhu Yilin highlights a traditional concern about foreign domination—this time in space. China's current dependence of foreign satellite services to meet many of its communications, remote sensing, and navigation needs are particularly worrisome.[14]

China uses international information exchanges and cooperation in the space area to enhance infrastructure, undertake espionage, and develop new technologies that have military and commercial uses, especially in the areas of space exploitation, laser development, military information processing, and automation. "China," writes Zhu, "shall further broaden its cooperation channels, utilize foreign funds, resources and technologies, and at the same time make great efforts to digest and absorb imported foreign technologies." This approach will shorten development times, increase overall quality, and permit China to "continue to act independently and rely on its own efforts."[15]

The Chinese space program is comprehensive, covering rockets, satellite development, and a network for telemetry, tracking, and control.[16] Although China's space programs are not as advanced as foreign programs, the persistent attention and significant share of China's resources allocated to satellite and launch development programs have provided it with important means for meeting national needs.

China, which made its first successful satellite launch in 1970, currently fields boosters for inserting payloads into low Earth orbit using the Long March 2C and 2D, into geosynchronous orbit using the Long March 3, 3A, 3B, and into sun-synchronous orbit using the Long March 4 and 4B. The Long March 4B, China's newest rocket, is said to be capable of launching multiple satellites, a capability that also demonstrates China's ability to deliver miniaturized multiple warheads on its intercontinental ballistic missiles.[17] The Chinese developed the Long March 2E, first launched in 1992, to compete commercially for a share of the business of launching payloads to geosynchronous orbit. The Long March 2E suffered failures in 1992 and 1995 and is at the center of the technology transfer controversy involving Hughes Space and Communications International (this controversy, the subject of the so-called Cox Committee, also focuses on the illegal transfer of technology subsequent to the failure of a Long March 3B, which was carrying a Space Systems/Loral–built satellite).

China's commercial launch business bounced back in 1999 from a disastrous string of failures (some of which resulted in deaths on the ground) in the early and mid-1990s. In a step away from its more pragmatic approach to space (with its focus chiefly on communications and remote sensing) China is also one of the few spacefaring nations that desires to put a man in space. China is expected to accomplish this feat sometime this decade under a program dubbed Project 921, probably using a Russian Soyuz-type spacecraft (rather than a space plane).[18]

China depends upon space-based telecommunications services to link into the remote regions of the country. Satellite services are particularly important, according to one noted analyst on the Chinese space program, because there are so few phones for public use outside of the hotels (which cater to foreigners).[19] The new mobile communications systems currently in operation and under development, although today rather expensive, ought to have great appeal among those in China who favor modernization.

Domestically developed Dong Fang Hong (DFH), or "East Is Red," satellites have serviced China's communications needs until recently. The more capable *DFH-3* satellite reportedly relies heavily on foreign technology and design. Other communications satellites servicing China include Sinosat (a Chinese-German joint venture to upgrade the *DFH-3*) and the Apstar and Asiasat series, all of which required heavy foreign technology involvement. China also plans to acquire a large, geosynchronous communications satellite with a very large antenna array that would provide a space-based signals intelligence collection capability, allowing it to eavesdrop on neighboring countries and monitor regional communications.[20]

There are also political reasons for unifying the country through space, which can provide rural peoples the means to stay in touch with the more populated regions. Satellite communications satisfy Beijing's ideological requirement for centralized control over the entire country. The positive implications of a global telecommunications capability, of equipping military units with high-technology communications equipment also have not been lost on Chinese security analysts.[21] China has acquired large numbers of Western-developed VSATs (very small aperture terminals), which will permit mobile units to transmit reliably voice, data, video, fax, and computer-to-computer communications, improving overall military command and control capabilities. A new military communications satellite and battle management system will enhance the ability of the PLA to coordinate and support its expanding air, sea, and land forces.[22]

China's remote-sensing program has progressed since the mid-1980s. Lacking a robust home-grown capability in this area, China has acquired data from the Landsat, SPOT, and ERS Earth observation satellites to assist with monitoring and cartographic requirements.[23] It will continue to rely on commercial satellites to provide high-resolution imagery. China may also play a heavy hand in distributing satellite imagery to potential adversaries of the United States and its allies. According to strategist Chou Kwan-wu, by selling commercial satellite images, "China can break the technological monopoly and political blockade of the West in this regard and help underdeveloped countries obtain essential photos to develop their economies or increase their defense capabilities."[24]

As part of its FSW program, in 1975 China began retrieving from orbit remote-sensing satellites carrying imaging packages for infrequent "quick-look" missions. Again demonstrating the dual-use quality of most space technology, these reconnaissance operations serviced Chinese development and security requirements. Because images are retrieved only after some time, China recognizes that its FSW Earth observation capability is of no utility when it comes to tracking mobile tactical and strategic targets. Thus, together with Brazil, China is developing the CBERS (or Ziyuan) series of satellites carrying multispectral cameras to compete in the commercial world. In addition to these commercial objectives, Chinese analysts believe the development of the Ziyuan satellites will not only help meet some of the more strenuous military demands (such as tracking mobile targets) but also help China acquire advanced space technologies.[25] Ziyuan satellites will have a real-time transmission capability, a marked improvement over China's recoverable capsule program.

China launched its first meteorological and oceanographic satellites in the late 1980s. The Feng Yun, or "Wind and Cloud," satellites are comparable to the U.S. NOAA satellites, but unlike NOAA's spacecraft, China's have had a spotty track record, which has forced it to depend in some cases on weather imagery data freely provided by NOAA.[26] China announced in 1996 that it was pursuing the development of a new radar satellite, probably with the assistance of Russia's radar satellite design bureau. Details about the resolution of this satellite are not publicly known, but even a ten-meter resolution would give it an all-weather imaging capability of military significance.

China also desires an independent satellite positioning and navigation capability for civil and military users. The Twin-Star communications satellites reportedly would have about a twenty-meter accuracy. The Chinese already use the U.S. GPS and Russian GLONASS systems for navigation and positioning purposes and to improve the precision of strategic missiles.[27]

Finally, China is believed to be developing, with the intention of deploying, space-based and ground-based antisatellite laser weapons (see chapter 4).[28] There has been a great deal of cooperation between Russian and China in this area, to include help in the area of radars for satellite tracking and imaging and possibly the development of orbiting nuclear reactors that may be used to pump lasers, which may be used to control space. China, of course, also has the capability to build direct-ascent antisatellite weapons.

Europe. European space endeavors are covered by a number of organizations spread out over several countries. Many of Europe's space technology development activities take place under the auspices of the European Space Agency. Other organizations, such as Eutelsat and Eumetsat, operate satellites to satisfy, respectively, Europe's telecommunications and meteorological needs. Multilateral European military interests may be worked through the ten-nation Western European

Union or the nineteen-member North Atlantic Treaty Organization. Europe, including the United Kingdom, also has a strong space industry. Although it may suffer from a lack of consensus on direction and resource allocation, and there does not appear to be a single agency with responsibility for formulating a European space policy and strategy (a potential role for the fifteen-nation European Union), "Europe" may be judged fairly as a space power in its own right.

Although ESA's mission is to undertake programs having only "peaceful" applications, its leadership recognizes that there is a fluid relationship between civil and military uses of space, a fact that will allow ESA to develop technologies that have military applications.[29] Earth observation and launch programs dominate ESA planning and spending. ESA also allocates a significant percentage of its budget (funded by the member states) to manned space flight (to include the establishment of a sixteen-man European astronaut corps for the *International Space Station*), scientific programs, telecommunications, and satellite navigation. A mainstay of Europe's space activities is the Ariane commercial launch program (led by ESA), with the latest in the Ariane series being the heavy-lift Ariane 5, which is managed and marketed by Arianespace (a commercial launch consortium). Ariane rockets launch out of Europe's Guiana Space Center in Kourou, French Guiana. Other future European rocket programs may include the French-Russian Starsem joint venture to launch medium-lift Soyuz boosters and the Italian-led Vega light launcher. In a further attempt to win greater autonomy for Europe in space, ESA also will begin looking at future launch design concepts, including those using reusable launcher technology.[30]

The European Union is the largest single purchaser of remote-sensing imagery in Europe, including French SPOT imagery. The European remote sensing satellite (*ERS-2* provides radar imagery of a few meters resolution) launched in 1995, which is a follow-on to *ERS-1*, launched in 1991, the proposed *Envisat*, scheduled for a late 2000 launch, and the recently approved *Cryosat* mission to view ice flows, represent ESA's latest interest in Earth observation. ESA recently revamped its imagery sales policy to become more competitive, which should make ERS radar imagery more readily available on the market.[31] In the interest of formulating a global strategy for Europe with respect to remote sensing, there are advocates within European industry and government for a "NIMA-like" agency (referring here to the National Imagery and Mapping Agency) in Europe to coordinate European Earth observation centers to improve distribution of satellite imagery having dual military-commercial applications.[32]

The European Meteorological Satellite convention established in 1986 meets European requirements for weather reporting and monitoring. Eumetsat gathers visible light imagery data every half hour from the Meteosat series of satellites, the first of which was launched in 1977. The processed pictures are then uplinked to the Meteosats for radio retransmission to national weather bureaus in Belgium, Denmark, Finland, France, Germany, Great Britain, Greece, Ireland, Italy, the Netherlands, Norway, Portugal, Spain, Sweden, Switzerland, and Turkey.[33] The Metop (meteorological operational) satellites series will provide Europe a capacity for monitoring weather and assisting search and rescue.

Europe's first non-experimental telecommunications satellite was launched in 1983. The forty-seven-member Eutelsat organization operates the communications satellites currently in orbit. With Eutelsat improving telecommunications for more than 280 million households in Europe, its next-generation satellites, called "Hot Birds" (Eutelsat 2s), have advanced switching capabilities, which will allow individual countries to uplink their signals separately from different ground stations (as opposed to channeling signals through a single ground facility).[34] The most recent generation Eutelsat 3s will provide improved business network communications. The European Space Agency is currently considering investment in broadband, multimedia satellite projects (SkyBridge, the West Project, PC Matrix, and EuroSkyWays) to build upon Eutelsat services providing high-speed Internet connections to personal computers.[35] Eutelsat currently services all of Europe, Central Asia, the Middle East, and North Africa, and it is expected to privatize, spinning off a new company from the organization's existing commercial activities.

Europe's most recent space undertaking is in the area of navigation. In 1999, the ESA, with the endorsement of the European Union, embarked on a major study to define a system to provide Europe with a civilian-run global satellite navigation system independent of the GPS system operated by the United States. The Galileo program, which the Russians view as a successor to their GLONASS navigational system, will be a joint industry-government venture to provide a highly reliable service over and above that offered by the GPS system. The Global Navigation Satellite System will comprise twenty-one or thirty-six medium-Earth-orbit satellites, which will be integrated with a ground-based wide-area augmentation system. The Europeans are looking to develop their own navigational system for strategic and economic reasons. Not only do Europeans worry about a U.S. decision to "turn off" GPS signals, they also lament the fact that most of the equipment in use today that utilizes GPS is produced in the United States.[36]

Individual countries like the United Kingdom, France, and Italy, have military space programs, some of which are supported and financed by Europe's defense arm, the WEU.[37] It is likely to become part of the European Union in the near future and will provide input and support for the next-generation radar and optical imaging military satellites developed within Europe. The WEU has a satellite processing center near Madrid, Spain, operational since 1997, which purchases commercial (from U.S. Landsat, French SPOT, Indian IRS, European ERS, and Russian SPIN-2 satellites) and French-led *Helios 1* "spy" satellite imagery.[38] European defense leaders also have access to NATO's communications satellites. Over the years France has sought unsuccessfully to invigorate a pan-European military space program, including in the areas of reconnaissance and signal interception.[39] On the whole, Europe's participation in military space programs is modest when compared to the United States and Russia.

Since the end of the cold war, Europe has taken a more independent stance toward the United States and NATO regarding the uses of space for peaceful military purposes. A "continental approach" to enhancing national/regional capabilities, therefore, has become increasingly popular. As Italian parliamentarian Francesco Aloiso notes, "We need our autonomy in space-based observation and navigation. It's a matter of strategic independence." Aloiso also believes that "Europe will be a better partner for America when Europeans are able to shoulder more of the load in regional crises like the one in the Balkans now." Pierre Ducout, president of France's Parliamentary Space Group, indicated that there is a growing consensus in Europe that a range of observation spacecraft will be needed for military and civilian operations, that "a viable security policy in the future means having an independent space capability."[40] That Europe will remain a major independent force in space is without question.

France. The management focus for France's space activities is the French space agency, the National Center for Space Studies, or CNES. The agency has striven to establish European autonomy in the areas of Earth observation and telecommunications, and Europe's success in the independent development of reusable launch vehicle technology will depend to some degree on the progress made by CNES in this area.

France launched its first SPOT remote-sensing satellite in 1986. Since then, it has successfully launched a series of SPOT spacecraft, the 1996 loss of *Spot 3* being an exception. *Spot 4* provides ten-meter resolution with its panchromatic cameras and has a multispectral and infrared resolution of twenty meters. France (CNES, SPOT Image, and Matra Marconi Space) is pushing to develop a high-resolution stereo imager for its *Spot 5* Earth observation satellite, which will assist French and allied defense forces by, among other things, making digital models of the terrain useful for missile guidance. A small imagery spacecraft with a two- to three-meter resolution, called 3S, is also under development.[41]

France, Italy, and Spain own the *Helios 1* optical imaging satellite, which gives Europeans a remote-sensing capability dedicated for military use, one of the drivers for this program having been the U.S. decision during the Persian Gulf War not to share all of its reconnaissance data with its European allies. The one- to five-meter-resolution *Helios 1* provided high-quality pictures to NATO planners during Operation Allied Force in 1999, assisting in battle damage assessment and evaluation of the refugee crisis. Helios was the only non-U.S. observation satellite used by the NATO countries.[42] Italy and Spain will join France in the development of the more advanced Helios 2 series of "spy" satellites, the first of which is expected to be launched in 2003, with a follow-on satellite launched in 2007 or 2008. France has considered selling Helios spacecraft to friendly countries on the condition that it have the ultimate right to switch off the satellite.[43]

Space-based military telecommunications is a budget priority in France. The French Syracuse military satellite communications system is not based on a dedicated spacecraft but shares the French Telecom 2 platforms with civil service providers. This system, replete with encryption and antijamming and anti-intercept capabilities, provides full telecommunications services to users within the satellite zone, although crises such as the one in Kosovo in 1999 severely tax the data-handling capacity of this system. Plans are underway to upgrade Syracuse (giving it a limited extra-high-frequency capacity and greater data-handling capacity) and to work with European neighbors to launch new communications satellites.[44] France is cooperating with Germany on next-generation satellite communication systems.[45] This program would work to deploy four satellites, all visible to Europe, to provide allied forces with a secure communications and a measure of interoperability for peacekeeping operations, crises, or wars in and around the European theater.

Germany. Germany's principal international partners have been the United States and Russia, but increasingly it has undertaken projects with India and Japan. Germany also is a member of ESA, where it is the second largest contributor to European space activities, and strongly supports Eutelsat and Eumetsat.

The German decision not to participate in the French-led Helios 2 reconnaissance satellite program led many to doubt that Germany considers space imaging to be a national priority. Lack of funding also compelled German lawmakers to abandon the Horus radar imagery satellite project, which would have provided a capability to identify vehicles and troops during nighttime or through dense cloud cover. In the eyes of some Germans, this was a mistake, as the country is now left entirely at the mercy of the United States and its European allies when it requires reconnaissance imagery. As a consequence, the German Ministry of Defense considered the feasibility of developing and operating mini-radar-reconnaissance satellites in order to establish a national assessment capability.[46]

Germany is involved with France in the development of advanced telecommunications satellites for military uses. Currently, German armed forces rely on links to Intelsat, Inmarsat, and the DFS (German telecommunications satellite) *Kopernikus* for out-of-area operations. *Kopernikus* transmitted data, faxes, and voice communications by German armed forces operating in Bosnia-Herzegovina and

Croatia for peacekeeping purposes. Germany also sanctions military uses of commercial satellites, "provided no direct control of weapon systems is undertaken."[47] To have true wide-area communications over the next several years for out-of-area contingencies, however, Germany will have to utilize under international agreements foreign military telecommunications satellite systems, such as Skynet (UK), Syracuse (France), or DSCS (United States), or systems operated by NATO.

India. For India, a country more than 1,222,000 square miles in area hosting a population approaching one billion, space offers unique vantage points for dealing with a number of national problems, including military operations and intelligence collection of activities in neighboring China and Pakistan, land use, resource management, weather forecasting, and communications. Indian officials acknowledge that the same satellite technologies used to improve the prosperity, education, and health of the Indian people will be available for military use.[48]

When in July 1980 it launched the *Rohini 1* satellite aboard its satellite launch vehicle (SLV), India became the seventh country capable of placing an object in orbit. The priorities of the Indian Space Research Organization (ISRO) are remote sensing and telecommunications, capabilities that will be developed to the greatest extent possible indigenously and through the exploitation of foreign technologies, personnel, and international research and development collaboration.

The Indian Resources Satellite (IRS) series of spacecraft images service national needs and are available commercially. India also has launched remote-sensing satellites to scan surrounding oceans to assist fishing, study temperature variations, and provide other oceanographic data. It plans to launch a next-generation series of imaging satellites with improved resolution and may develop whole new imaging satellite systems (Cartosats and ResourceSat). India's *IRS-1C* carries a panchromatic camera having a 5.8-meter resolution (versus SPOT's 10-meter resolution), and it, like the other imaging satellites, may be diverted to military missions, such as monitoring Pakistani and Chinese missile developments and movements.[49] The Indians have a very large commercial remote-sensing program and from time to time have sold imagery data to competing remote-sensing programs. India has more than a half-dozen ground imagery receiving stations and plans to establish additional stations in Argentina, Australia, Nigeria, and the Philippines.[50]

The backbone of India's communications capabilities is the Insat series of satellites, the first of which was built by Ford Aerospace. They also carry cameras for Earth observation. First launched in 1982, this versatile spacecraft not only delivers television signals, radio broadcasts, and telecommunications services, but may be used to assist in search-and-rescue missions and its optical and infrared sensors aid weather forecasters. India plans to upgrade its current Insat-2s and increase the number of transponders available with a next-generation Insat-3 series and the Gramsat. Indian industry also plans to invest heavily in mobile satellite communications.[51]

Satellite navigation capabilities, including their many possible military applications, have not been overlooked. Officials in India's armed forces have ambitious plans to integrate GPS into combat operations, and are building GPS receivers entirely in India.[52]

Most Indian satellite payloads are inserted into orbit by foreign launchers like the Delta, Ariane, and the space shuttle. Nevertheless, India continues on a course it set long ago to become self-reliant in this area, in part by leveraging private industry.[53] India's first major space launcher, which borrowed designs from the U.S.-developed Scout sounding rocket, was the four-stage SLV-3. Next in line was the augmented SLV, or ASLV, which was developed to validate the Polar, or PSLV, series used to place IRS satellites into sun-synchronous or polar orbit. Using third-stage cryogenic rocket technology obtained from Russia, India is also developing the geostationary satellite launch vehicle, or GSLV, which will be used to launch Insat and other geosynchronous payloads. In 1999, India gingerly stepped into the commercial launch business using its PSLV to launch payloads for Germany and South Korea. Plans are to make the PSLV a serious competitor in the international launch market. India will use its rocket technology and expertise not only to expand its nuclear missile arsenal (which now includes the development of the intercontinental Surya 1 and Surya 2), but also to develop a reusable, multirole space plane called Avatar, for (among other things) reconnoitering from space to improve national security.[54]

Indonesia. For a country comprising more than thirteen thousand tropical islands, space plays a major role in linking together Indonesia's more than two hundred million inhabitants. East to west, Indonesia's boundaries are greater in size than the continental United States. Communications is the major challenge, hence Jakarta's primary interest in securing the benefits of telecommunication satellite services.[55]

Indonesia embarked on its space program in 1975, which reached a major milestone with the launch of the *Palapa B1* communications satellite in 1983. Since then, India has launched additional Palapa B and Palapa C satellites using foreign boosters and announced in 1996 plans for a Palapa D to provide regional telecommunications and direct-to-home television services. IndoStar geostationary satellites already provide direct-to-home television and radio programming, and follow-on satellites are planned. The Indonesian National Aeronautics and Space Institute maintains ground stations for reception of remote-sensing imagery. It built a Landsat station in 1981 and acquired a SPOT receiving station in the mid-1990s.

Israel. Israeli leaders view their investments in advanced technologies, especially in the space area, as decisive counterweights to numerical advantages in population and conventional weapons held by hostile neighbors. Early warning intelligence is priceless to a country that is comparatively very tiny and whose principal benefactor, the United States, lies more than ten thousand miles away. Israel's leading generals and intelligence officials believe that space capabilities not only will enhance early warning, they also will contribute to deterrence and the overwhelming military force required to transfer land and air battles to the territory and sovereign air space of the enemy.[56]

The surprise Arab assaults that began the Yom Kippur War in 1973, and a U.S. decision to withhold critical intelligence information prior to the opening of the war, convinced national leaders that Israel needed an independent space reconnaissance capability. Israel launched *Offeq-1* and *Offeq-2* in 1988 and 1990, respectively. *Offeq-3,* launched in 1995, may have a resolution as high as 1.8 meters. It is reported to have an orbital maneuvering capability and a longer lifetime than its predecessors, which had life-spans measured in months. By the end of 2002, Israel is expected to have in orbit a constellation of eight one- to two-meter resolution commercial reconnaissance satellites called Eros to view Earth not only in the Middle East but also around the world. Eros will be partly owned by a company controlled by the Israeli government. Eros will further blur the lines between civil, commercial, and military/intelligence quality imagery satellites.[57]

As part of its antimissile early warning system to be used in conjunction with the Arrow antitactical ballistic missile system, Israel launched the *Amos-1* communications satellite in 1996. It experienced trouble attempting to get *Amos-2* off the ground in the late 1990s, although *Amos-1* is expected to remain in orbit ten to eleven years.[58]

The Shavit 1 rocket is nearly identical to Israel's Jericho 2 intermediate-range ballistic missile. Shavit 1 first flew in 1988 and successfully inserted *Offeq-3* into low Earth orbit in 1995, although this booster experienced failures in 1993 and in 1998. Shavit 2 is a follow-on launch system that may be launched from a mobile platform.[59]

Japan. Since 1969, Japan's underlying space policy held tight to the principle that Earth's orbits should only be used for "peaceful purposes." With the publication of a 1997 white paper that recognized the growing importance of "spy" satellites, and Tokyo's recent decision to develop space-based reconnaissance sensors, a decision prompted by North Korea's August 1998 launch of the Taepo-Dong 1 missile over the mainland, Japan shed one more layer of its pacifism and stepped into a new policy that emphasizes the role of space for defensive purposes.[60]

Japan entered the remote-sensing era when in 1977 it launched the first in a series of geostationary meteorological satellites (built by Hughes Communications). Japan discontinued the GMS series in the 1990s, but by 2000, Japan is expected to launch a new meteorological satellite (MTSat), which also will augment GPS navigation signals and assist air traffic control.

More home-grown satellite programs began in 1990 with Japan's launch of a marine observation satellite, and then in 1992 when it started collecting radar and optical images from the *Japan Earth Resources Satellite-1,* or *JERS-1.* The eighteen-meter resolution imagery of the *JERS-1* assisted Japan on several civil missions, including environmental monitoring, resource exploration, and agricultural and fishery studies. When in October 1998 it lost attitude and electrical power, *JERS-1* was shut down. Another earth observation satellite designed to provide weather and oceanic information, *ADEOS-1,* was launched in 1996 but failed in 1997. *ADEOS-2* will enhance studies of the ocean and atmosphere. Japan's ALOS (advanced land observation satellite) series, once launched in 2002, will have global mapping and environmental monitoring responsibilities.

Japan announced in November 1998 that, in cooperation with the U.S. Department of Defense, it would begin definition of a significant military space program in response to the North Korean missile threat (in addition to cooperating with the United States in the area of ballistic missile defense). Mitsubishi Electrical Corporation proposed to develop in cooperation with U.S. industry two pairs of satellites to be used for reconnaissance missions. Called the Multipurpose Intelligence Gathering Satellite Systems, the first satellite pair will carry optical sensors with one-meter resolution and the second synthetic aperture radars capable of providing imagery having a resolution of one to three meters. The purpose of the satellites, according high-level officials, is to "collect information the Government needs for crisis-management measures to deal with the defense and diplomatic issues as well as natural disasters." First launch is expected in 2003. This activity is being controlled by the office of the prime minister rather than by the defense establishment, purportedly to avoid the appearance that Japan is using space for military purposes.[61]

Japan has a number of advanced broadcasting and communications satellites and desires to contribute to the enhancement of a global communications infrastructure. Investments in high-speed satellite communication technologies, mobile communications capabilities, and digital multimedia broadcasting will help Japan meet this goal.[62]

The land of the rising sun also has progressed in the area of space transport systems. It has developed the M series of rockets used for medium-sized science-oriented satellites, and H-class rockets to meet national launch demands. The H-2A, which was expected to take the place of the less efficient and more costly H-2, was developed so that Japan can be a major contributor to the *International Space*

Station as a launch provider and developer of the station's Japan Experiment Module. Technical problems forced Japan's National Space Development Agency to kill the H-2. Japan has plans to develop by 2003 a new small launch vehicle called the Advanced Technology Proving Rocket, which will incorporate U.S. and Russian technologies. Japan cancelled plans to develop the HOPE-X unmanned minishuttle to service the space station.[63]

Russia. Once an esteemed space power, Russia today struggles to hold together degenerating capabilities, a saga perhaps best symbolized by its aging *Mir* manned station. As with *Mir,* the problems with Russia's space programs are systemwide. Lack of financing lies at the heart of this hapless development. Cash-starved Russians opened a cosmonaut training center in 1999 in Star City, where well-to-do space-oriented adventure seekers from all over the world pay thousands of dollars for training in space survival.[64] There also has been lack of agreement over general policy, which systems to develop, which elements of the ground infrastructure to leave unsupported in the effort to streamline operations, and how to develop a cooperative relationship between the Russian Space Agency and the armed forces.

In the meantime, Russian spacecraft are failing in orbit with little prospect of national space programs identified with the Soviet era being resuscitated. By June 1999, as much as 70 percent of Russia's 130 active satellites had outlived their guaranteed service lives. Lack of steady funding domestically and poor accountability in the management of funds it receives from abroad have raised international distrust of Moscow's commitment to the *International Space Station.* In the face of all this, Russia's leadership has urged on numerous occasions that everything possible should be done to ensure Russia's inviolable position as one of the world's leading space powers.[65]

Historically, Russia has had very strong Earth surveillance and reconnaissance programs. There currently are plans to continue the polar-orbiting Meteor series of weather satellites and to develop an Elektro satellite to meet weather monitoring requirements. Aiding in the meteorological mission are the Okean-O spacecraft, which provide oceanographic information, including data on ice flows, storms and cyclones, floods, and ocean surface phenomena. The Resurs-class of optical remote-sensing satellites, which take color imagery at resolutions from 45 to 170 meters, assists the national economy and other civil needs and will be available for sale on the open market. Like the military reconnaissance spacecraft, the Resurs satellites transmit images by ejecting a capsule containing exposed film. The *Almaz 1* synthetic aperture radar satellite, which did not last its full service life, transmitted its data digitally to ground stations. The Russians hope to launch a follow-on *Almaz 2* spacecraft. Earth imaging cameras also have been used aboard the *Mir* manned space station.

Russia's military reconnaissance satellite program also suffers. In June 1999, Russia's deputy premier deplored the country's inability to track and analyze events during the 1999 NATO bombing campaign against Belgrade. Russia, said Ilya Klebanov, "has turned out to be blind from the military point of view."[66] Operation Allied Force, watched as closely as possible by Russian analysts from satellites that had to be retasked to view the region on a regular basis, highlighted to Russians just how much national capabilities in the area of military reconnaissance have declined from just a decade earlier. Russia reportedly had only one military digital imaging satellite aloft during this operation. At one point in 1999, Russia reportedly lost all satellite photo reconnaissance capability. This dire situation has prompted the Russian government to make procurement of intelligence satellites a top priority.[67]

The Russians deployed in 1998 an advanced imaging satellite, the *Kosmos 2359,* estimated to be the size of a school bus, that resembles the Hubble Space Telescope. This reconnaissance satellite reportedly has a two- to five-meter resolution and is deployed higher in orbit than U.S. KH-11s in order to give it a wider field of view and a slower overflight time. Reports indicate, however, that the *Kosmos 2359* will not be among the military fleet at the time of this publication. To be sure, Russia orbits other reconnaissance spacecraft, such as the naval intelligence satellite *Kosmos 2349,* but it, like the others, has a short life expectancy.

Today's failures in space obscure the country's rather lustrous spy satellite history. Radar Ocean Reconnaissance Satellites (RORSATs) and Elint Ocean Reconnaissance Satellites (EORSATs) were developed to monitor, among other things, the activities and electronic transmissions of the U.S. Navy. While the nuclear-operated RORSATs were eventually left to expire in 1988, the solar-powered EORSATs reportedly still circle the globe.[68]

Russia currently relies upon approximately ten to thirteen aging Gorizont and Express satellites to provide basic telecommunication services across the country. Joining Gorizont in providing civil communications capabilities are the Yamal (GEO) and Gonets (LEO) series of satellites. A new constellation of Gonets satellites will enhance facsimile and electronic messaging services. Molniya spacecraft provide television, telephone, and telegraphic link, and given the highly eccentric orbits of these satellites (250 miles at its lowest point and nearly 25,000 miles at his its highest point above Earth), each satellite is able to provide eight hours of continuous service. A fleet of Raduga geostationary and Strela 3 low-Earth-orbit satellites provide military communications capabilities. Gaps in Russia's more primitive and aging satellite communications architecture have hobbled the transmission of messages to and from the *Mir* station and left many to ponder Russia's future in space. None of Russia's civil telecommunications satellites was designed to operate past 2003, and Russia is expected to turn to the West to buy the technology it needs to replace its spacecraft.[69]

Helping to meet the navigation needs of the country are two aging Parus ("sail") satellites. The GLONASS system of twenty-four navigation satellites, however, is the real workhorse. Originally designed for military use, Moscow approved a program for civil use of the GLONASS constellation as recently as October 1997, in large part to raise funds to keep the system alive. Russian military leaders believe that native space-based navigation and positioning capabilities are critical to the management and control of rockets, submarines, and aircraft.[70]

Russia deploys Cosmos and Oko satellites in elliptical orbits and a Prognoz satellite in GEO to watch for ballistic missile launches around the world. Yet there are a number of deficiencies in Russia's ground- and space-based early warning systems, which include ground-based radars (some of which were based in territories lost following the breakup of the Soviet Union). Russian satellites no longer provide twenty-four-hour coverage of U.S. missile fields, meaning that those in command of national nuclear missile forces are susceptible to receiving information that cannot be checked independently, increasing the chances that a nuclear strike inadvertently might be launched against the United States.

National launch capabilities, once impressive and without parallel, today are kept alive mainly by international commercial and noncommercial joint ventures.[71] Russian launches have declined steadily in number since the late 1980s, primarily because the operational lives of spacecraft are longer than in the past and the number of imaging, communications and weather satellites being deployed is much smaller. Given the great demand for boosters, several Russian and Ukrainian launchers are in active service, including the Zenit 3 (as part of Boeing's Sea Launch initiative), the massive Proton rocket used by Lockheed/Krunichev/Energia, the Soyuz launcher as part of the Starsem partnership (involving the Russian Space Agency, the Samara Space Center, Arianespace, and Aerospatiale), and the so-called Start booster, which is a modified road-mobile SS-25 ICBM. The SS-18 ICBM (renamed Cosmos-M) has been used primarily to launch military satellites, but it too, along with the adapted SS-N-23 SLBM, has been used to place foreign spacecraft in orbit.[72]

In the late 1990s, Russia opened up the Svobodnyy Cosmodrome in the far east Amur region for commercial liftoffs. The Plesetsk launch site, located in the north, remains active and under operational control of Russia's Strategic Missile Forces. The Baikonur Cosmodrome, currently the only site available from which to launch heavy boosters and by far the busiest of the spaceports, is leased from Kazakhstan. Officials in the Strategic Missile Forces, concerned about this dependency, have stressed that Russia needs to have launch sites on its own territory if it is to carry out an independent space policy.[73]

Russia, and to a lesser extent China, are the only foreign countries that have invested significantly in space-control and force-application capabilities. Under the Soviet government, Moscow devoted significant resources to the development of antisatellite weapons, space strike complexes or battle stations, manned military space shuttles and stations, antiballistic missile forces, and directed energy weapons. Although most of this research took place from the late 1960s to the late 1980s, like all knowledge and technological inventions, the weapon designs and concepts for military applications explored in that period cannot disappear or be uninvented. There is no evidence that Moscow is devoting significant resources to the development or even the maintenance of space weapons programs, although there are indications that Moscow has an ongoing antisatellite weapon program. For instance, Russia resumed testing of a high-altitude electromagnetic pulse weapon in April 1999. Given sufficient funds and appropriate policy motivation, one would have to assume that Russia one day could revive its space combat programs.[74]

United Kingdom. Collaboration with allies and friends in the space area has permitted the United Kingdom to reduce its net financial investment in space systems while deriving great utility from space. London invests heavily in ESA's Earth observation, telecommunications, and space science projects.[75] It does not deploy its own remote-sensing satellite either for civil or military purposes, not for want of technical capability, but rather because it is not currently viewed as affordable, given other domestic priorities and the availability of images commercially and from UK allies. The UK does not have a launch capability.

While London has supported extensively ESA's work in the telecommunications area, and currently benefits from satellites operated by Eutelsat, Intelsat, and Inmarsat, it also has developed a domestic capability. The United Kingdom became the first country to deploy successfully a GEO-based military communications satellite when *Skynet 1A* entered service in November 1969. Today Britain operates the Skynet 4 series satellites and plans to further upgrade the system with a 5 series. Skynet satellites provide secure, jam-resistant telecommunications with advanced signal processing to the British armed forces that are interoperable with NATO and U.S. systems. In August 1998 the UK backed out of a planned joint European communications system called Trimilsatcom when it determined that the system would not meet UK needs. The United Kingdom will rely instead on follow-on Skynet satellites and next-generation satellites developed within government and industry partnerships.

What Are the Military Implications of Proliferating Space Systems?

Sputnik was, simply stated, an unanticipated Russian technological advance. Americans were shocked that a country could be backward in the way it governs and in its living standards and at the same time awe the world with its sophistication in science and technology. Reinhold Niebuhr, American political philosopher, took note and concluded that "we were wrong to assume that a technical culture, requiring so many centuries to germinate in the West, could not be transplanted in much shorter time."[76]

Similarly, it would be a grave mistake to assume that other countries are incapable of challenging or competing with the United States in space simply because the United States leads in so many other areas of human activity. We would also be wrong to assume that only dedicated "military" systems can affect national security. What Billy Mitchell said about air power, that it is "the ability to do something in or through the air," is also true of space power and the space environment.[77] In the past, technological sophistication helped distinguish between military and civil spacecraft. Higher resolution intelligence spacecraft and advanced communications satellites capable of countering attempts to steal data or cause interference with the reliable flow of information now are entering into common usage. The "military," "civil," "commercial," even "scientific," and "educational" designators given to satellites are becoming less tenable and useful.

Commercial space activity over the past decade has radically changed the political composition of the space arena. Commerce is an inherently cosmopolitan and global activity that is not bounded by national goals and priorities. A commercial man is, by his very nature, a man of the world. Through the relentless workings of space business, the rise and steady growth of the market bring a wide variety of space capabilities to any country that can afford them, and in many cases it supplements services already provided by robust national systems.

Commercial products develop much faster than military products. The new-product cycle time for GPS-associated equipment and computers is less than a year. Commercial electronic products contain more recent and often higher technology than products developed by the armed forces. It is not inconceivable that future opponents may be able to acquire military systems more quickly than the United States or Russia historically could design, develop, and produce them. Since technology and expertise are widely available, they also will acquire greater understanding of how space systems work, which will lead to a better understanding of satellite countermeasures. Americans could face a situation where an adversary, despite being "underdeveloped" and considerably less vulnerable in space than is the United States, may have access to high-tech systems that will place significant stress on U.S. military operations.

According to a report by the 1997 National Defense Panel, in this day and age "we must anticipate that our enemies will seek to use commercial remote-sensing and communications satellites, along with space-based timing and navigation data, to target U.S. forces with high degrees of accuracy." We must, therefore, broaden our definition from the more traditional understanding that states must have national space programs in order to qualify as space powers. In evaluating who is a space power in the twenty-first century, one would be remiss to ignore the evolution of the global commercial space infrastructure and its increasingly international character. A space power is any entity that has the capacity to utilize effectively the space medium for commercial or military operations. The pieces of this increasingly complex puzzle will range from dedicated military systems to private commercial ones. Clever, resourceful space powers will be able to effectively utilize different combinations of those puzzle pieces.

Launch

Without launch, there can be no space activity. Barring international sanctions and embargoes that impose special constraints on a country's ability to work a deal with a foreign launch operation, one must assume that foreign access to launching services will be limited only by the cost and by the availability of boosters. The rise of commercial launch services makes space accessible to an increasing number of users, although the business of reaching orbit remains very much dependent on nationally developed and operated boosters. If the laws of the market are permitted to influence developments in this area, one also may expect the cost of space transportation to decline, launching efficiency to improve, and the number and quality of the services to rise.

While many civil and commercial satellite payloads will be multipurpose, commercial providers also may be used to launch disguised military payloads. Keeping the satellites true capabilities secret is not an easy matter, although the *use* of those capabilities can be more easily disguised. Even assuming the United States discovers prior to launch that a given payload has a military purpose, lack of U.S. control over launch operations means that it cannot be assured of preventing the insertion of hostile payloads into orbit short of engaging in preemptive attack to destroy or disable the booster and launch platform. The "train" to and from space will continue to roll, and the frequency of trips without question will continue to rise. "Getting there," though it is not being accomplished as routinely and reliably as one would hope, will gradually become less and less the problem.

Space-Based Imaging

Not only are there imaging services available to provide wide-area reconnaissance (at lower resolutions), but also services to provide highly detailed (one-meter resolution), photo-quality "pictures" of objects on Earth. "Department stores" offering space imagery are as close as the Internet and the fax machine.[78] Within three hours of the first release of spy-quality imagery data from Space Imaging's *Ikonos* satellite, inserted into orbit in September 1999, more than 386,000 people tried to download a high-resolution photograph of Washington, D.C.[79]

A variety of space imaging products and services is currently available. Whereas the long-lead times required in the acquisition, transmission, and processing of overhead imagery even just a decade ago meant that these satellite pictures could only serve national-level defense needs, developments in digital image acquisition, recording, transmission, and near-real-time processing will make it feasible for foreign armed forces to have more routine access to intelligence data.[80] Imaging sensors that use a much wider area of the electromagnetic spectrum, from ultraviolet to visible light to microwave, improve opportunities for viewing terrestrial objects (even in foul weather) and reduce susceptibility to countermeasures. Competition will ensure that these trends will continue to expand and improve in quality, coverage, and timeliness.

One of the few significant differences between what is now available on the remote-sensing market and the $1 billion intelligence satellites operated by the National Reconnaissance Office is that NRO systems have a much larger and more capable processing infrastructure to digest quickly the enormous quantities of information on specific targets.[81] Space images not obtained through commercial means may be acquired using intelligence-sharing arrangements with satellite owners and data distributors. The military utility of commercial remote-sensing satellites is indisputable, a selling point used by the French SPOT Image to promote its services around the world. An advertisement placed in *Defense News* in 1996 read, "SPOT: The right place at the right time. For global defense decision support, mission planning and rehearsal, peace keeping, to build your navigation databases, *Just tell us Where and When.*"[82] The United States used SPOT imagery to enhance military operations in Iraq, Rwanda, Haiti, Somalia, and Bosnia.

Given the spread of technology, the increase in the number of imagery sources, and the less restrictive regulations on space images in the United States and abroad, the United States can have little confidence that it can control through political means foreign access to pictures taken from space, be the requester friend or foe. While adversaries may not always have access to imagery in real time (many commercial space services offer archived imagery data and take orders for images to be delivered at an appointed time in the future), they will be able to enhance military operations by improving mission planning, route and target identification, intelligence preparation of the battlefield, and other intelligence capabilities. Image quality tradeoffs, lack of timely delivery, restrictions on placing orders, launch times (for film-capsule return systems like the one operated by SPIN-2), lack of tasking flexibility, inadequate processing and distribution capabilities, and unpredictable cloud cover will limit the military utility of the commercial systems.[83]

Image processing and pattern recognition technologies and software are improving, however, and promise to simplify this process.[84] In wartime, especially if the engagements occur over a matter of days and weeks, space imagery will be most useful if this procedure can be executed (received, processed, and acted upon) in a timely, reliable manner. The longer the lag in acquiring, processing, and distributing imagery data, the less useful it will be to campaign planners striving to degrade mobile targets or react to offensive enemy actions. Commercial ventures are improving ways to meet the technical challenges involved in organizing and analyzing this data.[85]

Many newer commercial imaging satellite systems will offer strategic reconnaissance capabilities with resolutions between one and three meters. Higher imagery resolution will permit general and sometimes precise identification of militarily significant objects, such as rockets and artillery, troop units, surface ships and aircraft, roads, and command and control headquarters, but their field of view will be very limited. Multispectral and hyperspectral imaging sensors will further assist in the precise identification and technical analysis of objects on Earth. Lower resolution systems, a description that fits most civil remote-sensing satellites, will provide a look over a wider area, assisting area surveillance requirements and weather monitoring. Sensors with low spatial resolution may provide warning of hostile activity and cue more capable reconnaissance systems.[86]

A number of countries have data receiving stations that are not controlled by satellite operators. Information on weather, critical to mission planning and execution, is generally openly distributed and, thus, will be available from a number of sources. Knowledge of cloud patterns and storm activity may be used to hide force movements from U.S. overhead reconnaissance assets or from an important element of an offensive operation. Any question as to whether meteorological satellites can provide hard intelligence on military activities was answered affirmatively when, during Desert Storm, a homemade receiver in the United Kingdom picked up signals from ESA's low-resolution *Meteosat 4* satellite that, when processed, displayed images of troop concentrations in the Persian Gulf area.[87]

In answer to the all-important "so what" question, most foreign imagery may be of little military consequence in the near term, even in wartime. Much of the information obtained from viewing Earth is probably available from other sources. Yet an honest answer concerning the significance of viewing capabilities from space is a situation-dependent answer. The ability of regional powers to view objects and activities on Earth is a source of some concern—so much so that U.S. policy makers have gone about the nearly impossible task of controlling domestic imagery distribution using "shutter control" measures and by trying to persuade our allies to do the same.[88]

Even in an era when other countries cannot hope to match the capabilities of the United States in space, there remains a need for the defense and policy making communities to be wary of what others can see from space. One "snapshot" from high above may compromise otherwise clandestine operations or secret activities below. So, from an operations security and intelligence-gathering point of view, space will have much to offer future adversaries. Space will help potential foes reconnoiter the battlespace prior to launching operations and to facilitate target planning well in advance of any conflict.

In the near-term, defense planners can remain fairly comfortable that the technology and space gaps between the United States and others will make it very difficult for other countries to use space as the United States uses space, to integrate it fully into operations, to deliver timely data to war fighters securely and reliably, to develop an integrated space systems approach to war fighting. Tactical superiority is only an advantage, however, if the enemy of the future chooses to fight like the forces of the United States and its NATO allies.

Communications

Extraordinary advances have been made in the communications area since the first geosynchronous communications satellites were launched. The industry has experienced dramatic changes just since 1990, when only sixteen countries or agencies were operating their own satellite communications satellites, Inmarsat launched its first dedicated spacecraft, and Eutelsat had only 4 satellites in orbit. By 1996, the GEO, fixed communications satellite market had grown to 174 operational spacecraft, involving twenty-three nations and agencies. Growth was particularly profound in the Middle East and Asian Pacific regions. Ownership over dedicated satellite communications gave countries control over the distribution of telecommunications services, greater independence, and a new-found source of prestige.[89]

Foreign ownership of private communications and the growing demand for bandwidth propelled by the growth in Internet services may affect the availability of satellite communications capacity to U.S. forces. Foreign ownership means that, should a severe disagreement occur over U.S. policy, U.S. forces may be denied access to systems. A shortage of transponder capacity may make utilization of a foreign owned system impossible in any case. U.S. policy—which states that U.S. forces only may use communications systems equipped with tracking, telemetry, and command links (at present only Intelsat, Inmarsat, Orion, and Panamsat fit this bill)—may further complicate efforts to gain access to foreign communications systems.[90]

Routine access to mobile communications services and the expansion of direct-to-home TV distinguishes the modern space telecommunications industry. With the advent of mobile satellite services allowing near-instantaneous point-to-point communications and portable very small aperture terminals, or VSATS, foreign military forces will be able to improve command and control over units in the field and send data of tactical importance to combatants. VSATs are small enough to be mounted on vans or trucks, contributing considerably to the mobility of armed forces, who would be capable of receiving intelligence and targeting instructions from distant headquarters. One VSAT terminal can be used to transmit increasingly large amounts of data directly to multiple users at any point on the globe. The U.S.-dominated VSAT market has rocketed in Europe and Latin America, and is gradually building in the Middle East, Africa, India, and China.[91]

Hand-held phones operated by commercial satellite constellations will take mobility to the furthest extreme by allowing individuals and special forces to range in all directions over the countryside. Large bandwidth capacities and advanced processing on new satellites will make it possible for a user to change frequencies using different satellites within the same satellite system. Multiple communication pathways, in addition to high-capacity systems, will complicate space-control tactics that rely on jamming and terminal/node destruction.

GPS: A Global Utility

For more than two decades now, the U.S. Department of Defense has operated a constellation of satellites that provides precise position, velocity, and timing data to anyone in the world with a receiver free of charge. Unanticipated civil and commercial interest in GPS satellites has spawned highly profitable businesses established to produce GPS receivers and associated hardware. The declining cost of exploiting the GPS signal has contributed to the ironic situation where the military user is now, by far, the minority user of the system. The United States discontinued the encrypted Selected Availability signal, making it available to all users in May 2000, six years ahead of schedule.[92]

Already governments and armed forces around the world are looking for ways to exploit and improve the accuracy of GPS signals. Even more accurate differential-GPS systems are in operation in Europe and are expected to be featured in other countries, including China, Poland, and South Africa, and whole regions such as the Middle East.[93] Because GPS is a global utility and universally accessible, the United States must expect that future adversaries will use its signals to pursue objectives hostile to U.S. interests.

Can We Counter Hostile Uses of Space?

National responses to threats posed by the hostile uses of space are very much within the purview of the space-control mission discussed in chapter 3. If there were a need to revisit this ground yet again, it would be to reemphasize two points.

First, practical solutions to adversarial situations posed by hostile spacecraft may be found along the entire space-control spectrum ranging from international agreement and negotiation to a resort to military "hardball" in the extreme situation.

Indeed, there are several levels of space control that may be incrementally or simultaneously applied. Consider the following. Level 1 activities are preventative in nature and may involve technology controls and cooperative agreements with other nations as part of a larger strategy to limit the spread of space capabilities. Its diplomacy and negotiation skills give a country a Level 2 space control, or preemption, capability once the threatening activity has been identified. Threats to implement economic sanctions against the offending parties and other forms of diplomatic leverage would be communicated. U.S. military forces would at the same time increase operational security and implement deception operations (such as reducing operational activities during satellite overpass).[94] Level 3 activities may involve attempts to control the offensive activity using whatever commercial leverage may be available. The U.S. government would ask domestic imagery providers, including commercial providers in allied and friendly countries, to exercise "shutter control." GPS signal transmissions may also be limited. Communications satellite operators may cooperate to control services regionally.[95]

Deliberate, nondestructive interference with hostile satellites constitutes Level 4 activities. Information warfare operations (including the use of software viruses) to attack a satellite system, jamming or distorting adversarial signals, spoofing, and special operations to cut satellite control and processing centers from their power sources would attempt to deny temporarily hostile satellite operations. Finally, at Level 5, a more permanent solution may be sought using deliberately destructive techniques. Special forces and attack aircraft may be used to destroy with explosives critical nodes and facilities. However, such operations would risk the lives of the troops undertaking the mission, the lives of military and civilian workers or residents in the localities under attack would be jeopardized. Destruction or capture (using the space shuttle, for example) of satellites in orbit would in most instances be considered the most militarily and possibly even politically effective choice.

Desert Storm was the first war in which a concerted effort was made to control space.[96] The allies took several steps ranging from diplomacy to destruction to cut off Saddam Hussein from orbital support. Well before the start of hostilities, leaders of the U.S.-led coalition negotiated with SPOT Image to keep vital imagery data out of Saddam's hands. The space-control campaign also included air strikes to knock out satellite communication stations and disrupt Iraqi command, control and communications assets. The coalition space denial campaign was rather nontaxing. Iraq had no satellites for the coalition to neutralize or jam, and this time at least, there was little concern of the use Saddam's forces would make of

GPS satellites. It is also important to note that Iraq did not have an antisatellite weapon to confound coalition space operations, nor did it elect to use its UHF jammers, which could have shut down up to 95 percent of U.S. Navy communications.

The coalition, nevertheless, failed to cut Saddam completely off from space. Saddam not only had access to satellite weather data, but CNN was available to Iraq throughout the war, bringing extensive coverage of the war to the Iraqi leadership, including live briefings and news conferences, weather reports, in-depth analyses of the war's progress, and public reaction in the United States. The impressions and responses of U.S. citizens are very important, since knowledge of the general squeamishness or vigorous support in the United States concerning a military action would allow a foreign leader to better determine the course of a war.

An array of choices is desirable, so one would presume that optimal balance in the protection of national security would mean that the defense leadership should want to preserve an ability to act along the entire length of this spectrum, depending on the political, diplomatic, and strategic requirements of the moment. Conversely, arbitrarily narrowing the list of remedies to just a few presumably would constrain unduly national command authorities.

The second noteworthy point is a derivative of the first. A combined approach to denying another country, or nongovernmental entity, the ability to use space is a superior approach to securing space interests. Demarches may have to be coupled with ASATs in order to achieve the desired outcome. As a carpenter prefers to have many tools in his tool box so that he can tackle the job with the right tool, many instruments of space control are preferable to just one.

If there is one lesson that the reader should take away from this chapter it is that the space-control mission, especially as it involves preventing adversaries from using that medium for hostile purposes, is an increasingly complicated one. It is bound to be a future source of profound and agonizing frustration. The policy implications may well be staggering. Some U.S. officials believe that interference with a foreign satellite operated in a hostile manner will cause a dangerous disturbance in the diplomatic world. Disruption of commercial services may lead to lawsuits in the United States or new legislation protecting commercial interests. The potential is there to damage alliance relations, to turn the friendly into the disgruntled.

In real estate, the key to success is "location, location, location." In political-military events, the key to success is "context, context, context." Describe the situation, and then a reasonable solution may be worked out. Space is a big common area, and it is not very well surveyed, which means that there is a lot of space to do many disagreeable things. Clear answers for how best to proceed in preventing Earth's orbits from being used against U.S. interests are not now at hand, but policy on this matter is not at all clear. Indeed, the negative international and domestic political and security implications of space combat may be exaggerated, especially if larger issues are at stake.

One would be hard pressed to argue that twenty-first-century security problems favor simple solutions. A look at the communications area will give one an appreciation for this thicket we are entering. From a telecommunications perspective, satellite constellations have become no different from the cellular or wire line networks. The links and nodes in space are elemental cogs in today's highly capable and vibrant communications and information infrastructures.

An embargo of a commercially available communications service is one possible avenue for dealing with hostile forces using satellite communications. Or is it? There are many potentially serious drawbacks to this approach. Any number of companies emerging to make space-based communications a part of our lives are likely to have an international profile. One of the ways the United States may want to deny satellite communications services to a future enemy is to try, through a combination of negotiation and strong-arm tactics, to shut down a communications system within a particular country or region. But international board members may not want to shut down. Ownerships are likely to be mixed, and future adversaries may even have representation on company boards. The United States may be a minority stakeholder in the corporate structure, or the corporate legal entity may be out of U.S. jurisdiction in the Cayman Islands. Market considerations also will be thrown into play. Will commercial satellite constellations shut down their services to its customers and renege on promises to shareholders, especially if there is significant business competition?

The merchant will argue, and with some justification, that in times of crisis and war basic telecommunications services are generally *not* cut off. Taking into account recent history and the crises in Iraq and Kosovo, the United States has not bothered to interrupt GPS signals even though Iraqis and Serbians had full use of the navigational and positional services. Jamming or spoofing the signal is always an option, but until a method for denying a true signal to the enemy while at the same time preserving U.S. and allied access to GPS data is discovered, such countermeasures also could degrade the U.S. ability to utilize GPS. It may be that the U.S. military will have to operate on the presumption that future foes will be able to communicate and command troops more efficiently than in the past. To be sure, if it is at all achievable, selective deniability may provide one way to reestablish for the U.S. armed forces a communications advantage that would be palatable from a commercial provider's standpoint.

The United States could not deny Iraq during Desert Storm all data from remote-sensing satellites without impacting the operations of the coalition. Since the United States does not control all satellite weather data (indeed, even U.S. military satellite data is publicly available), any effort to shut off weather transmissions would have required the onerous task of convincing a number of countries (including the Soviet Union and China) that the merits of this space-control strategy outweighed the obvious drawbacks.[97]

Right now, we have systems to take care of some foreign space threats. At some point in time, however, more conventional methods, such as special operations or air attack operations to destroy ground nodes and terminals, will no longer work very well. As space assets and suppliers grow in number, the ability to first identify and then target assets exploited by the enemy necessarily will decline. One must also not discount the international repercussions that most certainly would arise should innocent bystanders be killed in our attempts to control space by destroying objects on Earth. Satellites using encrypted signals or that are impervious to blinding, such as the old Russian RORSATs, may also survive attempts to temporarily interfere with its operations. Even short of having to deal with dedicated military satellites, however, we must anticipate that at some point in the future that there will be so much activity in space, that we will eventually move to that next level of shooting—in space. U.S. proficiency to reach this next level, from a technological standpoint, is not in question.[98]

In space, "nonlethal" or "nondestructive" weapons may well win the admiration of defense and policy officials. Critics of combat operations in space argue that ASAT attacks would inevitably produce orbital debris, creating hazardous conditions in space for a long time and sending fragments speeding in multiple directions, possibly on intercept paths with other friendly satellites. On the other hand, the international relations disasters that are predicted to result from the use of ASATs and the prospect that friendly space operations may be jeopardized by space debris are unlikely to be such weighty concerns when American lives and vital national interests are at stake. Context, context, context.

What Is the Bottom Line?

Space systems provide global coverage, unique vantage points, and exceptional promptness, all of which are advantages increasingly useful to foreign military forces and intelligence communities. We have learned in this chapter that the proliferation of space capabilities will give potential enemies of the United States improved "vision," a remarkable sense of direction and understanding of geographic position, and a reliable ability to tie together electronically military units geographically separated from commanders and from the political leadership.

We may presume with growing confidence, therefore, that space assets will grant the enemy improved intelligence capabilities in peacetime and enhanced situational awareness on the battlefield. One does not have to be prescient to believe that in the future these developments will compel officials responsible for the country's defense to begin developing appropriate countervailing strategies and doctrines, operational plans, and acquisition programs, to include countermeasures to the enemy's new-found prowess in space, which may range from employment of deception operations on Earth to destruction of offending satellites. One does not need a Rosetta stone to know that the United States can no longer assume it will remain unchallenged or unmatched in this area.

The growth of space capabilities and services through military, civil-sponsored, and commercial means is essentially boundless. Because so much cutting-edge technology is now available on the commercial market, the advantages offered by space will be freely available (but obviously for a price) to other countries. Many questions concerning what to do from a policy and legal standpoint will arise. These questions will be complicated by an irrepressible and nagging fact—space technologies serve multiple purposes and cannot be filed neatly away into purely military, civil, or commercial categories. The real challenge will be to come to grips with what can and should be denied the enemy, and when, and to fathom militarily effective and politically acceptable ways to counter hostile uses of space.

To be sure, the information made available courtesy of satellites to the United States' foes will be of little consequence if the hostile party lacks the hard assets of war—troops, weapons, logistical support—and the opportunities to use these assets to strategic effect. While space certainly will grant advantages to wartime activities that rely on persuasion and effective communication, such as diplomacy and psychological warfare, information alone will not carry the battle and it cannot reliably win the war. Indeed, according to the authors of *Joint Vision 2020,* U.S. forces "will not necessarily possess a wide technological advantage over our adversaries."[99] So when considering the implications to U.S. national security of adversarial uses of space to facilitate information gathering, handling, and distribution, it will be important to consider all sides of the picture.

In 1995, Russian major general Vyacheslav Georgiyevich Bezborodov observed that "the use of space systems presents unique opportunities for performing military missions. . . . Each state naturally pursues its own goals. A clash of different countries' interests is inevitable here considering the strategic importance of space, contradictions existing in the world and the danger of outbreak of conflicts and wars."[100] The Russian general's matter-of-fact statement is instructive on more than one level. Along with his self-interest, man tragically carries his torment and discord wherever he goes, and we ought to be aware of this fact as he journeys into space.

Bezborodov's comments on space and security also point out the urgent need for Americans to struggle diligently to understand foreign and strategic perspectives. The enemy of the future may not, indeed probably will not, operate according to the United States' peculiar political, diplomatic, strategic, and cultural paradigms. Failure to realize this, as we shall see in chapter 6, could blind us to possibly surprising and highly damaging uses of space by an innovative opponent, even a comparatively weak one.

6

The Pitfalls of Arrogance and the Limits of Military Power

How Might a Technologically Inferior Adversary Gain an Advantage?

Threats from enemies that think in unorthodox ways and employ unconventional weapons have surfaced lately in defense lexicons and analyses. Such threats, commonly referred to as "asymmetric," are as old as warfare itself. Said Sun Tzu in *The Art of War* (c. 500 B.C.), a manuscript discovered in the West only in the late eighteenth century, "As flowing water avoids the heights and hastens to the low lands, so an army avoids strength and strikes weakness." He also observed that "rapidity is the essence of war; take advantage of the enemy's unreadiness, make your way by unexpected routes, and attack unguarded spots."

Some of the most alarming threats, in other words, involve attacks against the blind side, the employment of indirect strategy, and attempts by an enemy to be as unconventional and as unpredictable as possible in order to upset (not necessarily defeat) a foe's superior armed forces—to disorient or weaken them just enough to achieve a political or strategic object.[1] This is a "thinking challenge" more than it is a tangible capabilities problem. The military requirements presented here represent more a tax on our military and strategic imaginations than on our resources and technological ingenuity. Solutions begin at the top and are fundamentally the business of policy makers, strategists, and defense planners.

The energetic and innovative fighter will use asymmetric tactics to circumvent the strengths of opposing forces, exploit vulnerabilities, or attack in ways that cannot, or will not, be matched. Present-day discussions of asymmetric tactics by U.S. defense officials is bald recognition that future wars are unlikely to emulate the experience of the 1991 Persian Gulf War.[2] As long as the United States is a great military power, its enemies probably will not rely on conventional forces, using massed armored formations and air and naval forces to impose their will—that is, they are not likely to rely on these forces in the same manner as the United States. Indeed, a cunning and bold enemy will not face the United States' overwhelming military superiority head-on. Rather, he will strive to force upon U.S. commanders alien terms of combat and deterrence, introducing early on a discontinuity that could wreck a canned or otherwise predictable military strategy.

The failure to allow for, if not entirely account for, such unexpected occurrences could have very serious military consequences and, ultimately, undermine national interests. Although by definition we cannot know the unexpected, we can expect the unexpected (if not its specific content) and make judgments about our weaknesses and the risks attendant in any given situation.

Finding and exploiting an enemy's weakness is not new to our age, nor does it necessarily require all-encompassing intelligence capabilities or "dominant battlespace knowledge." The United States' potential vulnerabilities, after all, are not too difficult to identify. The country and its allies, for example, invest in programs to improve defenses against nuclear, biological, and chemical weapons, as it is generally recognized that the use of these weapons could have tremendous physical and psychological consequences on and off the battlefield. An enemy may choose to employ ballistic or cruise missiles, taking advantage of the fact that in the early twenty-first century the United States will deploy only very shallow active defenses against short- and medium-range ballistic missiles and no homeland missile defense whatsoever. Information warfare techniques would exploit the United States' heavy reliance on computers by aiming to undermine segments of its vulnerable information, communications, and command infrastructures. A military adversary also may seek to degrade the U.S. advantage in long-range and precision strike weapon capabilities by fighting in urban centers or under the cover of jungle foliage.

Another approach involves the exploitation of foreign and commercial space capabilities by U.S. adversaries and disruptions in the flow of critical information available to U.S. and allied armed forces. Space, in other words, represents one avenue by which an enemy may surprise the United States. Americans rightly view their prowess in space to be unmatched by any other country. Yet even a half-hearted adversary will attempt to nullify or overcome his opponent's advantage. It may be that supreme confidence turns into overconfidence, as U.S. commanders are led to believe that the satellites they rely on are invulnerable and will always be available. Sever those ties to space even partially, and unwelcome surprises could visit upon U.S. military operators untold misfortune and hazard.

It is also important for Americans to understand that simply because something is "foreign" does not make it irrelevant or inconsequential. This may be a difficult concept for Americans (as well as for peoples of other polities and cultures) to understand, for U.S. culture, commerce, language, security strategy, and political influence are globally dominant. While yielding bountiful returns in the political, economic, and security arenas of power, such dominance by the United States nevertheless also may have the effect of insulating Americans intellectually and depriving the country's leaders of important international sensibilities.

Foreign views of the conditions of peace, of what is valuable and dispensable, of what is politically and strategically important, of weapons and their utility, and of tactics might not conform to U.S. perceptions and understanding. Foreign countries necessarily will not have similar security objectives. Their measures of prosperity may not be familiar to us, they may view and use tools of strategy (such as arms control) and diplomacy differently, and they might not place the same value on human life and freedom. Different political and cultural ideals will influence national decisions, including military decisions.

Why must we consider foreign and strategic perspectives in making our assessments of the future of U.S. space power? Because so much of the analysis that dominates decision making on this subject today is political and theoretical (that is, highly abstract) in character, leaving little room for complex strategic arguments or complicating considerations of foreign ideas. If you want to raise the neck hairs of many legislators, administration officials, or policy analysts today, talk to him or her about ASATs. Debates today over the military uses of space cannot occur without the interlocutors immediately shooting beyond the stratosphere of reason to the level of political slogan or unproved, but very clever and easily understood theoretical assumption ("antisatellite weapons are bad because they would cause nuclear instability" or "the sanctity and purity of space must never be violated by weapons"). What is usually missing from the debates are references to historical experience (drawing analogies from other geographic environments and the political past) and strategic reasoning. Indeed, the reliance on slogans and abstraction to argue critical issues is in most every respect a mark of intellectual laziness.

This subject is really grist for part 3; for now it is only important to recognize that serious defense discussions have been thoroughly politicized, bound by theoretical shackles placed there by those who believe that preparations for engagement with the enemy in space must necessarily lead to an "arms race," which itself is held to be an irrefutably "destabilizing" turn of events, something to be avoided at all costs. Highly politicized slogans about the militarization of space, belief in the central importance of arms control to the strategic dialogue on defense space issues, and down-on-both-knees deference before sacred arms-control treaties (most notably the 1972 ABM Treaty) have squeezed out honest and competent consideration of the strategic importance of space.

The reader should not take away from this the idea that all arms limitations in space are ipso facto undesirable. The point is really more basic: all space combat acquisition questions and all doctrinal considerations concerning space warfare *ought to be examined with a strategic eye*, guided ultimately by high policy considerations and only *tempered* by special political sensitivities. The slogans themselves are utterly irrelevant to deciding the important issues at hand—or, perhaps closer to the truth, they have become a significant distraction. Indeed, there may be good reasons for attempting to restrict specific weapons from space, but the country's leaders should come to that conclusion only after turning over in their minds the most important questions with some intellectual care and rigor. Very little of the current policy discussion is imbued with strategic meaning and insight; rather, it has been influenced quite harmfully by a political correctness that has influenced both perceptions and actions related to U.S. national security since the early 1960s, when theories of arms and instability were propelled into the policy limelight by Thomas C. Schelling's and Morton H. Halperin's *Strategy and Arms Control* (1961).

But sound answers to military space questions remain fundamentally reliant on rigorous strategic analysis. How can military operations in space serve national security and economic goals? How might the use of our capabilities in space facilitate military success, our own and our enemy's? Why is space control the governing concept of a strategy for space power? What is stability, and what contributes to or detracts from it? Why can space be considered a fourth military environment? What are the strategic and military dangers we must face as a nation in space? Thoughtful works exploring comprehensively a strategy for space power, defining key principles of space power, and elaborating on the strategic impact of military space operations remain to be written.[3]

This chapter is not by any means intended to develop a strategy for space power, nor does it constitute a holistic analysis of how space may impact the strategic ends of war. It explores some of the interactions between U.S. and foreign space operations and strategy with an eye to possible effects on policy, and it strives to reveal to the reader what is possible in military science, to show that threats to security may arise in unanticipated guises.

What Is Mirror-Imaging and How Is It Harmful?

Antisthenes, the fourth-century B.C. Greek cynic, warned that you must "pay attention to your enemies, for they are the first to discover your mistakes." Indeed, if one is alert to one's potential enemy, one not only stands a good chance of learning his mistakes, one also may come to understand how he thinks, how he works, how he reacts, and how he fights. The United States, arguably, is the most observed, most studied country in the world today, simply by virtue of its power and influence and the attractiveness of its ideas on liberty and government. But do the country's leaders and analysts pay sufficient attention to life outside the United States? Is the U.S. intelligence community appreciably mindful to the military ways and traditions, not to mention cultures and deeply held beliefs, of countries such as China, Iran, Russia, and North Korea? These aspects of human activity, in the end, affect the evolution of foreign military forces and development of operational concepts.

Awareness of the threat or assessments of military challenges, in other words, has as much to do with evaluating an enemy's capabilities as it does with thoroughly understanding him. So-called mirror-imaging represents an intellectual failure to assimilate important cultural, political, strategic, and military information about a potential enemy and to assume that the adversary with whom we are dealing thinks, acts, responds, and basically functions "like us." Lack of awareness of the other side is a self-imposed vulnerability.

Another word for this phenomenon is "ethnocentrism," although the use of this term is unfortunate in the case of the United States, a country containing a multiplicity of ethnic groups. Nevertheless, defined primarily as the view that one's own group or state is the center of all activity on Earth, "ethnocentrism" does have application in the case of the United States. It is another word for arrogance, or the rejection of other cultures and ways of life as inferior and the casual dismissal of foreign modes of thought and action. Ethnocentrism works particularly hard on national strategic thinking, which must necessarily adopt a nationalistic view of the world, in no small measure because strategic thinking tends to reflect a host of traditional ideas, beliefs, and prejudices. As Ken Booth has argued, "In most professions it is not difficult to brush aside alternative viewpoints and accumulations of contradictory knowledge. In the case of strategy, its practitioners have more than enough to think about, what with the speed of change in technology, the convolutions of strategic doctrine, the need to grapple with the problems of core concepts, and the burden of trying to keep abreast of current developments. Under such pressures it is only human that some corners are cut. The problem, and possibly tragedy, is that the corner is cut where strategists can least afford to engage in underthink."[4]

In their already extraordinarily complex line of work, strategists, analysts and defense planners all too frequently tolerate a lack of curiosity about the enemy. Ethnocentrism, therefore, is apt to distort strategic thinking and interfere with adequate defense planning, largely because of a general reluctance among planners and analysts to learn about the character of the enemy and allies. From a military perspective, such a lack of curiosity assumes that U.S. forces will always have the initiative.

The strong desire to establish and work within a conceptually and operationally familiar environment has led some strategists of the past to mold the image of the enemy to fit their preconceived strategies.[5] Like Procrustes, who according to Greek mythology seized unwary travelers and tied them to a bedstead, only to cut off or stretch his victims' legs to make them fit the frame, the strategist who is unaware of his enemy strives for narrow conformity and, in doing so, operates contrary to the nature of things.

Strategic thinking is hard work. Conscious or subconscious avoidance of reality, or purposeful ignorance of uncertainties, clouds one's assessments. Far from being typical "political units" or "rational actors" on the international stage, or simply "State A" and "State B," polities such as Serbia, Iraq, and North Korea actually are complex entities and better likened, in the words of former British prime minister Winston S. Churchill, to "heaving, thrusting, pulsating organisms which think and act with purpose."[6]

As creations of nature, states pulse with an energy of life generated by their national, political, economic and strategic objectives, their different purposes leading to the development of different beings. Polities aim at unique ends, for each makes its own assessment of what is just and what is advantageous. Political organization and ideas matter, so much so that "national interest" and outlook do not always coincide with the interest and outlook of the regime—Chaing Kai-shek differed radically from Mao Tse-tung, Nicholas II from Stalin, Streseman from Hitler, and the shah of Iran from Ayatollah Khomeini, despite their governance over the same peoples and within similar geostrategic contexts.

In some cases, the ends of different regimes will be radically different, alien, and practically unrecognizable to Americans. The ends of the U.S. government, for example, are tempered by the sacred principles of the Declaration of Independence and set out in the Preamble of the Constitution—they are justice, domestic stability and tranquillity, security from external injury, promotion of the general welfare, and the security of "the blessings of liberty to ourselves and our posterity." In the case of Iraq, to provide a contrast, the ends of Saddam Hussein's regime are his absolute personal security and the assurance of unity in his tyrannical rule, dominance of the Persian Gulf, the Arab world, and its oil reserves.

Political ideals affect every national decision, to include military decisions. This fundamental diversity has meaning for all aspects of statecraft. Not properly incorporating this truth about the ways of the world into key political and military decisions can lead tragically to situations where the real identities of foreign states are misperceived. Military operations are essentially functions of a country's foreign policy, and as such should be evaluated and assessed in relation to its policy and strategic thinking. We will explore this in greater detail below.

The people of the world are not everywhere "basically like" Americans. According to one observer of this phenomenon, the United States "has been led throughout its history to trust its own favored image of a unified mankind and to overlook the plurality of culturally diverse images, world views, and patterns of behavior."[7] It is, to be sure, a real analytical challenge to understand the histories and the political and cultural intricacies of many states of the developing world.

While the United States may have invested enormous intelligence resources to understand the Soviet Union, and now Russia, the wars of the past several decades have not been against Russia. Rather, recent conflicts or crises have involved such "backward" or "unsophisticated" states as Vietnam, Iran, Iraq, Somalia, Haiti, North Korea, Serbia, and East Timor. This is not the place to examine in detail these strategic-diplomatic encounters. It may nevertheless be said that a general lack of knowledge about the other side by U.S. policy makers, supplemented to some degree by the belief that the country's opponents possessed inferior martial skills and spirit, have contributed to the creation of what may be judged as bad policies and strategies toward these countries.

Key questions lie at the core of good policy and strategy making. What do we know about the political imperatives and long-term foreign policies of these states? How might their national histories and cultures influence present courses of action? Do we understand their strategic interests, ambitions, and internal problems? And how are they likely to engage superior U.S. forces in battle and achieve their goals without doing battle with the United States? To be sure, the country's enemies will not always fight as U.S. forces fight, nor happily accept U.S. terms of combat and deterrence. To think otherwise is an exercise in self-deception, a demonstration of willful ignorance of the moral and physical qualities that sustain the enemy, an ignorance that, in the end, may create for the United States military disadvantages.

Counterspace tactics that may be abhorred by U.S. officials in fact may represent perfectly acceptable military alternatives to the nation's adversaries. As a signatory to the 1967 Outer Space Treaty and the 1963 Limited Test Ban Treaty, current U.S. policy and political sensibilities adhere rather closely to the position that nuclear weapons must not be deployed or detonated in space. Based in part on its own experimental nuclear detonations in space in 1962 the United States eventually came to terminate its only viable antiballistic missile and antisatellite programs, which required a nuclear explosion to destroy the target. The tests confirmed what the United States recognized for a number of years already, that a large nuclear explosion above the stratosphere could cause a massive disruption of electronics and blackouts across a very large area in space, paralyzing many of its own space operations.[8]

Although analysts are in near-unanimous agreement about the disabling effects of the electromagnetic pulse on machines having electronic circuitry, many are quick to caution that the enemy's use of nuclear weapons in space is not realistic: "Why, after all, would the enemy use a weapon that would not discriminate between friendly and hostile spacecraft and effectively fry its own electronic operations?" And, "Are there not less costly, and easier ways to use nuclear weapons?"[9]

Consider the following. U.S. policy makers and defense planners regard the Chinese People's Liberation Army as technologically inferior to (which may translate in the minds of some as militarily less capable than) the more modern U.S. armed forces. And in head-to-head, like-on-like combat, this assessment could be correct. But as we shall explore in greater detail, China's technological disadvantage is really just a matter of perspective in the world of strategy—a much more interesting consideration is how Chinese forces may use available technology.

Consider also that China may one day attempt forcibly to claim Taiwan, a 230-mile-long island off the coast of mainland China, knowing full well that such a campaign could lead to war with the United States. While Taiwanese leaders have consistently opposed political independence from mainland China (and would seek unification only on condition that Beijing reject communism and institute democracy), Taipei walks a fine line between viewing itself as an independent state and as a province of China.

Of course, China could best secure its objective if intervention by the United States were avoidable. Ideally, in a hypothetical confrontation, the Chinese leadership could endeavor to utilize the advantage of surprise and press its initiative with urgency until its objectives were in hand. According to strategic analyst Chou Kuan-wu, "China boasts a large number of long-range antiship missiles and submarines. If reconnaissance satellites are used to coordinate a saturated strike of these missiles and submarines off the Chinese shores, neither U.S. nor Taiwanese naval forces could resist."[10] Lacking the required logistics capabilities and the command, control, and communications infrastructure, Chinese strategists naturally will look to delay a U.S. response to their aggression. It might do so, not by meeting U.S. military power toe-to-toe, but rather by implementing a counterspace campaign to attack the C^3 assets upon which the U.S. armed forces depend. In many cases, the communications satellites used for military purposes are commercial and not hardened against nuclear weapon radiation, and hence would be susceptible to disruption. U.S. war games have provided good reason to believe that attacks against satellites probably would occur early in a war and possibly cause other U.S. and allied military operations to grind to a halt.[11]

Were China to detonate a series of nuclear explosions in the air above the Taiwan Straits and in space, it would paralyze resistance activities by causing blackouts in Taiwan, and it would cripple unprotected space-based communications. A fifty-kiloton nuclear weapon detonated sixty-two miles above the earth would pump enough electrons into the Van Allen belts to cause havoc for all unprotected satellites.[12] Familiar with the effects of EMP, China could plan in advance for combat in an electronically disturbed environment. Should it move out quickly to destroy critical military targets with its missile and air forces and strive to control the strategic points in Taiwan with its ground forces, China might well achieve its war objectives long before Washington could mount a serious response.

Chinese military writers have expounded on the virtues of "wars of quick decision." They also believe that the high-technology approach relied upon by the United States ultimately makes it vulnerable to losing protracted wars.[13] In the event of resistance from the United States, and considering that nuclear weapons already would have been introduced in the battle, the U.S. Navy also may become vulnerable to nuclear attack. These are shock tactics, to be sure, but given the United States' lack of defenses against such assaults, they are not outside the realm of the possible.[14]

While the United States might not be materially harmed by this outcome, the end result, nevertheless, might be the defeat of U.S. policy and influence in Asia. Not only does this hypothetical example illustrate that one cannot judge the outcome of war simply by looking at the balance of forces, it also shows that there are assumptions about the enemy, and the tactics he might use, that are in need of careful examination. Much as the planners of World War II did not anticipate Japan's use of Kamikaze tactics toward the end of the war (who in their right mind could have believed such suicidal tactics were even possible?), the use of nuclear weapons for strategically supportable reasons may not be a source of disgrace for the opponent.

The future opponent may endeavor to conduct war by more "unsophisticated" methods, using every means other than electronic to communicate, which would free him up to cause massive disruption of space communications. There were numerous examples in the Vietnam conflict and the during the crisis in Somalia where a technologically simple and economically undeveloped people managed to construct a communications system that adequately serviced their ends. Our values are not necessarily their values. Our ways may not be their ways. Just because we would not do it, does not mean they would not do it. While we may expect that a regional opponent must fight his way through the heart of the U.S. armed forces before he could reach his goal, that same opponent likely will contemplate ways to reach his strategical ends without crossing swords with technologically superior U.S. forces.

The superiority of U.S. military power and the United States' strong adherence to a power projection strategy for protecting global interests will mean that a serious military adversary *must* consider indirect approaches. U.S. defense planners ought to presume that future foes will pursue anti-access strategies and design and develop forces (such as WMD-armed ballistic and cruise missiles or antisatellite weapons) principally to keep U.S. forces out of a particular theater. If the strategy adopted by an enemy is asymmetric, the kinds of support he will need from space will be very different from the kinds of support the United States will need. Washington may have to confront a country that does not own much in the way of space infrastructure, but nevertheless is able to exploit space for military ends.

If the enemy can buy imagery products and continues to have access to space during a protracted conflict, this may be a more difficult problem with which to deal than if he simply owned his own satellites and operated in space as Americans operate. The denial and negation of third-party satellites used for hostile purposes becomes a severe military *and* diplomatic challenge if the enemy can skillfully use international, commercial, globalized, networked space systems, thereby retrieving militarily useful products from different places, some through third parties, others under assumed names. The reader can be assured that Saddam Hussein's views of the utility of commercial satellite imagery, or the possible political and strategic implications of attacking those satellites, will not mirror the views held by U.S. policy makers. U.S. scruples in this and other matters most certainly will not be shared universally.

How Might the Weak Defeat the Strong?

Arrogance contributes to mirror-imaging, leading one to disassociate with the unfamiliar, which can cause one to make distorted assessments. Arrogance, by deflecting proper attention from the military and political sciences, also can make for bad strategy. There are, after all, approaches for achieving strategic ends that challenge more conventional notions that hold tightly to the belief that the apparently "stronger" will always defeat the apparently "weaker." If it were only that simple.

The Pitfalls of Arrogance

Hubris is a nasty vice. Excessive pride in its military and economic strength, in its diplomacy and technology, left the United States politically paralyzed and unable to exploit vastly superior military power to achieve victory in the conflict against North Vietnam (1965–73). Despite impressive technological superiority in air, on land, and at sea, despite its exotic weapons and advanced overhead sensors, and despite the day-in, day-out punishment inflicted on the North Vietnamese through air power, the United States was unable in the end to arrest enemy infiltration into South Vietnam. Following evacuation, Washington was unable to stop the regime it supported from crumbling under the crush of a conventional North Vietnamese assault some two years after the Paris peace agreement was signed.

Of course, who is strong and weak, who is more or less powerful, is relative. The stronger is clearly the winner, irrespective of brute strength. Put differently, what good is a country's military power unless it can achieve its objective? Strength on paper—who has more forces, who deploys more advanced weapons, who utilizes more efficient communications—is a good indicator of who *should* come out on top in a military contest. But then moral considerations tend to modify these initial estimates. We know that fighting spirit, élan, and morale are critical to the initiation of action and pursuit of an object, that the side without these qualities lacks life in its military purpose. And so it is that who appears to be the stronger at the outset of a martial contest can begin to cross that bridge over to the losing side. Great Britain clearly was militarily stronger in the war to suppress rebellion in its thirteen American colonies, but it lost nonetheless.

Now add to this increasingly complex question of "who is stronger" a potpourri of political considerations. Are the people of the stronger side in harmony with the leadership? Do they support military action? Does the leadership understand the political and strategic objectives of the enemy? Is the opposing side subject to domestic or international political pressures? What actions is the enemy capable of perpetrating at home to bolster his power? What might the enemy do to strengthen his position through coalition building? In other words, politics too can weaken one's hand by undermining support for, and complicating the execution of, war plans.

What then happens when we add to one side or the other the advantages offered by military science—that is, strategy? A good strategist will excel at turning a weakness into a strength, or turning the enemy's strength to work against him. He will understand whether a given political object is attainable through a proposed strategy. In the end, strategic considerations may make head-to-head battlefield strength irrelevant. According to Sun Tzu, "For to win one hundred victories in one hundred battles is not the acme of skill. To subdue the enemy without fighting is the acme of skill." Said differently by twentieth-century strategist B.H. Liddell-Hart, "In the case of a state that is seeking, not conquest, but the maintenance of its security, the aim is fulfilled if the threat be removed—if the enemy is led to abandon his purpose."[15]

Destruction, though an important part of battle, is not essential for a military decision. Dislocation is the aim of military strategy. Keep the opposing side off balance. Surprise him, use and move forces in unexpected ways, keep him under strain, respond imaginatively to changing circumstances. In such a way, who is the stronger on paper can be made to crumble in the course of battle.[16]

Thus, when considering warfare of any kind—including warfare involving space—Americans need to beware of the pitfalls of insolence and avoid complacency that has tended to dominate American political minds in the wake of the cold war and Desert Storm victories. No other nation today can materially compete with the United States on the battlefield, especially in the space arena, which

U.S. forces skillfully exploit to receive unparalleled tactical support. But does the fact that the United States is dominant across the board mean it cannot be seriously harmed, that the nation can disrespect foreign power? The Vietnam experience suggests that the answer to these questions are no. The downside of dominance is arrogance.[17]

Limited Wars and Indirect Strategy

In his *History of the Peloponnesian War,* Thucydides relates a famous dialogue between the mighty Athenians, who were powerful on the land and at sea in their fifth-century B.C. wars against Sparta, and the far weaker inhabitants of the tiny island of Melos. History records that the Athenians believed they were compelled to teach the Melians a lesson for their insolence and open hostility toward the great polity of Athens. The "diplomatic" expedition against Melos was intended to be impressive: thirty-eight naval vessels, sixteen hundred infantry, three hundred archers, twenty mounted archers, fifteen hundred heavy infantry. The Melian leaders, who did not consult their people on this important matter, in the end did not relent in their rejection of the Athenian entreaty to join their camp.

This story, of course, did not have a fairy-tale ending. The conference, having failed to persuade the Melians of the folly of their ways, led eventually to the complete reduction of their community. After surrendering to the discretion of the Athenians, according to Thucydides, the victors "put to death all the grown men whom they took, and sold the women and children for slaves, and subsequently sent out five hundred colonists and inhabited the place themselves."

Despite the unsettling nature of this event, one is not really all that surprised by it. Such is how imperious, domineering powers can behave. The Athenian understanding of international relations is one that is shared by many other powers: "the strong do what they can and the weak suffer what they must."

More accurate, though, is that the strong do what they can unless the weak can "outstrategize" the strong or the strong somehow fail of their own accord. History is full of examples of weaker, underresourced, numerically inferior powers prevailing over powers that are superior in number and capability. The record may be traced as far back as the Persian defeat at the hands of the weaker but determined Athenians in the Battle of Marathon (490 B.C.). Thanks to the energy of the young democracy's military leader, Miltiades, the Greeks met the forces of Darius upon their landing at Marathon. Following their hard-won victory, the Athenians doubled back rapidly to defend their city, much to the surprise of the attackers, who relented and departed without achieving their objective (the punishment of Athens for their role in the Ionia revolt). This battle is said to have broken forever the spell of Persian invincibility and generated a spirit that led to successive Greek victories against the powers of Asia.

The historical record of unexpected victories continues up to the strategically inept attempts in the 1990s to control tribal warfare in Somalia and render impotent Iraq's Saddam Hussein and relieve him of his weapons of mass destruction. U.S. leaders underestimated the resilience and persistence of the more primitive ragtag Somali forces, which were controlled by the warlords, and were so shocked by the combat deaths of nineteen elite U.S. troops that they immediately terminated the humanitarian mission in Somalia and beat a hasty retreat. Saddam Hussein, who for almost a decade has subverted attempts by the U.S.-led coalition to undermine his regime and reduce his military power, has survived extraordinary international economic sanctions and several bombing campaigns.

Making war involves not only sensible estimations of the balance of military power. The United States may have superior space information forces, be able to swarm the skies with attack and combat fighters, and provide its soldiers with the latest in portable communications, exotic weaponry, and night vision technology, but all of this may not matter unless there is a sound strategy for employing them. To presume that the United States will prevail in all contests because of its superior capabilities is to presume that battle is the only means to the strategical end. Not all victories will come upon completion of a "decisive battle" or a massive bombardment campaign.[18]

In his book *On War* (1832), Clausewitz postulated that victory will come upon the defeat of the enemy's armed forces. The first man to systematize the study of war defined war as "an act of violence to compel our opponent to do our will." One lesson he drew from the Napoleonic Wars was that, in its most ideal form, war approaches an absolute state, where the act of violence must be performed with all one's means and with all available will, fighting until the strength and energy of the opposing side is exhausted and the enemy's military arm is put down.

But later in his work, the Prussian theorist and general cautions that this principle, while perhaps most applicable in the early stages of a war, is really only the starting point.[19] A state may pursue what Clausewitz called a limited aim, such as the seizure of a piece of the enemy's territory, which may be used to pressure the opposing side into accepting defeat. There are, in other words, other ways to compel an enemy to do one's will.[20]

Clausewitz recognized in book 8 of his work that the theory of warfare having defeat of the enemy's forces as its sole aim was too simple and abstract, for it did not adequately explain all the wars that had gone before the Napoleonic Wars, nor, as it turned out, did it explain all those that have since followed. Wars, in other words, are never purely military endeavors aiming at the extremes of what is possible. Political considerations and the degree of national interest always modify military energy. Policy is the master and must never be suspended by military action.

Indeed, political considerations make possible other, more limited opportunities for exploiting the enemy's vulnerabilities. War is an international relation that differs only in method from other international relations. Thus, the dominant questions in war include these: What is the political object of the war, what are the political conditions, and how much does the issue at hand mean to each side (how much is each side willing to risk)?[21] It is not the case that every consideration in war must be subordinated to the aim of fighting. According to Liddell-Hart, "Strategy has not necessarily the simple object of seeking to overthrow the enemy's military power. When a government appreciates that the enemy has the military superiority, either in general or in a particular theater, it may wisely enjoin a strategy of limited aim." He goes on to note that such a policy of pursuing limited aims has "more support from history than military opinion hither to has recognized."[22]

One of the more famous expositors of the classical theory of war, and a contributor to British naval policy, Sir Julian Corbett, observed in this regard that "the elements of strength in limited war are closely analogous to those generally inherent in defence. That is to say, that as a correct use of defence will sometimes enable an inferior force to gain its end against a superior one, so are there instances in which the correct use of the limited force of war has enabled a weak military Power to attain success against a much stronger one, and these instances are too numerous to permit us to regard the results as accidental."[23]

An inferior army pursuing the defeat of his superior enemy's armed forces follows a strategy that borders on the suicidal. Iraq's Saddam Hussein took on the U.S.-led coalition in the Persian Gulf War on the presumption that the United States would not have the political will to continue once it started taking massive casualties, not on the presumption that he could defeat the coalition in a head-to-head battle. To be sure, Saddam miscalculated the international will to reverse his aggression against Kuwait, much as Japan during World War II misjudged the conviction of the Allies to reverse its ill-gotten gains in the Asian Pacific. But at one point leaders in Iraq and in the Japan of the 1930s and 1940s must have had cause to believe that they could succeed.

To move directly on an opponent consolidates his balance and increases his resisting power. If they are to win, inferior forces necessarily must pursue an unapparent route to victory. Strategy is not merely concerned with the deployment and movement of forces, but with the effect of those forces. It may well be that the best use of military forces is to withhold their entry into battle, or to employ them at the margins and await a change in the operational or strategic balance of power. Readers of Julius Caesar's *War Commentaries*, which describe his military conquests and campaigns in the Gallic wars, will appreciate the candor of this observation. It may happen that a protracted period of involvement will drain the enemy's will and material power. Pinpricks on his flanks, rather than risky blows against his main force, may be the best way to weaken him. Other ways to engage only partially may include raiding his supplies, luring him into unprofitable attacks, or causing him to disburse widely his force, all of which could have the effect of exhausting his moral and physical energy. Such a strategy will succeed, of course, if the drain on the enemy is disproportionately greater than on one's own forces.[24]

What Kinds of Strategic Effects Might Space Operations Have?

Although space operations change the character of war, the timeless principles of strategy apply in all wars. The space environment will compel the consideration of new variables and the discovery of new principles applicable to tactics (the employment of weapons), concepts of operations for weapons and space systems, and doctrine (which gives discipline to tactics). Operations in the space medium also will change the U.S. force structure and affect military education and training. However, the many unique features of that environment—the predictability and duration of orbit, the force of gravity, the uncanny speeds of orbiting objects, the limitless operational area and the global access it provides, the vacuum and extreme temperatures of space, the lack of a fixed geography, and the operating distances from Earth—do not change the fundamental principles of strategy.

Policy should guide the conduct of a war, and strategy (which orchestrates means to service ends) will always be the handmaiden of policy, whether strategy concerns itself with operations on Earth or in space. The political object on Earth remains the driving force behind any war or diplomacy. Space is simply another medium that may be exploited by military means to affect life on Earth. It is only because space is so unfamiliar to us that we may even consider that laws of nature and the basic rules of warfare somehow might not apply.

Space operations, including offensive and defensive combat in space, may be fully incorporated into a war of limited aim or a strategy that seeks to be roundabout and exploit surprise. A strategy for space age warfare will adhere closely to tried and true rules for military success. Approach your opponent with indirectness, so as to ensure his unreadiness to meet your move. According to Liddell-Hart, that "indirectness has usually been physical, and always psychological. In strategy, the longest way round is often the shortest way home."[25] There are clear advantages to fighting in an unorthodox manner—"unorthodox" being whatever the enemy does not expect.

In other words, though it may not appear that an adversary has the infrastructure and the capabilities to challenge the United States in space, the most dangerous military challenge is not necessarily the most obvious and direct. In the end, "the longest way around" to arriving at a point is still as effective, indeed usually is more effective, as the shorter way. Saddam Hussein surprised the world in the early 1990s with his clandestine effort to manufacture weapons-grade fissile material for nuclear weapons using calutrons and techniques that had long since been discarded by the major nuclear powers for more efficient and advanced processes, although they had been used by the United States in producing critical nuclear material for its first nuclear weapons. His was the "long way around," but it was frighteningly effective.

In war, the object is to minimize effective, militarily significant resistance by the enemy. Since the placement and use of physical objects (troops, weapons, etc.) affects the outcome of war, movement in the battlespace is a fundamental tool of strategy. In the physical or logistical sphere, dislocation or disorientation is produced by upsetting the enemy's disposition (the orderly arrangement of his forces), compelling a sudden change of a front (he may be concentrating in the wrong direction), or dislocating or separating the distribution and organization of his forces. These same effects may be achieved by endangering his supplies and his lines of communication. The larger the forces involved, and the more complex the operations are, the more command and control of those forces and operations will be challenged.

Space assets are now a fundamental part of the general disposition and the overall operational effectiveness of modern armed forces. As physical objects—whose predictability, reliability, and accessibility are vital to operational effectiveness—they too are vulnerable to attack or exploitation by an enemy who endeavors to keep U.S. forces off balance. Indeed, the complexity and information-intensive nature of modern military operations can work in favor of the enemy. Disruption of lines of information (earth-space links) important for navigation or communication may slow down the movement of forces, make synchronization of operations impossible, or alter the operational tempo, resulting in localized or even theaterwide dislocation of ground, sea, and air forces. The "high-tech" nature of space assets is very vulnerable to low-technology threats, such as jamming and nuclear weapons. The disorienting effect caused by such weapons could have lethal consequences at the levels of operations and strategy if the enemy is able to exploit in a timely manner the openings he has created.

The enemy also might seek to establish another "front," one overhead, by launching imaging satellites or aggressively exploiting commercial satellite services. The effect on U.S. operations may be to cause commanders to alter plans and become more concerned with using deception to disguise ground operations, until such time as U.S. antisatellite forces (assuming they have been developed and deployed) are set loose to take out the prying eyes. It also may cause forces to be deployed to locations to undertake operations that are less profitable or attention to be diverted to more active space-control operations, such as destroying ground receiving terminals on enemy territory using air power or special operations forces, which may be very risky politically. In any case, by interfering with decisively intended moves, a surprise use of space assets by the enemy may have the effect of confusing U.S. commanders, who would no longer be able to concentrate on those objectives they wish to pursue in order to bring about a favorable end to the conflict.

The other means for reducing the enemy's resistance lies in the realm of the psychological. The sense of being trapped, or outmaneuvered, or in imminent danger may be expected to have psychological repercussions that result in the dislocation of military operations. Menacing the route of retreat may have this effect. Seriously damaging one or both U.S. major space launch facilities or highly vulnerable space ground terminals also could cause psychological tremors, especially if U.S. armed forces were engaged in a protracted conflict and reconstitution of space forces was an essential element of its military strategy. Suspicion that GPS signals had been rendered inaccurate, or at least unreliable, also would act to unnerve and disorient U.S. forces by causing general mistrust of ordinary navigation and positioning tools. The general purpose of these actions by the enemy would be to shake the United States' confidence in the predictability, reliability, and accessibility of its space assets.

The enemy does not have to meet U.S. armed forces head-on in order to chalk up a victory. He may pursue what we have called a limited aim. State A's limited war against State B depends on the importance State B places on a limited object—that is, an object important but not absolutely vital to the survival of State B. Limited objects could be useful for forcing a favorable peace. In past wars, limited objects have included territories that are adjacent to one or both of the belligerents (such as Alsace-Lorraine between Germany and France), an overseas territory (such as Guam, the Philippines, or Puerto Rico in the case of the United States), or even a portion of the enemy's force, where the protection of that force becomes such a concern that it starts to equate in importance with the overriding war objective. The survival of the British Royal Air Force in the Battle of Britain in World War II, in the end, may have meant survival for the country. The Battle of Britain highlights how important protection of a capability can be to the overall war effort, although loss of that capability may not necessary translate into a lost cause.

The importance the United States attaches to space, coupled with the reality that its significant preoccupation and main political object in any war must necessarily always be on Earth (the seat of power and political decision), will make a strategy of limited war very attractive to its future enemies. The natural difficulties and costs associated with reaching space and establishing important lines of information only give impetus to the enemy to consider a strategy that strives to defeat the United States by pressing on the space front. Any operations that increase the difficulties on the space front, in effect, stress all other major military operations and increase significantly the likelihood that the United States will fail to reach its main political object.

We ought not to expect future enemies of the United States to undertake operations for permanent conquest of U.S. territories. Such an offensive would require tremendous military capabilities that are beyond the reach of every other major foreign power today. Future enemies will undertake operations having more limited aims, aims incorporating methods for disturbing U.S. war plans and foreign policy and strengthening their own positions.[26] The power of enemy counterspace operations will be in their ability to contain larger forces, and this might be all that is required for a foe to win favorable terms of peace.

What Is Being Said Abroad about Space and Warfare?

One might get the impression that much of the above discussion is academic in nature, only tenuously connected to the real world. In fact, U.S. policy and defense officials, analysts, and scholars are not alone in considering questions having to do with possible asymmetric relationships involving space and war. Indeed, the very insights presented to us by Sun Tzu, Clausewitz, Corbett, and Liddell-Hart may be discerned today in the plans and dissertations of military thinkers around the world.

Considerable time and energy are spent in foreign military headquarters, academies, and defense universities thinking about the newest "revolution" in military affairs, to include examinations of the strategic and doctrinal implications of spacecraft and information warfare techniques, among other advances in military technology. Let us consider here primarily the examples of China and Russia, whose military writers and analysts have provided us with a trove of strategic analysis concerning the meaning of warfare in the space age.

China

Chinese analysts believe that while the land frontier will continue to be the focal point for warfare, outer space is becoming a new arena for competition. The military uses of space have not gone unacclaimed among China's political leaders, military officials, economic planners, defense science and technology researchers, academics, and journalists. Twenty-first-century China has the facilities to research, design, develop, produce, and test satellites and launch vehicles, all of which have allowed the country to develop a strong communications industry. China has long-term cooperative programs with advanced countries such as Germany, the United States, Russia, Canada, France, and Brazil, and is seeking to extend those international ties to other countries. These relations have helped provide China with reliable meteorological and remote-sensing capabilities (including Landsat receiving stations, which China has used for intelligence and military purposes), which could be put to military use.[27] Consider the following passage written by Chou Kwan-wu:

> Although China's reconnaissance satellites cannot "see the whiskers," they still have irreplaceable and decisive military value. Saddam Husayn was almost utterly routed during the Gulf War partly because he did not have any means of strategic reconnaissance. . . . China may be poor, but it is a big country nonetheless and possesses satellite reconnaissance capability that even developed countries like Japan, Germany and Britain does [*sic*] not possess. If the Americans tried to interfere with China's internal affairs, such as over the Taiwan question, by military means, they will discover that the Chinese can read their global military moves like the back of their hand. Dealing with China is a lot harder than dealing with most other countries.[28]

China established in 1997 the first space tracking installation outside its national territory in Kiribati, Tarawa. Kiribati, located directly on the equator near the International Date Line in the South Pacific, is a prime location for satellite management and launching. The center will not only have useful commercial applications, but also a range of militarily strategic uses, from satellite control, data downlinks, to data intercepts. According to one industry expert, "China has done a careful lessons learned from the Gulf War, and has understood the importance of space control and space denial for future wars. China has gotten one of the truly choice pieces of real estate from the stand point of command of space."[29]

China, in fact, is modernizing on many fronts—despite its very limited per capita resources. Over time, China believes it will be in a position to benefit substantially from cutting-edge information technologies that will enable the PLA to enhance its war-fighting capabilities.[30] It will modernize its forces because, according to Chinese military analysts, "in coming high-tech local wars, we will face powerful enemies with high-tech weaponry, with our opponents likely to be space powers with space forces, as well as probably regional powers or groups with great-nation space force support. So we will be unavoidably threatened by hostile space forces. In all crucial military actions in future high-tech local wars, we will have to conscientiously consider the key factor of space forces."[31]

Space is now considered to be one of China's "strategic frontiers," along with the land, sea, and air frontiers. Its thinking about the future of space warfare is serious and holistic, complemented by official calls upon the United States and

Russia during the 1980s and 1990s to prevent the weaponization of space. According to senior Chinese military officers,

> As space forces are used ever more widely in wars, countermeasures aimed at space force systems will inevitably appear, including destructive actions aimed at firing, measurement and control, support, command, and user systems, various electronic strikes against military space-borne systems, and land-based anti-satellite weapons threats that are likely to evolve in the near term. As these will pose a greater threat to space forces in high-tech local wars, we need to pay them much attention. So land-based defense, anti-electronic strike, and anti-satellite weapon countermeasures for space forces in coming wars, are matters that need conscientious consideration in the building and use of space forces.[32]

The reason for attending to the maturation of space forces, according to Major General Wu, is simple: "To control future battlefields does not mean only control over land, air and sea, but also over electromagnetism, information, and even outer space."[33] The PLA general wrote that the control of these last three environments are vital to realizing dominance over the air, land, and sea. "Military satellites are now legitimate targets in war," according to the latest thinking among Chinese defense professionals, "and thus ASATs are legitimate weapons."[34] In the war zone where two sides compete for strategic initiative, "the entire world and even parts of outer space are fair game in a high-tech war."[35] According to another Chinese writer, "The U.S. `Star Wars' program showed that as long as there are wars, mankind will one day fight a space war to gain `space domination,' conduct space-based offensive and defensive wars, and launch space-to-ground and space-to-sea attacks as well."[36]

All of these developments have military strategic significance, especially when one considers the expanding dependence of the U.S. armed forces on space technology. Blind devotion to narrow technological or force superiority will make the country more susceptible to defeat. Victory for an opponent of the United States, as we have seen, may come easiest if the strategy adopted strives to avoid direct confrontation with the enemy and seeks to knock him off balance by striking him where he is weakest or assaulting an unprotected center of gravity. "We should not fight with the enemy in a way anticipated by the enemy, in a time and in a place that the enemy is expecting," write Senior Cols. Huang Xing and Zuo Quandian, reflecting points made by Sun Tzu. It is important to strike out at those targets that promise the best results "whenever and wherever it is most suitable," they note. "Only in this way will we be able to change inferiority into superiority, and passiveness into activeness, and thus win the initiative in conducting operations." Victory will not come through "frontal engagement and direct, regular competition in battle strength." Rather, "only by adopting some irregular operational forms and modes will we be able to constrain the enemy's combat strength."[37]

"Paralyze the enemy by attacking the weak link of his C^3I as if hitting his acupuncture point in *kungfu* combat," it is written in a 1997 Chinese Defense University publication. Others have argued similarly the virtues of "paralysis combat," where the military object is not the destruction of the opponent's armed forces. "If there is no unified command and control monitoring and early warning by the information-transformed C^3I system, then it is difficult to obtain timely, reliable intelligence," writes Ch'en Huan. By striking the enemy at the "vital point," he continues, assaulting units are blinded, troops and weapons amount to "nothing more than a pile of trash," and you effectively "paralyze the enemy and collapse his morale."[38]

Consider also these reflections by Senior Col. Shen Kuiguan, who wrote in 1994 that the "concept of defeating a powerful opponent with a weak force . . . has continually been put forward by people from the Yin, Zhou, Ming, and Qing Dynasties." A country may be backward in weaponry and economics, he stated, but may be able to exploit the political and geographical conditions of a given situation, take advantage of one's own strong points, and profit by the enemy's shortcomings. The World War II Battles of Britain, Stalingrad, and Moscow, as well as the Vietnam War and Israel's wars with its Arab neighbors, bear out this conclusion. The Persian Gulf War, he noted, cannot be fully explained by looking only at the role of high-tech weapons. The real causes of Saddam Hussein's defeat may be found in the political nature of the war, the nearly complete international isolation of Iraq, as well as his economic dependence, inflexible strategies, and passive defense tactics. "A comprehensive understanding and analysis of the Gulf War," Colonel Shen correctly points out, "are needed in order to avoid the erroneous conclusion that it is impossible for a weak force to defeat a powerful opponent in a high-tech war."[39]

"The traditional linear positional operation," wrote Major General Wu, "will hardly meet the demands of future military actions. It can be estimated that new mobile warfare, nonlinear operations, unbalanced dispositions, and mobile actions will be adopted by more and more armies of the world."[40] Space assets also can contribute to the execution of indirect strategy. Given the spread of information platforms to space, notes Ch'en Huan, the "battlelines of the past, like the Maginot Line and the Bar-Lev Line are no longer terrifying shields," as future wars will contain "nonlinear attacks on enemy objectives" and will not necessarily use a "one-by-one breakthrough tactic" of first going through the front lines to reach the rear.[41] Mobility and surprise, the Chinese believe, are key elements of successful military strategy. "We are a weak country," according to the coauthor, Wang

Xiansui, of recently published book called *Unrestricted War,* "so do we need to fight according to your rules? No."[42]

Recall that the use of force is not central to Sun Tzu's approach to victory; indeed, he only uses the word "force" nine times in his work. We should not be surprised, then, when the cultural descendants of Sun Tzu underscore critical operational concepts applicable to the exploitation of the Revolution in Military Affairs, concepts that incorporate some of the more classical postulates of warfare that hold that the difference between victory and defeat is as much psychological and moral as it is material.

Russia

The 1991 political collapse of the Soviet Union and the unremitting economic deterioration of Russia throughout the 1990s ensured a decline in investment in its space programs and forces. A domestic battle over resources ensued, and in due course the armed forces steadily lost out to the civilian sectors. The dramatic decline in Russian launch activity over the past ten years, for example, has come at the expense of the military. So there are those in Russia who have noted the sickly state of their space forces and, hence, grown concerned about Russia's military future and the ability of the state to remain a major, competitive space power. Lt. Gen. Stanislav Yermak, chief of staff of Aerospace Forces in 1996, believed that failure to regain its space potential would mean that "Russia will turn into a second-rate country and will be deprived of that key element by which it could emerge from the crisis."[43]

According to former Russian president Boris Yeltsin, "The government is determined to do everything necessary, as before, to preserve and increase the Russian space potential for the nation's sake and in global interests."[44] Yuriy Baturin, Yeltsin's national security advisor until his dismissal in August 1997, stated rather forthrightly in 1996 that "it is possible to perform scientific-technical and economic tasks on which both the prestige and the very destiny of our country depend. This is why mastery of space has been and remains one of the chief priorities for Russia." According to Baturin, "It is our space capacities (plus a number of other S&T [science and technology] achievements) that will permit Russia to retain great-power status."[45] Thus, according to one Russian army publication, "the country possessing space technology will feel confident in the future world."[46]

Soviet strategic thinkers were among the first to recognize what is now being called the Revolution in Military Affairs. Several viewed the Persian Gulf War as a "technological operation," a harbinger of wars to come. The character of new wars would be marked by rapid deployment capabilities, dominance of air power, and a higher degree of weapon automation exploiting advanced C^3I systems, including space systems. Future wars would see a massive use of technologies, as quality, speed, and precision of force come to overshadow quantity of force, and non-nuclear technologies achieve effects and objectives approaching those once left for nuclear weapons.[47]

The Russians, in other words, viewed the Persian Gulf War as a paradigm for future war, with emphasis on the use of high technology, which had become the new basis for analyzing combat power. The Russian perspective, at least in part, sees tremendous potential for space forces in the international conflicts of the coming decades. Officials in Russian political and military circles widely accept that space is central to future military operations and success on the battlefield. Baturin freely acknowledged the vital role space forces would play in future military operations. He indicated that the Gulf War ought to be pictured "as the struggle of two groupings approximately equal in numbers and arms, one of which (the United States and its allies) possessed excellent vision, hearing and coordination."

Moreover, the Gulf War has shown that "the use of space technologies increases the number of targets engaged (neutralized) and simultaneously reduces friendly losses by several times."[48] A Russian major general explained in no uncertain terms that the 1991 war against Iraq portended an evolution of warfare in space:

> Space assets provided exhaustive information about the enemy's status and measures he was taking. By thoroughly knowing the status of Iraqi troops, the MNF [Multinational Force] command paralyzed their operations and stunned them with the unexpectedness of steps being taken. In the future the role of space in war evidently will rise sharply, since the capabilities of strategic means of warfare are realized to the maximum extent in the aerospace sphere. It is presumed that in the not too distant future unavoidable strikes by precision weapons and weapons based on new physical principles can be delivered from space against any targets regardless of their degree of hardening. Thus, a country not having the capability to counter space weapons may turn out to be doomed.[49]

The gradual incorporation of information technologies into war fighting led some Russians several years ago to recognize that these new conditions of warfare would have an impact even at the "strategic," or nuclear, level of warfare. Some Russians see that the important military-technological tasks are achieving "global control over the strategic space zone" and "reliable functional destruction of space optical-electronic and radar intelligence means." Upon the advice of Professor Nicolai Mikhailov, Russia should develop its Strategic Nuclear Forces all the while keeping in mind "the objective necessity of increasing the role of space as the sphere of military activity that has an impact to bear on the effectiveness of basic operations of land, sea, and air forces."[50] Space-based reconnaissance information systems have led to the development of new offensive and defensive weapon systems, the most notable of these being weapons relying on precision guidance. This, according to Lt. Gen. Aleksandr Skvortsov and Maj. Gen. Nikolay Turko, will eventually lead "toward the development of weapons based on new physical principles, and toward a shift of the center of gravity of warfare from continental and ocean [theaters of operation] into the sphere of space."[51]

There are also those who argue that aerospace should be viewed as a unified sphere of possible military operations, that troops must learn to act under the conditions of the "expanded battlefield," which could mean having direct contact with the enemy as well as the absence of a clear front and rear. The ability to hit the enemy from afar before entering into direct contact with him, according to retired major general Vorobyev, "means that with other conditions being equal, success in such opposition will be on the side of the one who has the capabilities for conducting deep reconnaissance, has the advantage in long-range weapons and means of command and control, and is capable of making a decision in a sharply changing situation."[52] Each of these capabilities can be tremendously enhanced with assistance from space forces. Lt. Gen. Anatoliy Nogovitsyn and Maj. Gen. Anatoliy Panchenko believe that "in wartime the prevention of general aggression consists of inflicting a decisive defeat on the enemy in the aerospace sphere and not allowing him to win supremacy in the air and space."[53]

Despite the emphasis on the military-technological dimensions of warfare, the benefits of indirect conflict, of "informational-psychological opposition," are not lost on at least some of Russia's military thinkers. The essence of this approach, according to Vorobyev, is not the "physical destruction of each individual weapon"; rather, it is aimed "at the destruction of the state's information resource, command and control system, and navigation and guidance channels." He stresses that "the pressure of force is not excluded, but is to be used first indirectly," the objective being to "force him to surrender without a war (the ideal option)."[54]

Vorobyev juxtaposes this approach with the belief among "Western experts" that victory demands operations from a position of strength, that "it is necessary to have a decisive technological superiority over the enemy," as is evidenced by the reliance on electronics, robotization, computerization, space assets and information science. The Persian Gulf War has become the "prototype" for the "controlled war" sought by Western leaders.

Finally, not out of line with this brief discussion, it is worth highlighting the Russian inclination to resort to strategic maneuvering at the political level to overcome asymmetrical disadvantages. Russians have what one might call a selective penchant for legalisms (while the United States has a distinct vulnerability to legalisms!), and are inclined to view legal agreements, not as documents worthy in their own right, but as tools to be employed in situations where more effective tools are lacking.[55] As a instrument to prohibit the development, testing, and deployment of space-based strategic ABM weapons, Moscow also will use the ABM Treaty to construct political and diplomatic barriers to the deployment of any space weapon, even though it may not have a "strategic" function. Former Russian Federation deputy foreign minister Georgiy Mamedov stated that (in the words of analyst Dmitriy Gornostayev) "in every specific case of nonstrategic interceptors being put into orbit, Russia will be entitled to state that there has been a violation of the existing accords. Thus we will be able to force the Americans not to deploy these weapons in space." Regarding the development of new kinds of antimissile arms, Gornostayev also observed that, "while not having the industrial capacity to compete with Washington in the sphere of antimissile defense today, Moscow made a successful attempt to restrict its potential by diplomatic and legal methods."[56]

Although the treatment of this subject here is far from an in-depth look at the Russian strategic mind, one can discern a fairly dominant expectation that future wars will feature the use of high technology and space will be a central element in combat operations. As far as one can tell from available writings, there tends to be a marked difference between Russian and Chinese views on the future of combat, one that rests upon an assessment as to how prominent a role technology ought to play in warfare. Perhaps because China is a less advanced power unable to devote the resources required to engage in military-technological competition with the United States and Russia, it must think of victory more in purely strategic terms.

India

Indian scientists believe today that, by 2025, global power will be defined by a nation's access to information, and that it will be in the national interest to develop new technologies for launching satellites (such as hypersonic missile technology).[57] Indian air chief marshal Satish Kumar Sareen knows that "success in future wars will depend on the ability to deploy space-based resources for surveillance, battlefield management and communications."[58] Indeed, the integrity military satellites can provide to national C^3I architecture makes space indispensable to a strategy for deterrence.[59] Some Indians believe that it is also in their best interests to stay at the forefront of technological development, where new-era warfare integrates space and electronic warfare. Yet, wrote Lt. Comdr. V.W. Karve, "history teaches us that better technology alone does not necessarily lead to victory. Rather, victory goes to the side that uses technology better or who can deny the enemy the use of his own technology."[60]

Some Indians believe that the deployment of space weapons to defeat space threats is inevitable. Warfare, according to some analysts, is evolving to the point where it is imperative to neutralize a few relatively important targets quickly. Thus, as satellite capabilities increase, other states using them become more dependent on them. "Chain reaction in turn ends up in suspicion about intention, thus laying the basis for anti-satellite weapons," wrote Major General Madhok. General Madhok maintained that

> no one will disagree that a weapons-free Space can help serve a large number of activities for the welfare of mankind such as manufacture of medicines, drugs, and vaccines, electronic materials, electricity generation, mining of the Moon and nearby planets and perhaps in establishing our first contact with other civilizations. But this is at best a pipe dream. The history of mankind has never been free of wars. . . . The seeds of a space conflict are already in place. . . . Therefore those who talk of a peaceful space environment as a prerequisite, show a lack of historical perspective. The reality is that the support from the space systems to the armed forces on earth will be a threat by itself. And because it is so, there will be no option but to deny and if necessary destroy these threats.[61]

Madhok noted that with emerging technologies, "when more countries acquire space capability, the chances of a conflict can only increase and not diminish."[62]

Iraq

Not too surprisingly, Iraqi analysts also recognize the evolution of space warfare. Computers and command systems will be networked in the future to "special command, control, and communications spaceships," that will enable the high command to give immediate orders and instructions to subordinate units, wrote Maj. Gen. Harith Lutfi al-Wafiy. "Satellites have not been in use in a combined war until late this century," therefore modern combined war after the year 2000 "will deal with space commands, rather than just air commands." The command systems of the future will be "under constant enemy surveillance" and will need to be "strongly fortified against enemy attack."[63] Israel's investment in military reconnaissance satellites represents just such a threat, according to official Iraqi memos on the subject, presenting a situation that demands an Iraqi response in the form of offsetting investments in space technology and related sciences.[64]

Some Iraqi military analysts, reflecting on the engagements during the Persian Gulf War, hold that superior weapons technologies are no guarantee of victory. One analyst asked what would have been the outcome had Iraqi forces decided to stage battles inside the Kuwait City. "Could the enemy, with its weapons and tanks, have fought a street war without paying a heavy price?" Major Generals Husayn and Kan'an believe that "the answer would be clearer if we go back to the enemy's history in Vietnam, Somalia, and other countries." They observe that "the Western soldier is familiar with technology. This is, of course, not everything in modern military warfare. There is something called determination and endurance in the face of different battle conditions." In the future, they additionally noted, Iraq must have the capability to keep the "communications routes" open and secure.[65]

Israel

The unique defense concerns of Israel, a country only 20,700 square kilometers in area and having a population of less than six million, spring from the ever-pressing reality that it is a nation surrounded by enemies who have frequently made war against it. The security of the Jewish state drives its leaders' incessant pursuit of tactical and strategic advantages.

The prod of necessity led Israel to develop its own Offeq series of "spy" satellites and to enter into secret dealings with Russian intelligence agencies that would allow the Mosad, the Israeli intelligence service, to purchase hundreds of satellite pictures of Israel's hostile neighbors, including Iraq, Iran, and Syria.[66] Defense analyst Yosi Melman believes that the "main difference between the situation in the Gulf war and now is that Israel now has an Offeq satellite capable of taking high resolution pictures of two-meter objects [*sic*] in Iraq."[67]

This is strategically significant for Israel. According to air force commander Maj. Gen. Eitan Ben Eliahu, Israel must focus on achieving space supremacy as well as command of the air, if only because satellite reconnaissance could replace three squadrons of intelligence-gathering aircraft. Israel's defense chief, Yitzhaq Mordekhay, established in 1998 that Israeli Aircraft Industries should pursue future technologies, including satellite technologies. This was in part because Israel's view of the threat has evolved to the point where it is now seen to emanate from a "distant circle of enemies," a geostrategic situation that places a premium on early warning capabilities, deterrence, and an ability to transfer the war quickly to the enemy. The leader of the Israel Defense Force at the time, Maj. Gen. Ben-Yisra'el, believed that the future battlefield would be dominated by space elements. As such, Israeli officials acknowledge that it enemies also will have access to reconnaissance satellites, whether through indigenous means or purchase of imagery from outside their countries. This situation will compel Israel to develop new camouflage techniques for its vital targets.[68]

Ukraine

Ukrainian leaders remind their fellow citizens that, despite economic dislocation, their country "formally and in fact [is] a space power." Ukraine has a unique space industry and is proud of its "space eye," the *Sich-1* satellite. According to Valeriy Komarov, the general director of the National Space Agency of Ukraine, "If we are headed into the 21st century and want to be among the ranks of the leading countries of the world, if we want to make use of the achievements and the new and advanced technologies that we have, then we cannot get by without space, without global space programs."[69] Ukrainians too understand the growing importance of investing in capabilities to exploit the space medium. "Just as once those who owned the sea owned the entire world (Spain, Great Britain . . .)," writes Volodymir Shelopov, "today the masters of the space ocean are becoming the masters of life. They are reckoned with and, certainly, respected. Therefore, more and more world countries link their hopes for a better life of their people with outer space."[70]

For the near-term, according to the chief of Ukraine's Directorate of Missile and Space Weaponry of the Ministry of Defense, Maj. Gen. Valeriy Lytvynov, Ukraine will concentrate on civilian space projects: "This proves once again that Ukraine is a peaceloving nation, and our principal aims in space are scientific research and national-economic work." But Ukraine will not neglect the defense needs of the country. "The military," stated Lytvynov, "needs primarily information and reliable command and control for units, . . . reconnaissance and communications satellites. The task of passing the command and control of the troops through space to each soldier is being posed in the leading countries of the world."[71]

What Is the Bottom Line?

There is a belief within the U.S. Department of Defense, not entirely unjustified, that the more complex the United States can make warfare for the enemy in the future, the more the enemy will be deterred. For if the enemy has no hope of overcoming all of the complexities of modern warfare, he will be reluctant to initiate war, at least a war that will require the outright defeat of U.S. armed forces. Thus, moving the conflict into space, so the thinking goes, may be an effective means of deterrence in the future. There is something to be said, after all, for the United States' asymmetric war-fighting advantage.

This viewpoint, of course, assumes a number of things about combat and deterrence, most fundamentally that the enemy will fight obligingly on U.S. terms. Yet the enemy may not always be so willing to collaborate with the United States in these areas. Moreover, it must not be forgotten that the United States' diplomacy and impressive display of modern forces in the 1990s did not deter Hussein or Milosevic from pressing ahead with their strategic agendas.

Some of those responsible for the United States' defenses are not unaware of the asymmetric possibilities. According to the authors of the Joint Chiefs of Staff's *Joint Vision 2010*, "Our most vexing future adversary may be one who can use technology to make rapid improvements in its military capabilities that provide asymmetrical counters to U.S. military strengths, including information technologies."[72] Although bounded noticeably by strict technology considerations (and incorporating less well strategic concepts with which thinkers such as Liddell-Hart and Corbett were wont to wrestle), there is explicit recognition that asymmetric war-fighting capabilities would allow future adversaries to wage limited warfare against the United States and its allies rather than confronting superior conventional military capabilities head-on. Accordingly, "the arena of conflict in 2010 will include the US homeland, strategic LOCs [lines of communication], and global information infrastructure, including space, as US adversaries attempt to exploit nontraditional vulnerabilities. Even inferior powers could offer serious challenges to US military superiority in 2010. Adversaries will closely observe US capabilities and tactics in an effort to exploit weaknesses by asymmetric approaches."[73] Yet it is not entirely clear that a real strategic understanding of warfare incorporating unconventional and non-Western approaches to victory influences current military thinking, analysis, planning, and doctrine—especially as they seek to define and exploit the space environment.

Strategic thought must be open to the new and the different. "To be practical," writes Liddell-Hart, "any plan must take account of the enemy's power to frustrate it."[74] It must be sufficiently agile to assess objectively one's own plans and to consider the science and the art of fighting from the foreign perspective. Science is, of course, an activity shaped by unbiased, systematized inquiry, while art is the exhibition of special skill acquired by experience, study, and observation, and both appear to apply to developing an understanding of war. Space opens up new strategic frontiers that will increasingly influence military affairs, from weapons and tactics to high-level policy decision making. One ought to expect that space and counterspace operations also will offer new opportunities for the enemy to frustrate U.S. war plans.

The United States' policy and military leadership must not become complacent. Technological superiority in and of itself is of little consequence; what matters is how it is used. Assessments of "the threat" must include and then look beyond considerations of capabilities. To remain narrowly focused on the technological issues and to examine capabilities without considering and thoroughly understanding the political and strategic dimensions of warfare is not just ineffective but also counterproductive. The blinders Americans refuse to remove will establish the very conditions needed for a technologically inferior, space-aware enemy to achieve a militarily decisive advantage over the U.S. military giant early in this century.

Part 3 ▸ Confronting Janus

The divergent set of policy visions for space is the heart and soul of the national quandary over defense space matters, despite the impression one might glean from the "National" Space Policy. Janus, the two-faced god of Roman mythology, is a wonderfully apt metaphorical tool that may be used to illustrate our national dysfunction in space.

Much like Janus, the United States' defense space policy continues a tradition of looking in two different and, at times, opposite directions. In very general terms, two heads vie for the position of prima donna on the country's space policy stage. Those who regard space as "a medium like the land, sea, and air within which military activities shall be conducted" to defend and fight for U.S. interests represent the first head. Or, in the words of the Pentagon under President Clinton, "Space power is as important to the nation as land, sea, and air power."[1] Following this interpretation, space has the status of a full-blown "war-fighting environment," where there are unprecedented possibilities of a military nature, one in which combat operations may be expected one day to influence the course of battle on Earth. This viewpoint, expecting no revolution in human nature or state behavior, reflects no obvious reason to believe that outer space could be preserved from military use.

In sharp and dramatic contrast, Janus's other head regards space as a peaceful preserve, a sanctuary that man must not despoil, an arena where military activities may lead to unanticipated provocation and dreadful consequences for security and international affairs. According to officials of the same Clinton Pentagon who maintained that space power had grown in stature to be roughly on par with other forms of power, opting for destruction of satellites "is going to be highly confrontational to the international community" and thus must be avoided.[2]

The most reasonable upholders of this point of view will confess that military satellites performing information-handling duties are here to stay, and in any case, such operations can be a stabilizing factor in state relationships to the degree they contribute to "transparency"—an objective of modern-day arms controllers that conjures up images of omniscience referred to in Psalm 139: "Where can I flee from your presence?" What cannot be tolerated, however, is the notion that combat operations in space can perform a useful security role short of upsetting "fragile" international security and strategic relationships. Janus's other head often is manifested in official public rhetoric, federal funding priorities, acquisition strategies, international diplomacy, and treaty commitments, chiefly U.S. obligations in the 1972 ABM Treaty not to deploy antiballistic missile interceptors in space. This face of Janus has tended to be dominant and controlling over the course of the last four decades.

The space world is indeed a dynamic one, made so by increasingly complex political and economic choices motivating national and foreign space activities. Although it is not my intention to downplay the importance of everyday space policy issues, it is a fact that most decisions made in the commercial, civil, and governmental space sectors today are not weighed down by consideration of the most basic or radical question of our time: How should Americans relate the space medium to national security? The space-control and force-application missions speak directly to the national vision by assuming a specific political-military posture and direction with respect to national defense. Is it any wonder that the most potent, soul-searching political debates on outer space of the past forty years have revolved around the combat uses of space? Resolve the more fundamental issues dealing with vision and direction in the hearts and minds of Americans, reconcile them to the country's laws and policies, and the way will be clearer with respect to many other policy matters facing the national leadership. Knowing where one is going before setting out on a long journey makes for more auspicious travels.

7
NATIONAL DEFENSE SPACE POLICY

How Has Policy Evolved since Eisenhower?

The United States' defense space policy, handed down from president to president, from generation to generation, has exhibited remarkable stability over the past four decades, primarily with respect to national goals. Although there is much to be said for stability, balance, and continuity in a policy regime, one should not discount the degree to which familiarity and continuity breed a very unhelpful complacency. What we say by way of justifying U.S. defense space activities is so done by rote, reflex, and habit. Habits, by definition the products of repetition and settled tendencies, require minimal exercise of intelligence. In settled times, involuntary, routine activities are welcomed. Complacency, or excessive satisfaction with the way this country regards space and acts in space, is an outgrowth of the peaceful times in which we live. In more dynamic periods, however, the changes we encounter can make our habits look foolish and outmoded. Although there is much good to be said about the consistency of U.S. defense space policy since its inception, we ought not to make light of the possibility that perhaps the daily policy habits that guide the country's activities in space no longer are suitable for our age.

Not all that meets the eye is real. A careful look at the country's general written defense space policy dispels any notion that the civilian and military leadership understands fully the prominence of space. Fragmentary spending on military space programs, a decentralized defense space organization, and the unbalanced rhetoric of public officials on military space matters indicate that this country is deeply divided on this subject, the tough words in the National Space Policy notwithstanding. This is not to imply an elaborate smoke-and-mirrors game by U.S.

officials, only that it is time for Americans to pay attention to the entirety of the U.S. space policy regime.

What Is Policy?

Official policy is a collection of carefully chosen words, capabilities, and deeds, "deeds" being what is done with capabilities to support the words. One should even add to this list "inaction," for what we do not do also can speak volumes. Policy is an authoritative pronouncement on a specific course of action or direction. It directs those elected or appointed to policy-making positions who are responsible for executing the goals. The executive power of the country is vested in the office of the president, and it is the president, as the primary maker of policy, who is authorized to pursue the broad goals outlined in the Preamble of the Constitution. Since policy is made by those whom we elect to high offices, the leadership they appoint, and career bureaucrats, it reflects a degree of partisanship and vested interest related to the expressed vision and agenda of a particular presidential administration.

Congress also makes policy through the enactment of laws. Yet the House of Representatives and Senate have neither the energy nor the required focus over the long term to implement policy objectives to influence continually general political attitudes toward military space activities. It is usually left to the president to see to "the steady administration of the laws."[1] For these reasons, the presidency is central in our discussion of U.S. defense space policy.

Where do we find expressions of policy? In the case of space policy, its first public pronouncement arrived in the form of a law that outlined in the impassive language of legislation the principles, national goals, and basic rules for governing space activities. With the National Aeronautics and Space Act of 1958 and Eisenhower's issuance of the first "U.S. Policy on Outer Space" in 1960 as a backdrop, subsequent administrations since President Jimmy Carter have sought to codify their own objectives for acting in the space environment in the form of comprehensive presidential directives or decision memoranda developed in the National Security Council (NSC) and coordinated among numerous government agencies and departments. These written expressions of policy usually have been accompanied by press releases, or White House interpretations of declared policy. Typically, the White House also issues a classified version containing information deemed by the president's advisors to be sufficiently sensitive that its public release would be harmful to national security. The publicly issued document gives policy implementers and other interested parties outside government an understanding of the administration's formal direction and a sense of the president's overall approach to space.

Yet a written and appropriately labeled "policy" is only one way to set out general guidance. There are other no less formal, no less authoritative methods for pronouncing policy or taking official positions. Understanding its many origins is very important for assessing present-day U.S. defense space policy. The president and, to a lesser extent, members of Congress are able to set and revise the goals of the nation and communicate the United States' attitude and approach toward space in different ways. Official rhetoric, or the carefully considered public statements made in press conferences by White House officials or its agents (such as the Department of Defense), official speeches, and congressional hearings and debates all add to the public record of policy.

President Ronald Reagan's famous March 23, 1983, speech introducing the Strategic Defense Initiative to the American people is a wonderful example of this kind of policy making. This speech registered White House intent to pursue a comprehensive defense of the United States by using extensively the space environment to defeat Soviet ICBMs. It also set in motion significant doctrinal and organizational changes in the Pentagon (including the establishment of a Strategic Defense Initiative Organization). And although Reagan subsequently issued written direction that specifically outlined this long-term goal (National Security Decision Directive 85), the clearest and most influential expression of his SDI policy came from his vision, which was manifested in his address to the nation, the wellspring of all SDI-related events to follow.

Diplomacy, or the measured conduct of relations with foreign nations, and foreign policy, or the collective expression of overarching objectives governing our association with other states, also help establish an administration's policy record. When U.S. officials take positions in international forums or communicate with foreign capitals, they typically act as a spokesman for the president and may serve to elaborate presidential policy. Similarly, the United States' treaty commitments and proposals for the development of new international law or international arms-control agreements express presidential goals, intentions, and understandings.

Other expressions of policy include procurement programs. In fact, if one were to "follow the money," acquisition programs and established funding priorities are an unmistakable and telltale sign of national policy. Most weapons procurement programs represent the country's steadfast commitment to an overarching military strategy (which itself broadly reflects policy) to protect vital and major national interests. Weapon programs mirror individual service approaches to executing their respective missions, so that weapon systems such as the aircraft carrier and the tactical fighter have become national institutions or monuments to U.S. military prowess. Leaders make U.S. policy tangible by reaching out to industry, which in turn develops the implements of policy.

Organization is another very tangible expression of policy. The official hierarchy of responsibility in the decision-making chain may be exploited so that work gets accomplished efficiently, or not. Where decision makers and policy writers sit is important. The decision to make the National Reconnaissance Office a clandestine operation demonstrated the national commitment to reconnaissance satellites as well as program secrecy. President Reagan demonstrated his keen interest in ensuring that the nation will not easily divorce space considerations from defense matters when he supported the births of the unified and service component space commands.

The *National Security Strategy,* drafted annually by the president in accordance with the terms of the 1986 Goldwater-Nichols Act, serves as the basis for the *National Military Strategy, Defense Planning Guidance,* and other Defense Department policy documents. It is a high-level policy document that establishes short- and long-term objectives for the national security establishment and offers general implementation guidance. This publication presents the overarching strategy for enhancing security and responding to threats and crises, and it provides broad guidance for evolving the United States' defense capabilities to meet challenges in the years ahead. In this document, the White House reaffirms old policy positions and has the opportunity to bolster new ones. In the area of space and security, for example, President Bill Clinton emphatically stated in the October 1998 version of the strategy that his administration's "policy is to promote development of the full range of space-based capabilities in a manner that protects our vital security interests."[2]

With the stroke of a pen, the president also may sign an executive order. The president has an extraordinary power to write executive orders, which do not require congressional approval and, once signed, in some cases can become the law of the land. They are aimed at government agencies under the president's direction. The president, however, is constitutionally forbidden from spending unappropriated money. Often executive orders are written to declare a national emergency, which in turn may be used to justify such measures as economic sanctions.[3]

Congress also may have a hand in the policy-making business. The legislative branch may pass laws that affect defense activities, which are signed by the president. Given the steps that must be taken to turn a bill into a law, these acts of Congress indicate the achievement of a national consensus around particular policy issues. In the space arena, for example, Congress passed the 1998 Commercial Space Act in an effort, among other things, to provide a benign regulatory climate for space businesses, encourage government use of commercial services, and guide developers of regulations for the anticipated arrival of reusable launch vehicles.

Short of legislating a direction for the country, Congress also may register a collective policy position by passing nonbinding joint congressional resolutions. Failed laws also can register support for a specific policy direction. When the House passes a bill that does not have the support to carry the Senate (or vice versus), or both chambers of Congress pass a bill that the president then vetoes, the voting positions taken by the representatives of the American people become a matter of public record, and in this way may contribute, however marginally, to the evolution of policy. The legislative history during the 1990s regarding the deployment of national missile defenses provides a compelling example of this sort of policy making.

News releases and other official document publications may be used to establish publicly policy positions. Congress and the president routinely issue reports or studies that give a sense of the official position with respect to important defense issues. Such publications also may be used to expand the public's awareness or understanding of a critical national security issue or an administration's objectives. The secretary of defense also may issue reports, the most important being his annual report to the president and Congress.

Finally, and simply, actions mold official policy. What the country does is perhaps one of the clearest signs available for assessing the national interest and determining where the line is drawn with respect to national defense.

The purpose of this section is not to capture all available means for molding new policies or sustaining old ones. The reader need only understand at this juncture that policy making comes in many forms. Indeed, it is my contention that a true assessment of U.S. defense space policy requires a hard look beyond the gilded edges of the National Space Policy.

What Are the Origins of the "Freedom of Space" Principle?

The national purposes currently guiding U.S. national security and civilian space activities grew out of the institutions and high-level space policy established by President Dwight Eisenhower. The original motivation for designing and funding programs to take the country into space was not predominantly a military, economic, or scientific one—rather it was political. Space was a new frontier for the display of national prestige and power. The United States and the Soviet Union demonstrated during the 1950s their leadership to other nations by waging propaganda wars, exhibiting their technological and military superiority, and portraying the inherent greatness and excellence of their respective liberal democratic and communist regimes. This bipolar competition generated national enthusiasm in both countries for accelerating space flight capabilities and undertaking high-visibility space quests. One of the more inspirational and spectacular space achievements was the Kennedy administration's Apollo Project, the goal of which was to land a man on the Moon and return him safely to Earth before the end of the 1960s.

The NASA Act of 1958, one of the earliest space policy documents, posits "the preservation of the United States as a leader in aeronautical and space science technology and in the applications thereof to the conduct of peaceful activities within and outside the atmosphere." But before the United States could lead in space, it first had to "walk." With this ambitious end in mind, Eisenhower and the new law sought to bring the United States into the space age as quickly as possible by expanding human knowledge of space and the atmosphere, designing efficient aeronautical and space vehicles, developing vehicles capable of instrumenting the space environment, and undertaking studies of the peaceful, scientific and military purposes of space flight.

Yet a nation cannot lead in any arena unless it is at liberty to act. Since 1958, U.S. space policies have consistently recognized that the principle of "freedom of space" is indispensable to the ability of the government to maintain peace and protect national security. The principle may be viewed as the sturdy frame that adorns and holds together the varied and multihued mosaic of U.S. space program goals. Interest in freedom of space, however, is not nearly limited to defense officials. Although it fundamentally rests with the armed forces to ensure its execution and general vitality, space power is also a basis for national economic and foreign policies. Indeed, as the prosperity and security of the United States and other nations depend so critically on information flowing through space (and in the future may rely on the execution of combat missions from and in space), access to space must rank in importance with freedom on the high seas and in the air, especially in time of war or crisis.

Although the United States continues to affirm its belief that space is a common realm subject to use and exploitation by all nations for "peaceful purposes," those purposes do not restrict defense and intelligence-related activities serving national security and other goals. The most recent written space policy echoes the policies of the past when it "rejects any claims to sovereignty by any nation over outer space or celestial bodies, or any portion thereof, and rejects any limitations on the fundamental right of sovereign nations to acquire data from space."[4] A slightly more accurate restatement of this "fundamental right" is that the United States upholds the right of nations to acquire data from space *for peaceful purposes.* U.S. policy upholds no such right if that activity ultimately damages U.S. national security.

The principle of "freedom of space" has been the cornerstone of every presidential policy relating to space and national security since Eisenhower, even preceding the issuance of the first policy. James Killian's Technological Capabilities Panel, authorized by the president in 1955 to recommend ways to reduce the probability of military surprise, concluded that freedom of space would be instrumental to the defense of the United States. The panel assumed that satellites would be

Picture 1: Works of Jules Verne and other authors of science fiction had a profound effect on rocket pioneers Konstantin Tsiolkovsky and Robert Goddard. In this illustration depicting a scene from *The Earth to the Moon*, passengers experience weightlessness.

Picture 2: The American Robert H. Goddard helped transform the pseudoscience of rocketry into a respectable discipline of systematized knowledge and laid the technological foundation for today's satellite launch vehicles and intercontinental ballistic missiles. Unless noted otherwise, all images courtesy of NASA.

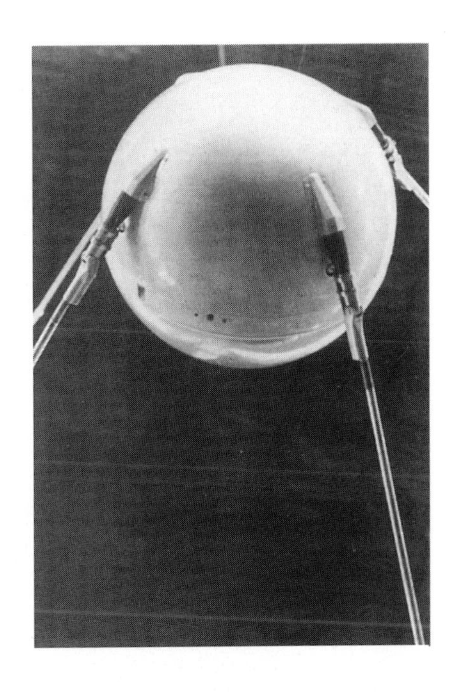

Picture 4: On October 4, 1957, the Soviet Union launched the 184 lbs. Sputnik satellite, the first to orbit the Earth.

Picture 3: The German government was the first to give significant support to rocket development. The V-2, developed by the Nazi regime as a "Weapon of Retribution," was the child of Wernher von Braun's engineering team and forerunner of 20th century rockets. Von Braun would go on to become an American hero.

Picture 5: From right to left, Von Braun, James van Allen, and William Pickering (director of the Jet Propulsion Laboratory) celebrate the January 31, 1958, launching of the first U.S. satellite, *Explorer 1,* which carried Van Allen's radiation detector. *Explorer 1* (the happy trio hold a full-scale replica) stayed in orbit for more than twelve years.

Picture 6: President Eisenhower poses on national television with a scale model of the Jupiter nose cone, the first to be recovered from outer space. (Courtesy of Army Aviation and Missile Command)

Picture 7: Posing for *Life Magazine.* A contemplative Hermann Oberth, front and center, dominates this portrait. Wernher von Braun is sitting on the table, and Eberhard Rees, his colleague from V-2 days in Germany, stands behind him. Standing off to the left is Major General Holgar N. Toftoy, who helped move Von Braun's team and equipment to the United States. Ernest Stuhlinger, ballistic missile research director at the Army Ballistic Missile Research Agency in Huntsville, Alabama, is seated in the chair behind Oberth. (Courtesy of Army Aviation Missile Command)

Picture 8: President John F. Kennedy and Wernher von Braun (center) discuss the Saturn Launch System in May 1963. Robert Seamans is to the left of von Braun.

Picture 9: A Delta 2 rocket.

Picture 10: U.S. Landsat imagery of the San Francisco Airport. (Courtesy of spaceimaging.com)

Picture 11: The U.S. Air Force Defense Meteorological Satellite Program satellite shown here teams up with NASA Geostationary Operational Environmental Satellites to monitor weather conditions across the globe. (Courtesy of Air Force Space Command)

Picture 12: The U.S. Defense Support Program satellites provide early warning data to U.S. Strategic Command and will remain the primary space-based early warning and cuing assets for U.S. ballistic missile defense systems until the constellation of DSP satellites are replaced later this decade by the Space Based Infrared System. (Courtesy of Air Force Space Command)

Picture 13: The Global Positioning System satellites have revolutionized modern navigation techniques using a highly accurate atomic clock to provide precise information to anyone, anywhere, at any time of the day regarding latitude, longitude, altitude, travel velocity and direction, and time. (Courtesy of Air Force Space Command)

Picture 14: U.S. Air Force Milstar satellites route sensitive military tactical and strategic message traffic and conversations. (Courtesy of Air Force Space Command)

Picture 15: Defense Satellite Communications System spacecraft provide the bulk of the Defense Department's long-haul, high priority satellite communications.

Picture 16: A Titan IVB/Centaur rocket launches the Cassini orbiter on its seven-year journey to Saturn. The Titan IV is a heavy space launch vehicle used to carry Defense Department payloads, such as Milstar and DSP satellites, and classified NRO satellites into space.

Picture 17: An Ariane 5 launch. The European Space Agency's Ariane commercial launch program relies on the Guiana Space Center in Kourou, French Guiana, to boost payloads into orbit.

Picture 19: The satellite's field of view is far wider than that offered by airborne cameras, as this image of the earth taken during an Apollo mission demonstrates.

Picture 18: In this December 1999 photo, NASA astronauts use the Space Shuttle as a garage in order to perform some "routine" maintenance on the Hubble Space Telescope.

Picture 20: The Space Shuttle, or Space Transportation System, is a semi-reusable aircraft-like vessel. In this picture, flight STS-1 makes history.

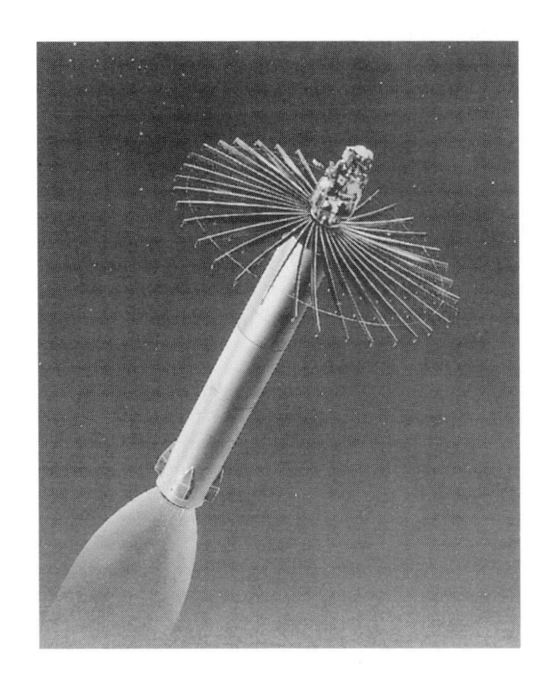

Picture 21: An early kinetic-kill interceptor developed by the U.S. Army successfully intercepted a target in space in the 1984 Homing Overlay Experiment. The kill stage in this experiment shown here weighed about 2,500 pounds and had a kill enhancer that measured 15-feet across. The U.S. National Missile Defense program developed in the 1990s an exo-atmosphere interceptor that weighs only 130 pounds and is capable of hitting the "sweet spot" on a target in space. (Courtesy of the Ballistic Missile Defense Organization)

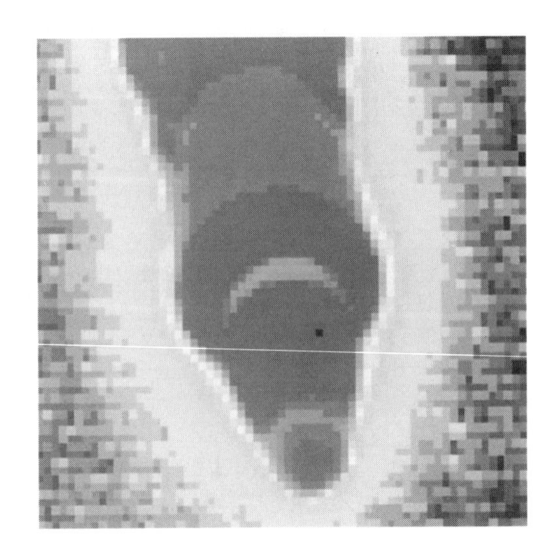

Picture 22: The BMDO's Theater High Altitude Area Defense system currently under development will intercept medium-range ballistic missiles in space. It must use infrared sensors to refine its path of collision with the target. This image was one of the last taken by the kill vehicle prior to a successful intercept in June 1999. (Courtesy of the Ballistic Missile Defense Organization)

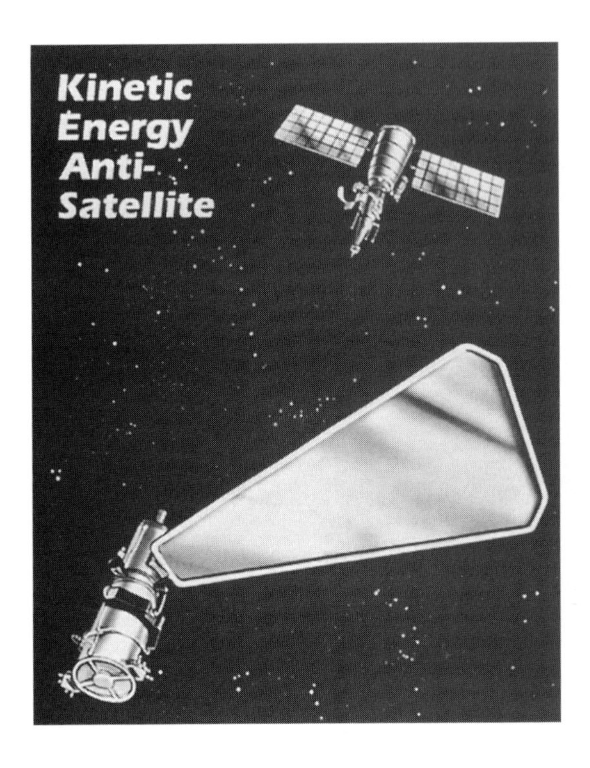

Pictures 23: The U.S. Army Space and Missile Defense Command is developing technology for the Kinetic Energy Anti-Satellite system, which would use visible light optical seekers to acquire and track satellite targets. The KEASAT would use a giant "fly swatter" to strike the satellite and knock it out of commission while minimizing orbital debris. (Courtesy of Space and Missile Defense Command)

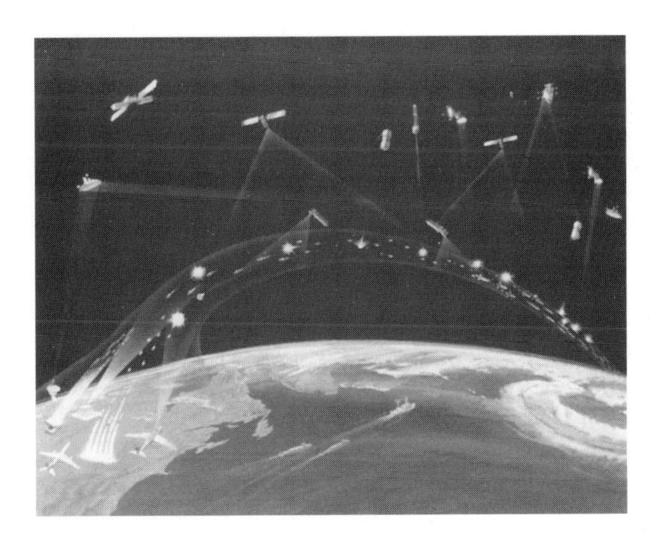

Picture 24: An artist's conception of how space-based force application weapons and sensors could contribute in a defense against a barrage of long-range missiles. (Courtesy of the Ballistic Missile Defense Organization) **Picture 25:** The Space-Based Infrared Satellite System (SBIRS Low) in this illustration is currently under development and would play a vital role in any ballistic missile defense architecture. SBIRS Low satellites will host multi-spectral sensors to observe, characterize, and report on ballistic missiles in all stages of their flight, especially the mid-course, or longest part of the flight path. SBIRS Low sensors will be capable of observing post-boost vehicle maneuvers and acquiring, tracking, and discriminating a complex target cluster, which might include reentry vehicles and BMD penetration aids such as balloon decoys. (Courtesy of the Ballistic Missile Defense Organization)

Picture 26: Orbiting Brilliant Pebbles interceptors were to be housed in "life jackets" that would provide "housekeeping" support until a missile attack was detected, at which time the interceptors would shed their jackets and intercept the ballistic missiles with ranges greater than 600 kilometers. BP interceptors could offer continuous world coverage. In this artist's rendering of a BP interceptor, it has just shed its jacket prior to attacking an approaching missile. (Courtesy of the Ballistic Missile Defense Organization) **Picture 27:** The Starfire Optical Range at Kirtland Air Force Base in New Mexico hosts the world's first laser beacon adaptive optics system (1.5 meters) capable of correcting distortions caused by the Earth's atmosphere. This low-level laser activity helps U.S. Space Command track and catalogue objects in orbit. (Courtesy of U.S. Air Force)

able to provide critical intelligence on activities in "denied areas," primarily the Soviet Union, so that steps could be taken to meet new threats and develop new strategies to exploit an adversary's specific weaknesses.

Having failed to establish an agreement on "open skies" with Moscow, which would have allowed the Soviet Union and the United States to monitor from the air strategic force developments within each other's territory, Eisenhower sought instead to establish international acceptance for the idea that the space environment, unlike the air environment, ought to be free and open. This freedom of space principle borrowed heavily from the long-established and internationally accepted maritime principle of freedom on the high seas. Nations customarily consider forceful interference with shipping outside national territorial waters an act of piracy or war. Eisenhower desired to apply this concept of environmental freedom of action to space. Supporting this objective, the White House National Security Council issued in June 1955 the following recommendation to the president: "*Freedom of Space.* The present possibility of launching a small artificial satellite into an orbit about the earth presents an early opportunity to establish a precedent for distinguishing between 'national air' and 'international space,' a distinction which could be to our advantage at some future date when we might employ larger satellites for intelligence purposes."[5]

The authors of "U.S. Scientific Satellite Programs," or NSC paper 5520, issued in May 1955, noted that U.S. scientific programs should be conducted so as to preserve U.S. freedom of action in space. It also stated that no actions should be taken in space research or international negotiations that would require the prior consent of other nations for U.S. space activities. These statements marked the birth of a central tenet in U.S. space policy.

Eisenhower's decision to participate in the International Geophysical Year was in part motivated by his desire to establish new principles and practices in international law to recognize free flight in space. But how was he to accomplish this? Simple declaration may have sufficed, yet it appears as though Eisenhower thought that the United States ought to lead by example and demonstrate its own intentions to the rest of the world to use the space medium for peaceful purposes. In the early phases of policy development, Eisenhower was concerned that U.S. activities not appear to be overly aggressive in the assertion of this freedom of space principle. He believed military activities incorporating enforcement options could lead to the formation of restrictive new laws that would make it very difficult for the United States to operate in space so as to improve national security.

The presidential effort to establish the custom of free passage through space, therefore, tended to manifest itself in U.S. foreign relations as a "space for peace" initiative. Eisenhower linked the United States' position on the control of space to the ongoing disarmament negotiations with the Soviet Union, which were taking place under the auspices of the United Nations. Eisenhower, in fact, became the first leader of any country to suggest limiting activities in space when in January 1957 he proposed a plan to the international community to "mutually control the outer space missile and satellite development." Until the fall of 1958, therefore, U.S. officials sought international supervision of activities in outer space, suggesting that they be limited to "peaceful and scientific purposes," prelaunch inspection of missiles and satellites, and the development of systems of international control.[6]

This linkage led Eisenhower, however, to make some unfortunate statements that clearly were not in the interest of the United States' long-term national security and obviously were contrary to U.S. plans to use information provided by satellites to support deterrence and war planning. Policy guidance produced by the NSC (November 21, 1956) stated that it is the purpose of the United States to seek to assure that the "production of objects designed for travel in or projection through outer space for military purposes shall be prohibited." The president even went so far as to propose to Soviet premier Bulganin on January 12, 1958, that "we agree that outer space should be used only for peaceful purposes. . . . Both the Soviet Union and the United States are now using outer space for the testing of missiles designed for military purposes. The time to stop is now. . . . Should not outer space be dedicated to the peaceful uses of mankind and denied to the purposes of war?"[7] An uncooperative Soviet Union forced U.S. officials to mature their own thinking about the relationship of "peaceful uses of space" to freedom of space and to separate these issues for good from disarmament. With an ever-increasing number of satellite launches by the United States and the Soviet Union, domestic requirements in both countries for flexibility in this new arena evolved to the point where the achievement of an early agreement on limiting activities in space became impossible.

Eisenhower's references to peaceful uses of space undoubtedly were intended to assuage international opinion, especially official opinion in Moscow, regarding U.S. intentions in space. Eager to avoid any challenge to the freedom of space principle, the administration sought to identify U.S. activities in space as "peaceful." Eisenhower believed a civilian administration of space would emphasize this to audiences at home and abroad.[8] By implication, his reference to "peaceful uses" and international concern about unrestricted arms competition in space, especially when coupled with his sharp distinction in his space policy between civil and military activities in space, came to be associated by some with "nonmilitary," an association that unfortunately has stuck in the minds of many who believe there is no military activity in space that can contribute to peaceful ends. It is clear, however, that Eisenhower intended no such thing; his program to deploy reconnaissance satellites belied the notion that he approved only "civilian" activities in space or believed that defense-related activities in space must be "militaristic." On the contrary, he sought to establish the principle of freedom of passage for his spy satellites.

The administration's push to establish international acceptance of freedom of space led it to make statements that diverged sharply from the viewpoint of the armed services, whose leaderships held that any space vehicle deemed inimical to national security should be considered hostile and countered accordingly, to include possibly its destruction. The U.S. Air Force in particular opposed all efforts by Eisenhower to limit military space projects, in effect restricting the services' traditional obligation to protect the United States. The military services worried that a "space-for-peace" policy would prematurely hinder efforts to establish U.S. control of space, even before progress may be made in negotiations to impose similar restrictions on other nations through international agreements. Although the government supported international cooperation and programs, it quickly backed off of the idea of establishing an agreement for the international control of space.

As originally conceived and interpreted by Donald Quarles, then deputy secretary of defense, Eisenhower's space-for-peace policy was sharply at odds with the views of uniformed officials. Quarles reflected the views of many in the Pentagon when in 1958 he suggested in his testimony before Congress that if the Soviet Union placed a satellite having reconnaissance capabilities in orbit, "we would consider that it was inoffensive in the sense that (they were) in outer space where (they) could do us no harm and we could not object to it."[9] U.S. policy eventually came to recognize that, Quarles's earlier statements notwithstanding, nations *could* do other nations harm from space, either by collecting information to support hostile activities or by using space to project force. In the end, the military services contributed significantly to the space policy debate with their early protests against the space-for-peace concept, thereby preserving the traditional role of armed forces as guardians of the national interest.

The 1958 "Preliminary U.S. Policy on Outer Space" (NSC 5814/1) showed early signs of the new thinking on the uses of space. It recommended that the United States consider a position that assumed the right of free transit in outer space of objects not equipped to interfere physically with legitimate activities of other nations or inflict injury or damage. Policy makers have never considered space to be a domain of anarchy in so far as it has recognized the applicability of the United Nations Charter and Statute of the International Court of Justice to space. The January 1960 "U.S. Policy on Outer Space" acknowledged that "principles and procedures developed in the past to govern the use of air space and also the sea may provide useful analogies."

In order to maintain maximum legal and political flexibility in the new environment, the NSC also advised against defining the boundary between air and space, a position eventually adopted by the authors of the first space policy. Similarly, and again for political and legal reasons, no attempt was made by the United States to define in international forums what it understood to be "peaceful uses" of space. This ambiguity has contributed to some confusion in current domestic debates on this subject concerning whether military or aggressive activities should be prohibited from space.

From Eisenhower's perspective, references to "peaceful uses," although intentionally never clearly defined, was a rhetorical tool intended to carve out maximum flexibility for the United States in space by stemming adverse international opinion. Eisenhower confessed his anxiety in this regard in his 1960 space policy when he emphasized "in so far as peaceful exploration and use of outer space are concerned, outer space is freely available for exploration and use by all." He followed up this declaration by stressing that "where the U.S. contemplates military applications of space vehicles and significant adverse international reaction is anticipated," the United States will "seek to develop measures designed to minimize or counteract such reaction."

Gradually, U.S. policy statements came to declare explicitly outer space to be a free domain so long as activities in that environment were peaceful, or nonaggressive (as opposed to nonmilitary). Secretary Quarles stated in early 1959 that it would be undesirable for the United States to make "any unilateral policy statement binding only on the United States and which might conceivably limit or hamper its own freedom of action." Quarles also noted the need to "ascertain the extent to which other nations may want to use space to the disadvantage of the United States." Eisenhower explained in his January 1960 policy that, notwithstanding the understanding that the full implications of the principle of freedom of space remained to be assessed: "It is possible that certain military applications of space vehicles may be accepted as peaceful or acquiesced in as non-interfering. On the other hand, it may be anticipated that states will not willingly acquiesce in unrestricted use of outer space for activities which may jeopardize or interfere with their national interests."

Thus, from Eisenhower's time forward, the United States would defend the right of *peaceful* transit through space, or the right of any nation to orbit satellites so long as those vehicles are not equipped to do damage to U.S. citizens, national territories, or national property (including national property in space).[10]

How Has U.S. Space Policy Evolved?

In this section, we will look primarily at presidential space policy, though congressional actions in this area will be noted when appropriate. And in keeping with the defense focus of this book, military space subjects shall be highlighted, and civil and commercial activities addressed as appropriate.

Eisenhower

The development of the first national space policy was the result of considerable teamwork between the legislative and executive branches of government. The National Aeronautics and Space Act of 1958 provided the basic authority and direction for U.S. policy. The act carried a declaration that the "activities in space should be devoted to peaceful purposes for the benefit of mankind," and it laid the basis for the separation (at least from administrative and organizational perspectives) of the civilian and military space worlds. Congressmen based the argument for NASA, a civilian-run agency, on the grounds that "national space policy is too important to leave exclusively to military authorities or to scientists alone."[11] There is nothing in H.R. 12575, however, to derogate the authority of the Department of Defense to conduct research and development of space-related weapons systems or to use space for defense purposes.

The executive branch developed NSC papers that, together with the new public law, embodied the U.S. vision and objectives with regard to outer space, including references to the importance of freedom of space, the intention of the United States to use space for peaceful uses, and the need to compete with the Soviet Union to enhance the international prestige of the United States and its allies. It has always been U.S. policy to seek bilateral and international cooperation in space and foster an exchange of information among scientists from allied countries to satisfy the space aspirations of "the Free World." Washington helped establish in December 1959 a permanent twenty-four-nation UN Committee on the Peaceful Uses of Outer Space in order to review the area of international cooperation, study programs for the peaceful uses of space that may be undertaken by the UN, and to study the nature of emerging legal problems. U.S. policy established the country's early opposition to the deployment of weapons of mass destruction in space.

While the administration publicly downplayed the launch of *Sputnik,* this world-class Soviet achievement clearly had an impact on the thought processes of a number of U.S. officials. The impressions of one of those officials, James R. Killian Jr., Eisenhower's special assistant for science and technology, had a particularly powerful impact on the formation of early national space policy. Eisenhower looked to Killian to survey post-Sputnik events from scientific and technological perspectives. The president urgently needed to assert a strategy for responding to Soviet achievements in space and countering Soviet military developments (especially advances in the areas of ballistic missiles and hydrogen bombs). Indeed, Killian observed that the American and world publics had a hard time separating Soviet scientific achievements from military progress. He noted, "The distinction between military and scientific implications is often not being sharply drawn and appears hardly to be drawn at all among the least informed. The USSR is diligently seeking to create the impression that in this field too a watershed has been reached, and that a re-evaluation of relative military strength and positions must follow."[12]

Killian also was behind Eisenhower's insistence that civilians direct the U.S. space program, and he set about working from an intellectual framework that posited two distinct objectives for the exploitation of space—exploration, primarily a scientific operation, and control, fundamentally a military operation.[13] He is, in some sense, the father of the organization we know as NASA and of the notion that only an aggressive, but civilian controlled, space program can assure true scientific progress and the protection of the United States' international prestige.[14]

On May 27, 1955, the NSC set out a national vision regarding space, establishing among other things the importance of space to the intelligence sector. NSC directive 5520 also took pains to distinguish civilian from military uses of space, suggesting for example that the United States should use a "civilian" launch vehicle to insert the first U.S. satellite into orbit in order to promote a peaceful image of America internationally, a guideline the administration revised upon the failure of the Vanguard program and the urgent need to launch *Explorer 1* on an army-developed Jupiter-C. The NSC recognized, however, that data derived from robotic spacecraft would have defense applications (for example, communications and missile booster research).

By August 1958, the administration decided not to incorporate ballistic missiles and antimissile systems into the nascent space policy in order to avoid confusion about what constituted a space vehicle (missiles may or may not traverse space). Preliminary U.S. policy on outer space also established Eisenhower's so-called conservative approach to the U.S. space program, emphasizing the importance of developing space technologies for meeting national needs at a measured and affordable pace. The president preferred this approach over any plan to match frantically Soviet space projects, which Moscow devised to show off the superiority of communism. The NSC also clarified the appropriateness of pursuing military uses of space essential to maintaining the overall deterrent capability of the United States.[15]

Eisenhower insisted that the U.S. space program pursue four different objectives. First, he acknowledged that man has a curiosity to go where he has not gone before, to explore the unknown. Second, he believed it was important to "be sure that space is not used to endanger our security. If space is to be used for military purposes, we must be prepared to use space to defend ourselves." Third, he noted the significant factor of national prestige. And fourth, space offers unique opportunities to expand our scientific horizons and add to our understanding of Earth and the solar system.[16] All subsequent U.S. space policies have retained these goals in one form or another.

With the issuance of NSC 5814/1, Eisenhower recognized that there would be a "psychological impact of outer space activities which is of broad significance to national prestige." He further recognized that the USSR surpassed the United States in scientific and technological accomplishments and warned that, left unaddressed, the Soviet Union would use its space superiority to undermine U.S. prestige. The overriding objective emanating from the Eisenhower policy, therefore, was to "carry out energetically a program for the exploration and use of outer space by the U.S."

Eisenhower's strong commitment to his new policy led to the creation of new governmental organizational structures and a commitment to expand the space program to achieve fundamental objectives spelled out in the 1958 NASA Act. He established the roles of the three armed services and other DoD agencies (such as the Advanced Research Projects Agency) in development of space projects, and minds in the military began to focus more sharply on ways to counter the Soviet satellite and ICBM threat and exploit space for national purposes. The services bandied about new doctrines for exploiting space. This was a time when heady ideas for developing space combat capabilities routinely bubbled to the surface. Congress became an important forum for discussing space and molding policy, and with the involvement of legislators, who often strongly supported the military view of the world, came assurances that the Defense Department would be able to engage in military-oriented space research.

The expansion and acceleration of the United States' space programs took place despite competition among the armed services for the lead space service position. Eisenhower did not look to the services, instead seeking to establish a separate procurement agency for his favored Corona satellite project. He secretly established the National Reconnaissance Organization in August 1960 to satisfy this objective and accommodate growing CIA interest in the utility of space for intelligence gathering. In any case, Eisenhower and his advisors were not inclined to give the military the keys to this kingdom, for he and others believed that the services would use this unique intelligence collection tool in a less than objective manner to find or even invent new threats, and establish expensive new programs to meet those threats. Eisenhower also expressed doubts about the efficacy and utility of many of the space weapon concepts brought before him, although he did sanction some space weapon research and supported the development of a satellite inspection system (SAINT) to monitor foreign space activities.[17] The president was more inclined in 1958 to emphasize the exploitation of space for the purposes of improving science, communications, reconnaissance, and weather forecasting.

Some might argue that because Eisenhower did not support a robust manned space program, his administration over space matters was "conservative" or "cautious." Indeed, he frequently wrung his hands over the possibility that space projects might break his budget, which lends some credence to this assessment of his overall conservatism. It is true that Eisenhower and Secretary of State John Foster Dulles objected to committing the United States to expenditures and efforts in outer space that were without practical limit and aimed primarily at achieving superiority over the Soviet Union. Rather, they believed that space capabilities should be developed to satisfy U.S. national purposes, not to outrace another country to achieve spectacular space "firsts."

One should not lose sight of the sensibility of this position and the enormous accomplishments of the Eisenhower administration, including the adoption of basic principles regarding activities in space observed to this day and the initiation of programs to provide the United States with a meaningful space capability. Indeed, during his last two years in office, numerous reconnaissance, meteorological, communication, and scientific satellites were launched successfully. There was considerable public support for Eisenhower's space program (at least the publicized part) and for the measured and well-conceived approach he took to prepare the United States for the space age. To be sure, compared to the frenzied space plans of John F. Kennedy, Dwight Eisenhower looked overly calculating and cautious, but in my eyes, Eisenhower was one of the more energetic and forward-looking presidents to contribute to the development of the United States' space capabilities.

Kennedy

President John F. Kennedy's administration accepted the elaborate space policy framework constructed by his predecessor and sought to strengthen many of its core elements. Although Kennedy came into office without a particular space agenda, he quickly understood the role the vast outer realm would play in enhancing security and bolstering the nation's international prestige. The space race with the Soviet Union became the centerpiece of his policy, a race he eventually sought to win by taking the country on one of the greatest, most dangerous, and spectacular adventures ever undertaken by any people at any time in history.

The decision to land a man on the Moon and return him safely to Earth defined Kennedy's national space agenda. Kennedy and his advisors aimed to overtake the Russians, and they were keen to do so in a convincing and dramatic way. The Mercury, Gemini, and Apollo Projects amounted to a decade-long commitment to garner unparalleled prestige for the nation, prestige Kennedy felt it had lost when the Soviet Union's Yuri Gagarin became the first man in space—only the latest in a series of "space firsts" for the communist regime. The Soviet achievement caused many Americans to lose confidence in the nation's educational system and in their collective scientific and engineering skills.

Thus, on May 25, 1961, President Kennedy asked the American people before a joint session of Congress to support his program, stating that it was "time for this nation to take a clearly leading role in space achievement." Kennedy put the country on a crash program to fully exploit its broad and rich economic, technological, and scientific bases and to achieve the ancient dream of space travel. Kennedy believed that the United States must occupy a position of preeminence in the space area, that the conditions for war and peace in the future depended on it.[18] By the time the decade ended, the country's generous funding of its manned space program had built up considerably its space technology base and capability to exploit Earth's orbits.

Kennedy, ably assisted by his vice president, Lyndon B. Johnson, sought early in his term to establish the new administration's plans and goals for space, although Kennedy never formally issued his own "U.S. Space Policy." Eisenhower's directives did not finally solve the issue of which of the armed services would take the lead in space. Kennedy appointed the air force the lead service for space. Air force space activities subsequently grew, and the department floated numerous proposals for new investments in programs to give the country military supremacy in space, to include deploying military astronauts. Kennedy's Defense Department established the first antisatellite weapon programs to address possible Soviet offensive uses of space and to defend peaceful uses of space. The United States needed to compete in missile and space technology, and the Pentagon took advantage of the opportunities it had to conduct research.

Shortly after assuming office, Kennedy pressed boldly ahead with the reconnaissance program, both to advance and legitimize it. On December 3, 1962, his national security advisors drafted this statement, which the administration used in negotiations with Moscow: "It is the view of the United States that Outer Space should be used for peaceful—that is, non-aggressive and beneficial—purposes. The question of military activities in space cannot be divorced from the question of military activities on earth.... Observation from space is consistent with international law, just as observation from the high seas."[19]

The government's position on the use of space remained essentially the same as it was under Eisenhower.

Believing that communications satellites would have a significant impact on security, Kennedy continued direction provided by Eisenhower in January 1961 "to encourage private industry to apply its resources toward the earliest practicable utilization of space technology for commercial civil communication requirements." On July 24, 1961, he asked that the organizational framework be created for all nations to participate in a global communications satellite system "in the interests of world peace and closer brotherhood among peoples throughout the world."[20] Kennedy stressed the positive foreign policy implications of establishing a satellite communications system for peaceful purposes and of making it available to developed and undeveloped countries to increase the global capacity for exchanging information cheaply and reliably. NASA and the Pentagon also continued research and development in the communications field.

The administration continued work within the UN against defining limits in outer space or restricting military uses of space. Kennedy emphasized the rights of passage in space, choosing to highlight definite parallels between space and the open oceans, which are not under the sovereignty of any nation. International registration of objects in orbit and establishment of a regime of international space law remained fundamental goals of the United States. Arguments for banning weapons of mass destruction and nuclear explosions from space, provided there were adequate inspection and control, also gained momentum under Kennedy.

Kennedy's March 23, 1962, decision to classify all U.S. space activities set his administration's policy somewhat at odds with Eisenhower's policy of openness. His famous "blackout" directive was aimed at improving national security by limiting disclosure of information to the public regarding all military space programs and space flights. With this directive, which given the nature of the competition underway with the Soviet Union at the time was not wholly unfounded, there began a long tradition in the United States of mystifying and distorting the country's defense space activities and hiding them from the general public, publicly distinguishing and sorting them apart from categories of military activities taking place in the other environments.

Johnson

President Lyndon B. Johnson further refined established space policy once he unexpectedly assumed office in November 1963. The goal of sending man to the Moon clearly dominated his political and budgetary horizon, despite the fact that his commitment to the lunar landings drew steady criticism from Republicans who believed his programs did not adequately meet military objectives in space.[21] This charge is full of irony, given Johnson's famous January 7, 1958, statement that it must be the goal of the United States "and of all free men" to control space in order to ensure national safety.[22] During his nearly six years in office, Johnson actively affirmed a number of principles and goals pursued by his predecessors in space. Making great strides in the areas of communications, meteorology, navigation and geodesy, space operations became more routine under his administration.

By the time Johnson left office in January 1969, international customary law and a freshly negotiated treaty had affirmed the freedom of space principle. Johnson carried through Kennedy's initiative to ban weapons of mass destruction from space by negotiating the multilateral 1967 Outer Space Treaty (it is worth noting here too that the United States never intended to deploy such weapons in space in any case). Hand in hand with his goal of establishing the rule of law in space, Johnson sought to broaden opportunities for international cooperation in that environment in the "hope that the conquest of space can contribute to the establishment of peace."[23] By early 1968, eighty-four nations had participated in cooperative space activities with the United States. The minimal cooperation that took place with the Soviet Union centered on communications satellite experiments and on sharing data from weather satellites.

From the early 1960s to the 1970s, Soviet leaders vigorously pursued military space activities. Aside from work on antisatellite weaponry, Secretary McNamara publicly disclosed in November 1967 that the Soviet Union had been working for several years on a fractional orbital bombardment system (FOBS), having tested at least twelve rocket carriers for the new weapon system. A FOBS would carry a nuclear weapon, much like a ballistic missile, the main difference being that once it was launched, it would threaten the United States from only about one hundred miles above Earth. In concept, it could have been called down on a target prior to completing one orbit and in a manner that would have sharply cut warning time (for presumably its final destination could not be known until the rocket engines required to propel it downward kicked in). A ballistic missile, on the other hand comes down out of a predictable trajectory from 700 miles above Earth, providing significantly more warning time ("significant" being a relative term).

In the fall of 1964, Johnson openly declared that United States was pursuing an antisatellite capability, one of the reasons being to defeat bomb-carrying satellites like the Soviet FOBS, although this was not publicly acknowledged at the time. Johnson also sought to enhance U.S. security against the FOBS and ICBM threats by improving early warning radars and approving development of the Nike-X (Sentinel) antiballistic missile system.

Secretary of Defense Robert S. McNamara drew up plans for a separate military communications systems on the grounds that the use of newly established public COMSAT facilities would preclude necessary military secrecy and compromise national security. Hence, Johnson announced in August 1964 that the air force would develop the independent Defense Satellite Communications System to carry critical military communications, a system still in use today.

The air force experimental manned space station begun in 1967, the Manned Orbiting Laboratory, was to be deployed in the early 1970s. The MOL would reconnoiter the earth from space for targets and militarily significant activities as well as inspect or destroy, if necessary, foreign satellites. However, increasingly proficient robotic reconnaissance craft would do this job more efficiently and a great deal more cheaply. The emergence of a superior way to gather intelligence from space caused President Nixon to cancel the program early in his first term.

Nixon

President Richard Nixon determined U.S. space priorities after Apollo, and he decided in the end that the country could no longer afford major investments in large new space projects. When Nixon outlined his objectives in a March 7, 1970, speech, he announced that the country would continue to explore the stars, but it would do so with unmanned vehicles (he rejected Vice President Spiro T. Agnew's recommendation that the United States initiate a manned Martian expedition by the turn of the century). He did, however, pledge to finish out the Apollo missions (although the final three missions were canceled) and to support the *Skylab* project to undertake further biological experiments in Earth's orbits. Cost efficiency drove policy. When in January 1972 he announced the start of a semireusable Space Transportation System (STS), also known as the space shuttle, he did so on the basis that this novel way for placing payloads in orbit would save the country money.[24] Nixon's message was loud and clear—the United States' space program would be "bold yet balanced."

Under Nixon military space programs matured. The United States deployed military satellites to last longer in orbit, which in turn reduced demand for space launches. Big Bird, the KH-9 series of strategic reconnaissance satellites, became operational in 1971, and improvements occurred in navigation, communications, and meteorological satellites. The Pentagon, seeing the possibilities of routine access to space for military purposes, expressed its interest in the STS following development approval.

Military space, in fact, was not a topic of great interest during the Nixon administration. Vietnam and Watergate voraciously consumed the attention of the president and his aides. Nixon managed, however, to make a significant impact in the area of arms control with the Soviet Union. The 1972 ABM Treaty, which banned or limited the deployment of space-based ABM components, to this day has political and legal implications for the military uses of space. Negotiators of the SALT agreement also inserted provisions to prohibit interference (although "interference" was never clearly defined) with the peaceful operations of National Technical Means of either party, an unofficial recognition of the role reconnaissance satellites played in monitoring arms agreements.

Ford

Soviet antisatellites and satellite survivability concepts did not excite much interest in the Nixon administration. During Gerald Ford's short term in office, however, the Soviet space threat reemerged as a presidential level issue and directly impacted U.S. defense space policy. The Soviet Union, right along side the United

States, had been gradually maturing as a space power through deployment of satellites to enhance national security. Hence, U.S. official interest in ASATs naturally grew. Concern also arose regarding the dependency of the United States on satellites for basic national security requirements (especially communications), and many in the Ford administration began to question whether U.S. satellites were adequately protected against deliberate attack by the Soviet Union.

The Ford administration commissioned a Defense Science Board study, which concluded that the ground segments of U.S. space systems were indeed vulnerable in wartime.[25] This got the attention of the president. Interest in space defense within the U.S. Air Force and the Pentagon, however, lagged behind the White House, enough so to prompt Ford to issue several national security decision memorandums (NSDMs) in an effort to implement his space policy. NSDM 333 addressed the vulnerability of U.S. military satellites and directed that the air force allocate funds to accomplish work in the space surveillance and space defense areas. It also led to the establishment of the System Program Office and the Space and Missile Systems Organization (SAMSO) within the air force. Another directive addressed relaxed attitudes in the Pentagon toward the development of an ASAT weapon. Nixon and Ford gradually terminated the U.S. nuclear ASAT program begun in the 1960s. However, in an effort not to give up the initiative in this area, Ford's NSDM 345 directed the Pentagon to develop a miniature homing vehicle ASAT weapon to be launched by an airborne F-15 to counter Soviet ocean reconnaissance capabilities and to study arms-control options. One of the last actions Ford took before leaving office was to approve the development of the new ASAT project.

Carter

Soon after assuming office, President Jimmy Carter directed his NSC Policy Review Committee to "thoroughly review existing policy and formulate overall principles which should guide our space activities."[26] He oversaw the first comprehensive review of the country's space policy since Eisenhower. The review captured civil and national security space programs (although Carter later published a presidential directive to address exclusively civil space programs).

The reviewers did not challenge the basic principles of U.S. defense space policy. Rather, management problems involving interactions among the major space stakeholders in government drove the review. The reexamination sought ways to reduce fragmentation and redundancy in the space program and improve coordination among the DoD, NASA, NOAA, the intelligence community, and private space organizations. For the first time, a broad interagency steering group was established (which included the Departments of Defense, State, Agriculture,

Commerce, and Interior, as well as the Arms Control and Disarmament Agency, CIA, Joint Chiefs of Staff, National Security Council, NASA, NOAA, and the White House Office of Science and Technology). Policy by interagency review had begun.

Based upon the recommendation of the National Security Council, President Carter signed Presidential Directive (PD) 37 on May 11, 1978, reaffirming broad objectives and basic principles of the U.S. national security space program. Policy continued to maintain that space systems are national property with the right of passage through and operation in space without interference, and that purposeful interference would be viewed as an infringement upon sovereign rights. Longstanding satellite programs were maintained, and development continued on the Rockwell International NAVSTAR GPS and the space shuttle. The shuttle was the object of some attention within the air force by officials who believed this semireusable spacecraft would offer more defense space program options, although there remained strong support within this service department to retain a robust expendable launch vehicle force.

The policy explicitly underscored the national commitment to operate on a global basis remote-sensing operations in support of national objectives. The country also continued "to pursue activities in space in support of its right of self-defense and thereby strengthen national security, deterrence of attack, and arms control agreements." The Carter policy was the first to encourage domestic commercial exploitation of space capabilities in order to bolster the country's technological position and economy. However, by stating that the U.S. government must authorize and regulate all privately funded Earth-oriented remote-sensing activities, Carter also put a damper on this incentive.[27]

PD-37 is an important policy statement in the history of U.S. military space programs insofar as it depicted space to be more than an environment for enhancing land, sea, and air power. While this directive did not set a course for the full integration of space weapons into the national military strategy, it did continue the research and development of ASAT systems begun under Ford. PD-37 expanded the government's view of military space by officially recognizing the potential for space to become a future combat medium (something that was also understood by earlier presidents).[28]

Carter pursued his primary objective of arms control in parallel with the MHV ASAT weapon as part of his "two-track" approach. The administration announced its arms-control objectives in March 1977 and subsequently undertook the first negotiation with Moscow on ASAT weapons to seek "an effective and adequately verifiable ban on anti-satellite systems."[29] Indeed, the pace of Soviet coorbital ASAT testing had increased during Carter's term in office. Carter viewed ASATs more as a bargaining chip than a war-fighting tool: they were a hedge or deterrent against the failure of the administration's negotiation strategy. The president sought comprehensive limits on ASATs, although his negotiators never clarified what devices and activities ought to be covered by a future agreement.[30] The president eventually initiated talks with Moscow when he became impatient with the lack of progress of his working group, which he set up to develop a U.S. position.

In an effort to defend plans to enter into a comprehensive strategic arms-control agreement with the Soviet Union, President Carter publicly acknowledge for the first time since 1962 that the United States had a strategic photoreconnaissance program. The president revealed in an October 1, 1978, speech at Kennedy Space Center that "photoreconnaissance satellites have become an important stabilizing factor in world affairs in the monitoring of arms control agreements." He reaffirmed that "we shall continue to develop them."[31] It was widely believed that Carter's revelation on "the fact of existence" of such systems was intended to improve congressional acceptance of his arms-control agenda by underscoring U.S. access to data from space that would uncover hostile activity by foreign countries.[32]

Reagan

Following a ten-month comprehensive evaluation of U.S. space programs, President Ronald Reagan issued National Security Decision Directive (NSDD) 42. The directive further developed PD-37's position that space is a potential war-fighting medium, where the United States could pursue activities "in support of its right of self-defense." The Reagan policy stressed the importance of providing space systems more assured survivability and endurance and highlighted the fact that the United States would continue the development of an ASAT weapon begun under Ford and carried out under Carter. Reagan also underscored that the United States would "maintain an integrated attack warning, notification, verification, and contingency reaction capability which can effectively detect and react to threats to United States space systems."[33] Although not fundamentally different from the policy adopted by the Carter administration, the Reagan policy received some heated criticism from certain quarters as being too "militarily oriented."[34]

Consistent with the broadened understanding of the national security implications of space activities, the 1982 Reagan policy sharpened the official understanding of what was meant by the "use of space by all nations for peaceful purposes." NSDD 42 followed the expression of this now-familiar principle underlying U.S. conduct in space with this simple statement, which was subsequently applied in the Bush and Clinton national space policies: "'Peaceful purposes' allow activities in pursuit of national security goals." This was the clearest policy statement to date clarifying that "peaceful uses" or "peaceful purposes" were more synonymous with "nonaggressive" rather than "nonmilitary" purposes.

Reagan continued to support the STS and set the goal of making the space shuttle the primary launch system for both national security and civil missions. He also sought to make the shuttle available to authorized commercial users. Eventually, according to the directive, expendable launch vehicle operations would be phased out if and when the capabilities of the STS were sufficient to meet all national needs and obligations.

This national launch strategy backfired when the shuttle *Challenger* exploded in January 1986, a mistake duly recognized and corrected in Reagan's 1988 space policy. That said, the Reagan administration's May 1983 initiative to support growing interest in the private sector to operate commercial ELVs sought, at least in some measure, to ensure a flexible and robust U.S. launch posture in order to maintain U.S. leadership in space.[35] Reagan also initiated in 1986 the joint DoD-NASA national aerospace plane research program to explore an entirely new family of aerospace vehicles, a program his successor eventually terminated.

Political opposition in the Democratic controlled House and Senate decried the "weapons race in space" taking place under Reagan. In his space directive, Reagan asserted that his administration would continue to study space arms-control options: "The United States will consider verifiable and equitable arms control measures that would ban or otherwise limit testing and deployment of specific weapons systems, should those measures be compatible with United States national security." By 1983, however, disappointed that the administration had not yet entered into ASAT negotiations with Moscow, many congressmen and senators sought to restrict unilaterally U.S. ASAT activities with amendments to appropriations bills. Although these attempts to restrict the ability of the United States to deny through the use of force the hostile uses of space generally never made it into legislation, Congress did pass an amendment withholding procurement funds for antisatellite weapons unless the president certified that his administration was actively assessing the possibility of negotiating with the Soviets and that the proposed ASAT tests were necessary.[36]

The administration's response to the requirement was to issue a report dated March 31, 1984, on U.S. policy on ASAT arms control, which essentially fleshed out the NSDD 42 statement on this subject. The report reviewed national security requirements for ASATs and the problems and possibilities for ASAT arms control. The many challenges of arms control in space identified by the authors of the report are summarized for the reader in chapter 4. Suffice it to say here that, though the administration's policy was to "search for viable arms control opportunities in the ASAT area," the report's conclusions and the administration's activities underscored Reagan's deep skepticism about negotiating with the Soviets for greater security in space. The report found that, despite efforts under Reagan and Carter to consider arms-control options "in light of whether they support our overall deterrence posture and are effectively verifiable, no way has yet been found to design a comprehensive ASAT ban that meets these criteria."[37]

In sharp contrast to Carter, however, Reagan, who had different judgments about Soviet behavior, never viewed the U.S. ASAT program as an arms-control bargaining chip. Reagan's policy was to deploy an operational ASAT for deterrence and defense purposes at the earliest possible date. While the president supported the continuation of the F-15 MHV ASAT program, Congress leveraged sufficient opposition to the program to compel the air force to kill it, after which time the air force unloaded all ground-based ASAT work onto the army.

Reagan's vision for space well exceeded the vision of his predecessors, as he believed the outer realm could be exploited to provide exceptional defenses for the country. NSDD-85, dated March 25, 1983, directed the DoD to define a multiyear research and development program (within limits of the 1972 ABM Treaty) to achieve a long-term objective for the United States—the elimination of the threat to the nation from nuclear-armed ballistic missiles through the deployment of strategic missile defenses. He directed that studies be completed on a "priority basis" to assess the role ballistic missile defense (BMD) would play in a future security strategy. The resulting major new initiative, the Strategic Defense Initiative, came to incorporate novel space-based ballistic missile interceptors, satellites for battle management command, control, and communications, and other supporting space-based BMD components.

Reagan signed the directive two days after his now famous televised speech to the nation outlining a new vision for making Soviet nuclear ICBMs "obsolete." Reagan's SDI speech, according to his science advisor, George A. Keyworth II, was not a result of interagency coordination and it did not originate with his staff: "This was not a speech that came up; it was a top down speech[,] . . . a speech that came from the President's heart."[38] The speech itself was a prime example of the president using the "bully pulpit" to appeal directly to the American people for their support to set the country on a new course.

In order to reaffirm the administration's basic goals in space and craft a policy that more clearly reflected the realities of national space activities, President Reagan issued his second national space policy in the form of a presidential directive on January 5, 1988. While the written policies might not have looked all that different, by 1983 it was clear that President Reagan had expanded Carter's vision to establish more fully a national war-fighting capability in space. Reagan's 1988 national space policy reaffirmed the basic tenets of space control and space-based force application established in PD-37, NSDD-42, and NSDD-85.

The 1988 national space policy was the first to spell out explicitly the four basic DoD missions in space, namely, space support, force enhancement, space control, and force application. A February 4, 1987, Defense Space Policy issued by

Secretary of Defense Caspar Weinberger, which featured the four mission areas, gave guidance to all DoD organizations with respect to space activities. Most novel in this array was the mission to project force in and from space. While previous policies dealt with the first three missions in one form or another, the new directive provided a policy basis for the Defense Department, "consistent with treaty obligations," to "conduct research, development, and planning to be prepared to acquire and deploy space weapons systems for strategic defense should national security conditions dictate." It was the boldest affirmation to date of an emerging understanding in several quarters in government that space really was like the land, sea, and air war-fighting environments, and that combat operations in space ought to be pursued to defend national interests and improve national security.

Commercial space activities were recognized for the first time in the February 1988 policy directive as comprising one of three national space sectors. Over the years, Congress passed several laws to facilitate the growth of commercial space activities to support national security and the economy. Reagan took the first significant step to provide comprehensive guidance to foster growth of commercial use of space and direct commercially available goods and services be used to the greatest extent feasible by the national security and civil sectors.

Although he may be faulted for his failure to implement his visions within the long-term defense planning process, the programs Reagan sponsored and fought for and the new space organization he supported demonstrated his commitment to strengthening American space power. On Reagan's watch, the United States established the U.S. Air Force Space Command to consolidate space activities and provide a link between space-related research and development and operational users. The army and navy soon followed suit with their own space commands and by 1985 the unified U.S. Space Command came into being. The air force also established a Space Technology Center at Kirtland Air Force Base, New Mexico. Other steps taken towards consolidating U.S. space organization included the formation of a DoD Space Operations Committee; elevation of the commander in chief of NORAD to a four-star position and broadening of his space portfolio (indeed, CINC NORAD was given two more "hats," as he became commander of both U.S. and U.S. Air Force Space Commands); the creation of a separate Space Division in the air force and establishing a deputy commander for Space Operations; construction of a Consolidated Space Operations Center; formation of a Directorate for Space Operations within the office of the deputy chief of staff/plans and operations in air force headquarters, and, among other things, the establishment of a course in space operations at the Air Force Institute of Technology.[39] Reagan's restructuring of national military space organization and programs also led to the establishment of the Strategic Defense Initiative Organization (later called the Ballistic Missile Defense Organization) in 1984 to plan, organize, coordinate, direct, and enhance research and testing of strategic defense technologies.

Bush

President George Bush issued National Security Directive (NSD) 30 in November 1989, a national space policy that essentially reaffirmed the national security space goals outlined by Reagan. The Bush policy, updated by an addendum in September 1990, further codified the existence of the three space sectors and reaffirmed the centrality in national space policy of the space-control and force-application missions. The United States must be prepared, stated the directive, to assure that hostile forces cannot prevent U.S. use of space and to negate, if necessary, hostile space systems. Although the F-15 MHV ASAT had been canceled by the Pentagon during the final years of the Reagan administration, it remained the policy of the United States to develop and deploy an ASAT capability "at the earliest possible date."

Bush also carried forward Reagan's ideas for the use of space combat systems to improve national security conditions. Bush's most notable contribution in this area was his restructuring of the SDI program away from providing a comprehensive national shield against ballistic missiles and towards a program to provide the U.S. homeland, allies, and troops "protection from limited ballistic missile strikes, whatever their sources." The Bush administration pushed this program in the face of mounting congressional criticism and dwindling appropriations for strategic defenses. Bush's system employing hundreds of space-based interceptors providing global protection against limited strikes (GPALS) would not survive after his administration.

Clinton

Promotion of the domestic remote-sensing industry began in the 1980s when Congress passed the 1984 Land Remote Sensing Commercialization Act and President Reagan removed many government restrictions on the spatial resolution of U.S. commercial satellites. When foreign suppliers of imagery began dominating the markets in the 1990s, Congress passed the Land Remote Sensing Policy Act in 1992, and President Clinton issued new policy guidelines and licensing procedures on March 10, 1994, to encourage the development and deployment of commercial satellites having a resolution no higher than one meter in order to increase U.S. industrial competitiveness.

President Clinton's 1994 policy on remote sensing opened wide the throttle for the commercial imagery business. The policy supported development of U.S. remote-sensing industry to compete with the growing number of commercial ventures worldwide, citing important national benefits such as national security and the promotion of regional stability. Also indicative of the push to nurture a new industry, the Clinton administration gave the pro-business Commerce Department jurisdiction in 1996 over satellite exports, a move that Congress reversed in 1998 following the political fallout stemming from public outcry over the transfer in 1995 and 1996 of sensitive satellite information by U.S. companies to China.[40]

President Clinton issued a "National Space Transportation Policy" on August 5, 1994, to establish guidelines and implement actions to sustain and revitalize national space launch capabilities. The policy underscored that the space program is "critical to achieving U.S. national security, scientific, technical, commercial, and foreign policy goals" and highlighted the requirements for assured, reliable, and affordable access to space. The policy named the Department of Defense the lead agency for the development of expendable launch vehicles and associated technologies and NASA the lead agency for the improvements to the space shuttle and development of next-generation reusable systems. The Departments of Commerce and Transportation were to promote domestic commercial space launch activities.

In March 1996, the Clinton White House released its "U.S. Global Positioning System Policy" to deal with the rapid increase in the use of the GPS signal worldwide for civil, scientific, and commercial applications. Under Clinton, the goal of U.S. policy was to support and enhance U.S. economic competitiveness and productivity while protecting national security and foreign policy interests.[41] The release of the new GPS policy amounted to an attempt to set guidelines for handling the system's uncontrollably rapid and expanding roles and for addressing the reality that GPS has become an international resource offering services even to the nation's enemies. The policy also attempted to keep the United States and U.S. industry in a dominant position internationally by ensuring that the GPS system remained the standard positioning and navigation system.[42]

President Clinton published his "National Space Policy" (Presidential Decision Directive 7) on September 19, 1996. Capturing and elaborating on policies from the Reagan and Bush eras, the 1996 policy once again affirmed the principles that drive U.S. space programs and recast familiar goals. Clinton's expansion of intersector guidelines was perhaps one of the more significant contributions to policy. Indeed, the convergence of some defense space programs with civil space programs, and the requirement for the government sectors to support commercial sector development and promote the Pentagon's exploitation of commercial services represent significant steps forward. There is an ongoing transition from government investment in the development of infrastructure to the exploitation of commercial services and products, which today are developing at a faster pace than what is found in the public sector. The underlying reason for the shift to commercial services and products (especially in the communications area) is that they can offer more efficiency and possibly even cost savings, although the jury is still out as to whether commercial space services can do the job for less money.[43]

Another notable achievement of the September 1996 policy is that it officially acknowledged the well-known fact of the existence of U.S. satellite photoreconnaissance systems having a near-real-time capability useful for intelligence collection, defense planning, and military operations. It also revealed that the United States relies on "overhead signals intelligence collection" and other forms of space-based intelligence collection. The 1996 policy officially declassified the existence of the NRO and senior officer positions, a process begun in 1992 under Bush. These steps represent progress toward the elimination of the excessive classification of the country's military space programs and organizations.

The national security space policy moved positively to direct that the defense and intelligence space sectors work more closely together. In an unprecedented effort to breakdown major institutional barriers, and impediments between the "white" and "black" space worlds, the secretary of defense was authorized to propose modifications to intelligence-gathering satellites and develop and operate, as necessary, DoD satellites "in the event that intelligence space systems cannot provide the necessary intelligence support." The director of central intelligence, in turn, must ensure that the intelligence space sector provides timely information and data to support foreign, defense and economic policies and military operations in addition to its more traditional missions of providing indications and warning and information on treaty compliance. The National Space Policy embodies many of the lessons learned from Desert Storm, namely, that national space systems had turned the corner, so to speak, and that, henceforth, satellites would be a critical tool of the war fighters.

Late in the second term of the Clinton administration, Secretary of Defense William Cohen signed a DoD directive on space policy to replace the 1987 version. Major changes in the international system and the issuance of a new National Space Policy in 1996 underlay the reasons for having issued the July 9, 1999, DoD Space Policy. The new directive did not depart significantly from existing practices and policies, and it reaffirmed the important role space plays in U.S. military strategy, the critical requirement of achieving information superiority, and the significance of intersector cooperation to national security. In an effort to ensure the continued integration of space into military operations, the directive stated that "space shall be considered as a medium for conducting any operation where mission success and effectiveness would be enhanced relative to other media."

The directive's bold statement that the United States reserves the right to use force to protect its space systems affirms a longstanding national position that interference with U.S. satellites will be construed as an infringement on sovereign rights. The explicit reference to the "use of force" in this context is a novel expression in policy, but it is not a new piece in the architecture. One feature of the 1999 defense space policy that did not get much attention in the press, at least not as much as the "use of force" addition, was the subject of military personnel in space. The policy stated that if it can be demonstrated that military roles for humans in space can make "unique and cost-effective contributions to operational missions," the Pentagon may explore the possibility of a military astronaut program. This "if," however, is a major hurdle to overcome given that most manned space missions probably can be done by satellites or other robotic means *and* given the historically proven fact that manned space missions are significantly more costly than unmanned. This policy reflected, ironically, opposition in the Clinton administration to dedicated military astronauts, a penchant first made evident in Clinton's policy directing the Defense Department to stay away from the development of reusable launch vehicle technology (which may be used in a military space plane) and later by the Clinton 1997 line-item veto of funding for the U.S. Air Force military space plane program.

All considered, the Clinton administration held space matters in relatively low priority. Defense space subjects were simply not part of the president's lexicon—he rarely, if ever, spoke to military space matters. Military space budgets were relatively low and flat under his administration and there was no progress in centralizing space policy making and unifying military space visions. President Clinton also terminated the National Space Council, which under the Bush administration ensured that space received executive-level visibility, replacing it with the National Science and Technology Council chartered to consider space along with a host of other science and technology policy issues.

There also was no settled organizational structure for space within the Pentagon. While there was for a short time a deputy undersecretary of defense for space to coordinate all space policy and procurement matters, Secretary Cohen dissolved that office soon after his arrival in the Pentagon and mixed it into the elephantine C^3I office of the Pentagon. The office of the national space architect was repackaged into the national security space architect. By the end of the Clinton administration, the president and defense secretary had effectively reduced the profile of defense space by shuffling it into a huge bureaucracy and retaining many of the cold war barriers between "white world" and "black world" space. Defense space policies and programs necessarily flounder in such an environment.

What Is the Bottom Line?

There is, in fact, sufficient vagueness in the written space policy to allow a variance in interpretation regarding what direction the United States must take with respect to military activities in space. With the exception of a few sentences here and there, all of the presidents over the past forty-two years effectively could have signed up to the same basic national space policy, despite the radical differences that appeared in priorities and program activities.

Yet these radical differences in approaches only serve to highlight that the country lacks a unified, coherent approach to space, especially with respect to the space-control and force-application mission areas. Congress can raise military space issues in a very visible and highly charged forum and fund the odd military space program, but the attention span of this body ultimately is too short to lead the charge in one direction or the other. The president and his activities, therefore, must be the focus of any policy discussion involving the state of U.S. military preparedness in space. Bearing in mind that policy is a living idea that finds expression in the country's program activities, laws, directives, as well as its deeds and inaction, let us now look at fundamental problem areas in current U.S. defense space policy.

8
Freedom of Space and the Defect of Present Policy

Why Is the United States Unprepared for Its Military Future in Space?

Freedom of space, a principle seated deeply in the psyche of the American people, remains a critical element of the national security strategy. Decades of prosperity and security at home cause many Americans to take for granted their freedom to explore and move about the world, to engage the rest of the world at all levels of interaction—economic, commercial, diplomatic, and military—on the land and the oceans, in the air, and now in space.

Principles and goals underlying the United States' declared space policy are essentially unadulterated from Eisenhower's time, notwithstanding the political tumult that often arises during national debates on defense space subjects. That said, the organizational structures, program commitments, funding priorities, and implementation guidelines laid out to execute U.S. space programs have changed considerably over the past forty years, reflecting vastly different interpretations from administration to administration of how best to protect U.S. freedom of space. Both Presidents Reagan and Clinton touted the requirement to defend U.S. interests and freedom of action in space, but could there have been a more widely divergent set of military space programs?

There is much to be learned about present policy deficiencies by simply examining recent space policy directives, not for what they address, however, but for what they ignore. Notable lapses include inadequate explanation of possible threats and ambiguous guidelines, especially in the areas of space control and force application. The policies leave each of the services to go their separate way in the development of space doctrines and visions at a time when unification of strategic thinking and consolidation of resources clearly is in order. In service parlance, the country has been "stovepiping" its visions. Declared policies leave untouched radical inconsistencies that hamstring the entire space policy regime and do not reconcile the country's treaty commitments and funding priorities to the now predictably tough public rhetoric about the importance of space for national security.

A common refrain among critics of President Clinton's military space policy, especially of its neglect of the space-control mission area, is that although the 1996 directive itself is fine, the policy's implementation is flawed. Current written policy is good, in other words, it is just not being acted upon. I contend otherwise. Implementation is not the crux of the problem. It is a question, rather, of *what it is* that is to be implemented.[1]

Is Space a Battlefield?

The National Space Policy is a flexible document, especially with respect to the two most politically controversial space defense missions, control and force application. "Flexible" is a positive-sounding attribute—indeed, it is the next best thing if one's preferred approach does not appear in official policy. "At least," it may be said, "the policy does not definitively preclude what we believe needs to be done." And this is true. No policy really is better than bad policy. But it is also true that no policy is . . . no policy.

The "Fourth" War-Fighting Environment

Don't believe it for a minute! It is not my intention to denigrate the progress made by the United States armed forces since Desert Storm in the defense space arena. Chapter 3 is testimony to America's grand achievements there. Much has been accomplished to make space more relevant to the needs of U.S. military forces. Yet despite the rhetoric that the armed forces must fully "integrate" space into military strategy and operational planning and that space is like the land, sea, and air environments, where the country's interests must be protected, outer space has not yet made the grade.

There is, in fact, clear policy support for the Defense Department to undertake aggressively and confidently only 50 percent of its assigned space missions! Mainly as a result of defective policy (U.S. technological capabilities to perform all assigned missions are not in doubt), the U.S. armed forces are fully prepared to execute just two of their four mission areas.

The space support and force enhancement missions sustained by policy are, in fact, undergirded by detailed policy implementation guidelines in each of the three recognized space sectors. The achievements in these two mission areas have been significant, although many in the defense community will maintain that, despite having ample authoritative policy direction, the DoD does not even perform the space support mission to satisfaction, that the United States still has neither reliable nor quick access to Earth orbit.

Contrast these missions areas, however, with the space-control and force-application missions—space defense's terrible twins, or those two missions that require engagement with the enemy. They are left to wither on the policy vine, waiting to be nurtured by clear program guidance—waiting, in fact, for the political justification they require to become fully a part of the space defense equation.[2]

If space is the necessary fourth warfare medium, "like" the land, sea, and air, *how* precisely is it like the land, sea and air? Or perhaps we should ask *how much* is space like the other environments? As we shall see, space often is treated differently by different parties, even within the Pentagon. Yet if space is "vital" to national security, why do we not have major programs in the space-control and force-application areas to provide needed protection and extend U.S. military power? Why do officials draw such a tight circle around U.S. freedom of action in space—in this case, freedom to undertake operations as required to secure national interests in space and on Earth? The United States maintains significant peacetime land, sea, and air forces for the very reason that leading officials recognize that world affairs can turn nasty. Why are we not doing the same in space? There is a great deal of talk about the "fourth" medium, but little in the way of policy, organization, and long-term programs to support that vision.

The Vanishing Mission

The force-application mission made sense under Reagan and Bush, when there existed supporting philosophies and funded programs to defeat enemy ballistic missiles using interceptors based in space as well as serious investigations into the viability of offensive and defensive combat options to influence the course of war from space. Yet despite the wording in the 1996 policy implying high-level "direction" to the DoD to maintain the capability to execute this mission, the force-application mission is a true orphan within the framework of the Clinton ballistic missile defense and space policy regime.

The nation's chief space commanders, not surprisingly, have recognized consistently this rather striking incoherence in top-level thinking, mainly because it is their job to advocate and plan for this mission and conduct supporting operations. Talk about grasping at a straw! "There's been no national action on this, other than to assign responsibility for it," Gen. Richard Myers, the United States' top space commander from 1998 to 2000, lamented in 1999 about the force-application mission. The part the U.S. Space Command must play in our national Kabuki dance with respect to this activity, therefore, must be to spin its wheels investigating concepts of operations and basic technologies applicable to the force-application mission, that is "*if* we're tasked by the national command authorities to go do that."[3] The presence of Janus here is unmistakable.

The doublespeak and confusion in policy even has Defense Department officials befuddled and at a loss for how to deal with the mandate to maintain capabilities to execute the force-application mission. Former secretary of defense William Cohen stated in one section of the 1999 Defense Space Policy that "capabilities necessary to conduct the . . . force application mission" area "shall be assured and integrated into an operational space force structure that is sufficiently robust, ready, secure, survivable, resilient and interoperable to meet the needs of the NCA, Combatant Commanders." In a different section of the same policy, however, he directed department services and agencies to explore force-application concepts, doctrines, and technologies "consistent with Presidential policy as well as U.S. and applicable international law." Again, statements about having force application *capabilities* "assured and integrated into an *operational* force structure" do not align very neatly with direction specifying that these options will be explored, and only when such activities are consistent with what is in effect a higher policy, a policy that is nowhere spelled out.

The 2000, 1999, and 1998 *Annual Report*s issued by the secretary to the president and Congress literally do not even mention space force application as a viable DoD mission area. Whereas space support, force enhancement and even space control get honorable mentions, there is nary a whisper about the role space might play to defend actively U.S. territory and interests or project U.S. power in space and to other regions on Earth. We lost a space defense mission somewhere!

The last time the force-application mission carried any status in the *Annual Report* was in the 1997 issue, signed by the newly appointed secretary of defense, William Cohen—the same William Cohen that dropped this mission in later reports. The 1997 report stated that "research in this area is aimed at developing treaty compliant advanced follow-on technologies offering promise for improved performance in both tactical and strategic defenses as insurance against possible future threats." The paragraph concluded, "At this time, the DoD space force structure does not include any capabilities for power projection." This brief section on force application actually parrots an equally ever so brief section on this "mission" prepared by former secretary of defense William J. Perry in 1996, who, in turn, borrowed from language used by Clinton's first secretary of defense, the late Les Aspin, who prepared the 1994 *Annual Report*. What is bizarre is that force application mysteriously disappeared from Secretary Perry's view in the *Annual Report* for 1995. Is it or is it not a critical mission area?

To be sure, one must look to the chief executive to place blame for this confusion. Evidence that the 1996 National Space Policy does not at all provide the controlling guidelines for what the United States does in space with respect to this mission is observable in many other places. Although he never explicitly stated so publicly, former President Clinton did not support active military measures in space, for offensive or defensive purposes, and was not given to challenge or renegotiate provisions in the 1972 ABM Treaty that ban the deployment of space-based interceptors in a ballistic missile defense system. His choice of programs and determination to keep space force-application research efforts small and deeply buried in the basements of U.S. research and development establishments plainly reveal this policy choice.

Fleeting Space Control

The ambiguity of current U.S. defense space policy is only slightly less pronounced in the space-control area. One must give the July 1999 Defense Department directive credit for its forthrightness, for it clearly admits that it does not provide the definitive guidance with respect to this controversial mission. Despite assurances that capabilities to conduct the space-control mission (like capabilities for force application) "shall be assured and integrated into an operational space force structure," the directive goes on to state that space-control capabilities will be provided only on the condition that they are "consistent with Presidential policy as well as U.S. and applicable international law." A restatement of "presidential policy" and international law applicable to the space-control mission would have been helpful, but the "directive" does not recite this higher level guidance. When one turns to the National Space Policy for an expanded reading of the published "presidential policy" relating to space control, one finds only the canned statement that U.S. space systems must be protected and hostile uses of space denied—nothing more. Also lacking are specific citations of international legal restrictions on U.S. space-control activities.

The testimony of John Hamre, former deputy secretary of defense, given before the chairman of the Armed Services Subcommittee on Strategic Forces on March 22, 1999, in fact, provided the most authoritative summary of U.S. space-control policy available, a fact that is actually more of a commentary on how little attention the executive branch had paid to space defense matters. Again, it is interesting to note that the deputy secretary's testimony outlining U.S. policy on space control and his other public statements on this subject is *not* reflected very well in the DoD Space Policy.

By way of illustrating this policy through congressional testimony and in subsequent public remarks on this subject, the Clinton Pentagon left the impression that the space-control mission was well in hand. Yet when the then–commander in chief of U.S. Space Command, the official responsible for actually planning for this mission, addressed this very issue, he did not appear to be all that sanguine. General Myers explained that there are four vital parts to the space-control mission, namely, surveillance (detecting and characterizing what is "out there"), prevention (denying an enemy the use of space through diplomacy and/or other nonmilitary means), protection (detecting and protecting against interference with U.S. satellite systems), and negation (putting hostile satellite systems out of commission, to include the use of force). The general maintained that the United States only does the surveillance mission well (one out of four—again, not a praiseworthy ratio). "The other three," he stated, "need much more work."[4]

The Pentagon's proposed initiatives in recent years for improving space-control capabilities have amounted to a few drops in an already near empty bucket. The department announced in 1999 the administration's decision to "pursue enabling research and technology development" in order to "implement the President's policy guidance" found in the 1996 National Space Policy. The money appropriated by Congress for fiscal year 1999, a total of $7.5 million, was spent to "posture" the United States to "address challenges posed by the spread of space systems, technology, and information" and "to protect U.S. and friendly forces against hostile use of space systems and services."[5] This compares in Pentagon budget requests to $500 million requested to buy Titan 4Bs and refurbish Titan 2s, $1,000 million for Milstar, $65 million toward to the joint transition to NPOESS, and $67 million for the final launch of a DMSP satellite. Space is not cheap, that is, unless one is talking about space control.

Yet there is an even more revealing story behind the drafting of the president's fiscal year 2000 space-control budget, and indeed, the Pentagon's reference to posturing seems to have been very appropriate in this context. A program budget decision (PBD) issued by the Pentagon comptroller in December 1998 to address space-control funding candidly confessed that the $7.5 million was requested in order to prevent legislators from forcing upon the Pentagon space-control initiatives the administration did not want to fund.

Only one month earlier, in an address delivered in Cambridge, Massachusetts, Senator Robert Smith (New Hampshire) lambasted the U.S. Air Force for not fully embracing space power. By January 1999, one would have to assume that the Pentagon had received word of congressional dismay over the vetoes and of Senator Smith's plan to hold a hearing on U.S. defense space programs in March of that same year.[6] Hence, the appearance of a PBD that lays out a politically motivated plan designed to let the Clinton administration off the hook in the eyes of Congress, rather than one that focuses on threats posed by the proliferation of foreign and commercial satellites, requirements for ASATs, or the vulnerabilities of U.S. satellite systems to justify the initiatives. According to the PBD, "The reluctance with which we have executed these congressional adds and our unwillingness to budget for space control have caused several key members of Congress to ignore the Department and to turn to outside special interests groups for ideas on how the Department should address the space control mission area."[7]

Part of this space-control deal between Congress and the administration was an assurance that the Pentagon would add money to continue the program, according to the PBD, "to address some very real problems that exist in the space control mission area."[8] Yet in the grand scheme of things, $10 million a year is small potatoes and represents more the administration's desire to ignore this mission, not a desire to "ensure an integrated approach for space control."

The Line-Item Vetoes. In truth, Congress has been largely responsible for ensuring even the minimal funding provided for a pitifully small selection of service programs intended to improve U.S. space-control capabilities. The legislative branch has kept the U.S. Army kinetic energy ASAT on life-support for several years, supported the air force's rationale for a military space plane, and insisted on funding for Clementine 2 to test acquisition tracking and intercept technologies against asteroids, technologies that may be used to enhance a BMD system.

President Clinton sent an unambiguous message regarding his intentions with respect to antisatellite systems when he vetoed funding for a series of military space programs in October 1997: the KEASAT, military space plane, and Clementine 2.[9] Using for the first time the newly granted power of the line-item veto, the president's choice of weapon systems to be cut from the fiscal year 1998 budget reflected policy differences with several Republican members of Congress. In any event, the Supreme Court found the president's line-item vetoes unconstitutional and returned the money to the Defense Department.

According to the official tapped by the president to answer questions about the vetoes, Robert Bell, then senior director for defense policy in the National Security Council, "We simply do not believe that this ASAT capability is required, at least based on the threat as it now exists and is projected to evolve over the next decade or two."[10] The administration also raised concerns regarding ABM Treaty compliance and "the militarization of space," in effect agreeing with arms-control advocates that the programs were "provocative" and violated the sanctity of space.[11] There is some evidence that the administration fundamentally opposed the idea of sending military man to space, despite draft requirements for such a role issued by U.S. Space Command and the U.S. Air Force.[12] The Clinton White House believed that there already existed adequate alternatives to ASAT weapons for protecting space assets, such as disabling enemy satellite systems either by jamming their signals or targeting their ground nodes and data linkages.[13]

The vetoes provided military space issues more visibility than the president might have wished. In January 1998, in the wake of the October vetoes, forty-three retired senior military leaders sent the president an open letter decrying the state of U.S. defense space programs. The focal point of their concern was the ability of the country to deny the enemy the hostile use of space and assure the dominance of U.S. forces in that "theater of operations" by making sure there are adequate means at their disposal to achieve space-control unilaterally. Diplomacy is often raised up as the space-control weapon of choice—and it is, in some limited circumstances. Proponents of the ASAT program noted, however, that the limits of diplomacy are reached when the country with which we are dealing refuses to cooperate.[14]

SBL and MIRACL. Past decisions by the Clinton administration only add to a somewhat confusing picture of White House motives and policies with regard to the military uses of outer space. In contrast to his strong objections to continuing a program that could have given the United States an antisatellite capability within a few years, Clinton made a couple of decisions that appeared to be contrary to his otherwise undisguised opposition to space weapons. He did not touch, for example, the air force space-based laser program, even as he vetoed other military space programs. While this program had the strong support of Senate Majority Leader Trent Lott, it was also true that the SBL program of the late 1990s was in the early phases of concept exploration. It may be that Clinton sought to avoid a high profile fight with the Republicans over a program that would not produce a weapon until well into the next century. A technology demonstration for SBL is not envisioned until early in the second decade of this century, while an operational capability is not projected until sometime late in the third decade. Moreover, although SBL survives as a research and development program, Clinton's negotiators laid formidable obstacles to deploying space-based theater missile defense systems in agreements with Moscow on ABM Treaty demarcation. Although research on weapons like SBL apparently was acceptable to the White House, the Clinton administration never intended to establish conditions that would allow future leaders to procure this space weapon.

Also around the same time as the October 1997 vetoes, the administration decided to permit the testing of the ground-based midwave infrared chemical laser against an air force Miniature Sensor Technology Integration-3 (MSTI-3) satellite at the White Sands high-energy laser static testing facility in New Mexico. The army and TRW developed MIRACL in the 1980s as a research project for the Strategic Defense Initiative. The test involved firing the MIRACL laser to illuminate, or "lase," a dying MSTI-3. According to the Department of Defense, the primary purpose of this test was not to destroy the satellite, but rather to compare the data collected with computer models that are used to develop methods and technologies for protecting U.S. satellites from such damaging interference. Since lasers are available worldwide, defense planners strongly believed the United States should identify vulnerabilities in U.S. satellite sensors and develop appropriate passive defenses and operational countermeasures.[15] The secretary of defense let the experiment to test the defense capabilities of U.S. satellites go forward, while the president remained publicly out of the decision loop to authorize the MIRACL–MSTI-3 test.

Saying No to Negation. In a public interview, former deputy secretary of defense John Hamre summarized the Pentagon's philosophy for the negation mission, stressing that opting for destruction "is going to be highly confrontational to the international community. . . . If we're going to concentrate on a program of destroying satellites in space—as a preferred method of space control—it will be much, much harder for us to get international cooperation" on issues such as frequency and orbital slot allocation.[16] Indeed, the administration often registered its concerns over Russian reactions to U.S. space defense developments.[17] But there was also an explicit concern that U.S. defense activities in space, specifically, "concentrating on a program of destroying satellites," might be viewed by other countries, including U.S. allies, as "provocative." The administration, in effect, imposed upon itself an international veto over how aggressively and efficiently the country could control space.

The country's political leaders have been keenly sensitive to international reaction to its space activities since the Eisenhower years. It is understandable why, in the late 1950s and early 1960s, there was profound concern over foreign responses to space overflight. But this issue was resolved well over thirty years ago. Gone also are the days when the United States and the Soviet Union faced off against each other, with their "itchy fingers" on the nuclear "buttons," a time when Washington feared that satellite warfare would take out critical sensor platforms and communications links both sides needed to verify that a nuclear attack was imminent or under way.

The former Pentagon official believed correctly that "preventing bad guys from using space against us is fully authorized under international law." However, he underscored, the United States should strive "to take steps and actions that don't create instability in the world. This [space-control] area is, frankly, on the edge, and we do not want to take steps that are precipitous and could create greater problems for the U.S." Hamre maintained that, should the United States ever procure the highly expensive space destruction capabilities it might need, "our leaders would never authorize us to use [them], for fear of international [backlash]."[18] He raised this concern before the March 1999 Senate subcommittee hearing by appealing to the innate American interest in commerce. Since commercial space assets are increasingly being used for a wide range of defense applications, "terrestrial-oriented negation measures may be more consistent with long-term American interests." He maintained that "physical destruction could undercut U.S. commercial interests that depend on global cooperation."

If the deputy secretary meant by "precipitous" either "hasty" or "rash," who would argue with him? But if he believed that it might be necessary someday to prevent "bad guys from using space against us," why is it hasty or rash to deal decisively with this problem in space? We have reached a mysterious conceptual void, where we apply different rules to space. U.S. officials routinely think and act through security guarantees with friendly countries and allies—except with respect to space. The country's leaders accept the reality that there are international ramifications to everything the United States does—except with respect to space. Global commercial relationships (where profit is king) also appear to take care of themselves in the hurly-burly caused by international politics and warfare. The world fully expects the United States to throw its weight around when its interests are threatened—except, apparently, with respect to space.

Critics of ASATs frequently maintain that satellite destruction would be "escalatory," implying that such actions will take the confrontation to new levels of horror and cause the United States' enemies to exploit its vulnerabilities in space. This depiction of satellite warfare was reinforced in the Clinton Pentagon by conscious references to "tactical" space control, implying that the deliberate destruction of satellites escalates the level of engagement to "strategic." We have all grown accustomed, after all, to understanding the dire consequences of "strategic confrontation." Yet ASAT weapons do not approach strategic nuclear weapons in their destructive power or indiscriminate effects. Stigmatizing ASATs ("ASATs = Armageddon") obscures the fact that counterspace operations can be very precise, highly discriminate, and locally nonlethal. Far from being "precipitous," satellite destruction could prove to be the most rational and militarily effective course of action for our leaders to take in some situations.

The United States continues to deploy intercontinental nuclear forces having the power to utterly devastate foreign lands. Yet few maintain that this force-in-being is "provocative," and many, especially in the arms-control community, will argue vigorously that this mutual assured destruction capability is actually stabilizing. Indeed, the United States continues to carry on routinely with its international relationships in a multitude of forums despite fielding and exercising with forces capable of atomizing its enemies. The logic of deterrence and the disciplined, responsible display and use of U.S. military power is accepted internationally and even praised and courted by allies—except as U.S. power might be manifested in space. Or so it is often portrayed.

The oft expressed fear by critics of ASATs is that the world's leading space power will be shut out of negotiations for frequency and orbital slot allocation, and the like. Yet is it not an exaggeration to claim that a nation acting according to its right of self-defense, a right recognized in Article 51 of the UN Charter (which also contemplates *anticipatory* self-defense as a basis for justifying unilateral preemptive uses of force), will be shut out of important international regulatory activities? As it would do in any other instance as a leading defender of international law and order, if and when it might occur, the U.S. government would make an appropriate public case for its use of force in space. Life would carry on.[19]

An international anti–United States campaign, having the requisite consensus to be effective, might be mounted should the United States ever behave as a heedless warrior in space, paying no mind to international customary practices and recklessly interfering with foreign and commercial satellite operations—"skeet shooting" in space, as some have characterized it.[20] Indeed, the description of the negation mission as "a program concentrating on destroying satellites" (as "the preferred method," no less) is rather like saying that the United States today has "a program concentrating on destroying aircraft," or ships, or tanks or, for that matter, people. It is an inelegant presentation of the space negation mission, one which conveys the sense that the armed services, once the chains restraining them to Earth have been cut, will bound out into the "wild frontiers" of Earth's orbits, breaking and smashing anything in their path.

Context is everything. In the heat of battle, when vital interests are at stake, should it prove necessary to negate enemy capabilities by disabling their satellites, the American people and friends and allies abroad will not judge what has happened apart from the circumstances. When U.S. and allied lives are jeopardized and there are bold and critical steps to take in the interest of national security, U.S. military actions and counteractions should not be guided by concerns about what a shapeless, nameless "international community" might think, say, or do. A hostile act is a hostile act, and there is a voluminous record of accepted practice in the international community for dealing with threatening activities. This is how it is on the land, in the air, or at sea—and this is how it will be in space. There are rules of the road and conventions at sea and in the air that do not preclude the force option. It should be no different in space.

At present, the United States does not deploy a capability to negate enemy satellites. Given U.S. technological capabilities and progress in the area of ground-based hit-to-kill antiballistic missile systems, I do not doubt the ability of the country to "kludge" a system in a desperate attempt to deal with some limited intolerable situations in space. But assembling on-the-fly a makeshift solution and fielding reliable, effective capabilities, to involve the development of policy and doctrine and consideration of strategy, are two dramatically different things. The country today is limited in its capability to enforce its stated objective of keeping space free for peaceful transit.

"Tactical" space control, the approach pushed by the Clinton administration, would involve taking temporary, or reversible, actions, where the focus would be on nonlethal effects, data link jamming, blinding, spoofing, and other ways of disrupting information flow.[21] Judging from past practices of the Bush administration during Desert Storm, and the Clinton administration in crises involving Baghdad and Belgrade, U.S. space-control activities also may include the destruction of enemy satellite ground terminals and nodes using air or special operations forces.

Clinton officials publicly maintained that the United States could defeat any hostile satellite during a conflict through electronic jamming or interference with communications. "We need to not be victim to 'old think,'" according to Robert Bell. "The old think cold war mentality was that we envisioned space control as ASAT, and we equated ASAT with a dedicated system that went up and destroyed something." Bell looked instead to the "revolutionary advances in technology, particularly in the area of information operations," so that all we need to do is "widen our horizon" beyond reliance on ASAT systems to protect U.S. interests in space. Bell indicated that there were "a range of alternatives being explored," and further "if we were in classified session I could say more, but I can't."[22]

The Limits of Temporary "Control." Administration assurances aside, every commander in chief of the U.S. Space Command since the command was established has agreed in one forum or another with General Myers's assessment that the country still has much work to do in the space-control mission area. From a military perspective, one would have to ask whether disrupting data from the ground would always be sufficient. Would a commander in the field always view this rather restrictive set of disruption options as enabling the best possible approach for taking out insolent eyes in orbit, whether operated by foreign or commercial entities? Logistically the U.S. armed forces are limited by geography to the extent they can do direct jamming or blinding (insofar as line-of-sight positions are required). Could a commander be confident that the actions he authorized were effective? In a protracted crisis or war, are not temporary and reversible measures counterproductive? Acceptable, or "tactical," space-control approaches may be too fleeting in effect and monopolize too many resources, as assets are diverted frequently in order to "revisit" again and again an old problem.

Indeed, the question may fairly be asked, why be consumed about pursuing temporary, disabling tactics in the space environment when there is little compunction to use similar tactics in the land, sea, and air environments? Although the Pentagon chooses not to highlight it in its policy, it has exercised terrestrial negation and ground destruction options in recent conflicts, combat actions that, ironically, do not inspire protest in the arms-control community despite the fact that they produce collateral damage and even death on foreign soil.[23]

Yet there are severe limitations and numerous points of possible failure in a space-control strategy that relies on the destruction of ground space facilities (mobile and fixed), nodes, and terminals located on enemy territory.[24] Target lists in *all regions* where conflicts involving the United States may occur need to be updated and expanded regularly. To complicate things, satellite terminals are proliferating. Mobile targets will provide their own unique challenges. This space-control strategy also requires air superiority—a condition that U.S. forces cannot take for granted. Moreover, attacks on satellites nodes and terminals will divert precious and finite air combat resources, in effect, granting a limited reprieve to enemy tank formations, ballistic missile transporters, and weapons manufacturing and storage facilities, targets that affect directly the course of the war. Intensive bombing campaigns also will rely heavily on in-theater basing—and the United States deploys a comparatively "thin" long-range bomber force and does not have operational space-to-ground strike weapons to provide global reach. Finally, this space-control strategy may be entirely ineffective and fall well short of its objective in a very short war (because attacking ground targets takes time). Although a ground-focused space-control strategy is plausible in a prolonged conflict, it ultimately may become too draining.

There is also a political cost to bending over backward to avoid knocking out enemy satellites. Not only might it make sense to attack satellites directly for reasons of military efficiency (as perhaps the surest way for cutting the enemy off from space) and economy in the execution of tactics (hitting a satellite with one "bullet" is more economical than expending thousands of "bullets" to impede his ability to use that satellite), but keeping negation missions on the ground runs a high risk of collateral damage. The political costs associated with killing foreign citizens, civilian or military, destroying economic infrastructure, sending U.S. military personnel and weapons platforms onto the sovereign territories of other countries, and jeopardizing American lives all but prevents the United States from exercising this type of space control in critical situations short of war. Why would one destroy satellite ground stations on the sovereign territory of another country, in times or crisis and war, when it could be avoided? The answer is that we may have to—we have now no option to do the job in or from space.

What Has Congress Said about the "New Battlefield"?

Congress, too, in its sometimes desultory style of engagement on defense matters, has voiced over the years a variety of opinions about what role U.S. military forces should play in space. It does so mainly through its interactions and policy debates with the executive branch. Congress too can contribute to bad policy. Since the collapse of the Soviet Union, Pentagon and congressional interest in space weapons (antisatellite and antimissile systems) waned, but continued to receive some attention and a small amount of funding. Although ASAT weaponry remains controversial, the political climate in the United States may be warming to the subject.[25]

MIRACL and "Mysti"

In 1990, congressional Democrats won a legal prohibition on test firings of the MIRACL laser against satellites. In 1995, the Republican-led Congress let the ban expire. Opponents of the decision to test MIRACL appealed to President Clinton to deny the U.S. Army Space and Missile Defense Command's request to execute a test planned for October 1997. They rested their case on arguments that are quite familiar to anybody who followed the space weapon debates of the 1980s. Most of the arguments were raised in a letter drafted by three members of Congress to President Clinton imploring his intervention.

Representatives Dick Gephardt (Missouri), John Spratt (South Carolina) and Ron Dellums (California) drafted a letter to President Clinton, Secretary of Defense William Cohen, Secretary of State Madeline Albright, and Director John Holum of the Arms Control and Disarmament Agency requesting a delay in the army's MIRACL test. According to the letter from the congressmen, the main problem with the test is that "a fully evolved policy" regarding "the issue of anti-satellite warfare" has not been developed—a point with which I could not agree more.

The congressmen argued that the country needed to debate this issue. There were many questions—many very predictable questions—that needed to be answered. "For example," they wrote, "given the strong U.S. national security and commercial interest in the use of space, should we begin the race to the development of an ASAT capability?" And "what are the likely international reactions, and specifically the likely Russian reaction, to the test of the MIRACL laser?" They noted that Russia has been notified of the test and was expected to object. Indeed, before and following the test, Russia did send strong letters of protest to President Clinton. Further, there was a need to review all arms-control and foreign policy implications of the test and to consult thoroughly with Congress.

The test amounted to provocation, pure and simple. These three opponents of the test believed that the army's argument that a laser test must happen soon, before the life of the air force satellite expired, was cover for the real motive—to conduct the test before congressional debate could take place.[26] Each of these arguments received a great deal more attention in the press in the weeks and days leading up to the experiment. The critics' major point of opposition to the MIRACL test was that, because the United States was back in the antisatellite business, the nation must deal with the risks that accompany an "arms race" in space. For this reason, a leading critic of the test, Senator Tom Harkin (Iowa), indicated he would oppose the test and use all means at his disposal to prevent the slide toward an ASAT arms race. "The United States," he said, "has the most to lose from the international development of ASATs due to our high dependency on satellites to support our diplomatic and military capabilities. Additionally, potential adversaries are unlikely to acquire intelligence satellites in the near future."[27]

The leading argument of those in Congress who supported the experiment against MSTI-3 was that this test would permit the United States to learn of inherent vulnerabilities in U.S. satellite sensors to laser-induced dazzle and damage. Indeed, the debate over this test provided an opportunity for many in Congress to voice their support for a number of Pentagon laser programs, allowing them to make the case that tests such as the MIRACL–MSTI-3 experiment are aimed at learning if the low-power lasers operated now by up to thirty nations could do serious damage to U.S. satellites.[28] According to Senators Pete Domenici and Jeff Bingaman and Representative Joe Skeen, all of whom represent New Mexico, the MIRACL laser would give the United States a capability that "has the potential to force overhead reconnaissance satellites, which we do not control, to close their shutters."[29] Possession of a capability demonstrated by MIRACL could provide the United States with a new and effective tool for executing its foreign policy and military missions and for protecting U.S. forces. In the end, the MIRACL test went off as planned, and despite the catcalls and dire warnings from critics, U.S. commercial, diplomatic, and strategic relations with other countries did not falter.

KEASAT

As the army's kinetic energy ASAT neared the point where it could become an acquisition program, the tempo and intensity of criticism against the national deployment of ground-based ASAT weapons rose. This technology was designed, developed, manufactured, and integrated under the kinetic energy ASAT Demonstration Validation Program from 1990 to 1993 and was making steady progress, much to the chagrin of ASAT opponents. Once again, many of the opposing arguments played out in the press surrounding the laser shot against the air force satellite were paraded in lengthy debates on Capitol Hill over the KEASAT program. Among these arguments, that ASATs will "militarize" the space environment and bring about a costly and destabilizing arms race received the most attention. Moreover, it was said, the use of an ASAT weapon would pollute the space environment with "lethal" debris that would harm U.S. interests in space.[30]

Proponents of the KEASAT rested their case on the emerging threat to U.S. forces posed by enemy satellites. Congressional supporters of KEASAT cited the fact that the administration had stated a need for a system to counter the space threat.[31] Congress had funded the KEASAT program for several years, citing a growing military necessity to protect our forces deployed abroad and U.S. satellites. Many legislators cited the ASAT program as a hedge against this danger. Senator Robert Smith argued that there are several countries with a capability that could harm U.S. interests by using such things as jammers, lasers, and direct-ascent ASATs carrying a nuclear weapons. Indeed, said Smith, "the capability is out there."[32] Senator Richard Shelby (Alabama) also noted the proliferation of satellites and the "obligation" to continue to strive to counter this threat.[33] The need to preserve an ASAT industrial base also was used by proponents to save the KEASAT program.[34]

Without strong congressional support for the KEASAT program, it would have died long ago as a result of active opposition from the Clinton White House.[35] In 1996 Senator Strom Thurmond (South Carolina) noted that "Congress has authorized and/or appropriated funds for the kinetic energy antisatellite technology program since 1985. For the past three years the administration has not complied with the law and obligated the funds for the program. Every year, as a result, we have to take actions to force the Department [of Defense] to comply with legislation to compel them to obligate the funds for this particular program."[36] Again, the two faces of Janus are set in full relief.

Space-Based Lasers

Congressional opponents of the space-based laser center on the financial, legal, and technological challenges facing this program. One of the technological "Doubting Thomases" is Senator Harkin, who has argued that the problem of firing a laser "is complicated enormously by the size and the multiplicity of targets using this big mirror in space. Millions of computer code must be written. One little mistake and that would spell the end."[37] As Representative Ron Dellums (California) explained, "The deployment of this system will one day, lead to the weakening or the abrogation of the [ABM] treaty, thus presenting us with the prospect of another arms race, this time probably in space."[38] Others in Congress believed that SBL is a system in search of a threat. According to Senator Dorgan (North Dakota), who exploited some of the arguments used in the Pentagon against SBL, "There is no validated military requirement for space-based laser."[39]

Supporters of space-based interceptors argued that such weapon systems could help deter and, if necessary, defeat ballistic missile attacks against forward-deployed troops, power projection forces, and U.S. friends and allies, thereby strengthening national security and global stability.[40] SBL backers believe the threat from ballistic missiles is more serious than that which the opposition portrays. According to Senator Inhofe (Oklahoma), "This euphoria that we seem to enjoy around here that there is no threat is one that is of more concern to me than anything else we talk about." The fact that Russian missiles can be retargeted in a matter of minutes (and we lack verification of it), he reminded his colleagues, is not comforting.[41] Even those who are not ardent supporters of the SBL program understand the irresponsibility of not making preparations to deal with the future ballistic missile challenge. Senator Inouye (Hawaii) noted, "I just hope that we will not have to use this in warfare. . . . We have not arrived at the millennium, so sadly we must prepare ourselves that if such a time should come, we are prepared."[42] According to Senator Thurmond, any reductions in the space-based laser program would be disastrous: "It would be short-sighted for the United States to constantly abandon this development effort at a time when the . . . missile threat is growing."[43]

Senator Smith supported the future constellation of space-based lasers and observed that it was making important technological progress. He believed this was an important program because of the expansion in the number of countries possessing ballistic missiles and WMD. Using a space-based system, "we are able to get these missiles in their boost phase and make the debris from those missiles fall back on the aggressor or the firer of the missile." Senator Kyl (Arizona) thought it was premature to tell whether the SBL would be cost-effective. But, he indicated, "we all know that a boost phase intercept is the way to go." To many of the senators, including the majority leader, SBL was "the national missile defense option of choice."[44]

Congressional interest in space weapon systems is usually tied to local politics, like any other military program. These factors, together with issues dealing with operational, technological, and financial effectiveness of the systems at issue, will always be in play. Space-based laser, for example, is being developed in Trent Lott's state of Mississippi, and the Senate majority leader is one of the chief champions of this space weapon. The MIRACL laser program is carried out under the watchful eyes of Senators Domenici and Bingaman and Representative Skeen, all from New Mexico, the site of the HELSTF (High-Energy Laser Static Test Facility). Occasionally, there are notable exceptions to this rule of local politics. Senator Smith's ardent support for the KEASAT program is noteworthy, because none of the development efforts took place in his home state of New Hampshire.

Solitary Voices

Much of what happens in Congress with respect to defense space issues has tended to be "staff-driven." Members are pulled in so many different directions by the torrid competition for their attention and generally are so lacking in interest in the defense space subject, that it really is newsworthy when a senator or representative takes the time to speak out on the future of U.S. space power. As counterweights to arguments of former Senator Charles S. Robb (Virginia), which will be presented in some detail below, a few legislators have voiced concern about shifting U.S. visions for space.

The year 1998 was a banner one for collecting congressional remarks on U.S. military space-control programs. Although Clinton's October 1997 line-item vetoes did not create much outrage in the halls of Congress, like an electrical surge after which energy rapidly dissipates, the president's action gave life to a brief coalescing of ideas on military space matters. A few congressmen reiterated their unwavering support for U.S. military space-control initiatives and talked a little further about their visions for space. Representative Curt Weldon of Pennsylvania, ardent advocate for ballistic missile defenses to protect the U.S. homeland, lamented the Clinton vetoes, suggesting that the president's dogged adherence to arms-control ideology guided his decision.[45]

A few senators also have entered the foray—if only for a brief moment. Senator Bingaman actively inserted himself into the remote-sensing debate of the mid-1990s. The 1994 policy issued by Clinton on high-resolution commercial imagery satellites stipulated no restrictions on the collection of imagery, save in times of national security emergency. According to Senator Bingaman, who published some of his arguments in the weekly *Space News,* the Clinton administration, whose policy favored U.S. industry, did a poor job of balancing security and economic interests in the remote-sensing policy. After all, he pointed out, U.S. policy did not contemplate how to prevent the nation's foes from scrutinizing U.S. forces from space, nor did it address how to deal with concerns of allied or friendly countries who do not want their territories imaged by adversaries. In 1995, President Clinton had not yet issued his National Space Policy, but rather was operating at that time under the Bush policy of 1990. Yet the senator could not help but notice back in 1995 that "neither a negotiating strategy nor an ASAT capability appear to be part of the United States' current policy."[46]

Senator Bingaman's twofold solution for improving U.S. remote-sensing policy was to construct a "stable international regime in which U.S. firms can compete and U.S. national security interests and the interests of U.S. allies are protected." The first part of his solution was to anticipate "foreign policy or national security problems" and increase the degree of government "shutter control," to involve slight digital degrading or a delay in timeliness of delivery of satellite imagery data, which the senator argued would be of no consequence to civil users. Countries on the "terrorism list" would receive great scrutiny, he argued. An obvious failing in the plan to preempt problems in the remote-sensing area is that U.S. laws and regulations only bind U.S. companies. So the senator proposed that Washington undertake multilateral talks with France, Russia, and possibly Israel and Japan, in order to universalize restrictions.

Senator Bingaman rightly recognized the difficulties of negotiating such controls and thus recommended a "fail-safe." The United States, he wrote, "needs to have an anti-satellite (ASAT) capability against low Earth orbit imagery satellites." Although one can legitimately pan the senator's suggestion that controls may be regulated and negotiated without harming the U.S. remote-sensing industry (and indeed there was significant debate on this issue in 1996), he is to be praised for calling attention to a major policy vacuum in the defense space area and for pointing out that the policy under Clinton on remote sensing raised more questions than it answered.[47]

Senator Joseph Lieberman of Connecticut appeared in February 1998 before the Association of the U.S. Army on future readiness and combat capabilities. He spoke at length to the recommendations of the National Defense Panel (NDP), a group of outside experts whose mission it was to assess current and future defense challenges and recommend changes to U.S. military forces. The NDP, which issued its report in December 1997, had the foresight to talk about space as being increasingly high priority, which in turn provided the senator a platform from which to make some uncharacteristically bold remarks on this subject.

Senator Lieberman noted the "relentless" growth in commercial and military satellites, reminding his audience that while there were roughly five hundred satellites in orbit in 1998, more than two thousand may be expected to be in orbit early next century. Space, he said, has become "an inseparable and vital part of our personal and national lives." The implications for security are critical. Our dependence on space, he continued, "will change warfare itself just as the rise of sea borne trade centuries ago expanded warfare from land to the seas. . . . We are rapidly creating new 'sea lanes'—space and cyberspace lanes—that can be attacked and that must in turn be defended."

He then concluded from this, probably by way of responding to the administration's plans to abandon all efforts in this area, that "it is inevitable that capabilities to attack and defend them [space lanes] will continue to be built and deployed." Reminding his audience that we are in a time of "revolutionary change," Lieberman concluded with this stunning *coup de main:*

> While we do have a national space policy, we do not now have a national military space policy. We do not have a joint space doctrine, space is not an assigned area of responsibility, and therefore there is no single US policy and program administrator for space. We need to direct the Pentagon to address these gaps. But we must go beyond these and address the elements of the national space policy directly as they relate to the militarization and weaponization of space. Space has long been effectively militarized. Weaponized space is the ultimate high ground. It is certain that someone will attempt, and perhaps succeed, in taking that high ground. To a large degree our current national space policy is based on a patchwork of treaties dating back to the 1960s. It is time to review these treaties to determine whether

> they continue to protect our long term national interests in the rapidly changing space environment.[48]

Forceful, blunt, and on the mark. The only exception I would take is to the senator's assertion that we have a National Space Policy that is sufficiently intelligible to inform U.S. defense space policy. But even here Lieberman concedes that we must go beyond "elements of the national space policy directly as they relate to militarization and weaponization of space." As we have noted, the Pentagon is impotent, not only to make the necessary changes in policy, but also with respect to undertaking important space combat missions without more direct guidance from the president. But like two trains passing on parallel tracks, the conductors having exchanged friendly waves, the senator appears to have come and gone on this issue and has not been heard from since.

Senator Bob Smith drew some attention for a forward-looking speech he delivered in November 1998 before the Fletcher School of Law and Diplomacy and Institute for Foreign Policy Analysis in Cambridge, Massachusetts.[49] The senator denounced the Pentagon's budget focus on traditional programs at the expense of significant spending on national military space programs. In this forceful oration, Smith laid out three frank assertions: "*First,* America's future security and prosperity depend on our constant supremacy in space; *second,* while we are ahead of any potential rival in exploiting space, we are not unchallenged, and our future dominance is by no means assured; and *third,* to achieve true dominance we must combine expansive thinking with a sustained and substantial commitment of resources, and vest them in a dedicated, politically powerful, independent advocate for space power."

Smith berated the air force for desperately hanging on to its traditional air power missions, and for refusing to allocate the necessary funds to grow national space power. The air force also has restrained its space activities to information-handling operations, to the maintenance and improvement of information systems to enhance force operations on Earth. "This is not space warfare," he chided them. If we limit our sights to simply achieving information superiority, "we will not have fully utilized spacepower." The air force, he stated, is not organized, trained, or equipped to lead the country in space. He threatened that, if the air force should fail to embrace space power, then perhaps it was time to strip the space mission from that service and create from whole cloth a new space force, which would then advocate vigorously for space power and build the necessary political and budgetary support it ultimately requires.

But his remarks also aimed a little higher. Senator Smith called upon Americans to "dramatically restructure our institutional approach to this ultimate strategic theater." The country must foster a "spacepower culture," maximize cooperation among the three space sectors (working "aggressively" with private industry), and "give our space warriors the tools they need."

Senator Smith finished his remarks by noting that "the future of millions of unborn depends on our willingness to master space." Space control and the ability to influence the course of a war from space, in other words, will give the country new possibilities of a military nature that may be expected to provide Americans enhanced security in this coming century.

Like Lieberman, Smith presumed that the policies issued by the White House, citing assertions in the 1998 National Security Strategy, provided the basis for our national commitment to space power. The air force, he stated, has the support of the "national command authority," which had "established the policy foundations for such a transition." I maintain, however, that there is ample evidence to the contrary, that what you see is not always what you get. The national indecision over how to act in space is perhaps the country's greatest weakness.

Is Space a Sanctuary?

Yearnings for eternal peace have found their way into modern political discussions on space. Such sentiments contemplate the mollification of the state of war in that environment and the disciplining role that man's activities must assume in order to establish an unprecedented harmony among states—a perpetual peace in space.[50] In modern times, there prevails an idea that we should "master the evil principle" within and make politics bend to what is moral and right. And what is moral and right in the eyes of many is preventing man from taking his conflicts into space.[51] Religious sentiments bolster the idea that space is a sanctuary, a place truly and almost divinely apart from Earth, an undefiled arena for men to behave toward one another as they ought. In the eyes of those of Judeo-Christian heritage, for example, outer space is God's "canopy" and often equates with Heaven ("Heaven is my throne, and the earth is my footstool").[52]

The U.S. defense space policy regime still carries within it an imprudent ambivalence with respect to the question: is space a sanctuary? Space sanctuary activists argue vehemently against any national strategy or defense plan that assumes space combat operations can contribute to peace and security. Official statements from successive chief executives routinely portray space as a future war-fighting environment. Yet as we have seen, many other official actions of the president and Congress betray an adherence to contrary conclusions about space and warfare, which has fostered our national ambivalence on this matter.

While many ideas about space sanctuary are rooted deeply in the sentiments discussed above, more "hard-nosed" arguments for "safe heavens" focus on strategic justifications. Even these strategic arguments, however, get an occasional moral or emotional boost from the more sentimental view of space. In the end, the strategic arguments fail because they harbor at their core an unwarranted optimism about the nature of man and rely upon an all too charitable view of world politics. From the standpoint of policy development, it is necessary to address and test the proposition that space can and must be kept free of man's wars. I am confident that history already proves this proposition to be an illusion. In the meantime, U.S. defense planners need guidance consisting of more than unrealizable dreams.

The Militarization of Space

The military use of space is not "new."[53] Most people today will concede that space already is "militarized" to some degree, since nations already deploy satellites that perform military support functions, such as communications, navigation, surveillance, and early warning, and contribute to the "transparency" required for global stability. Satellite systems support U.S. strategic missile and many other forces. Ballistic missiles traverse the space medium on the way to their targets. Indeed, in 1999, the United States Ballistic Missile Defense program smashed three ballistic missile targets in the space environment to test the accuracy and effectiveness of ground-based missile intercept systems. The real questions are: Should the space battle medium be extended by the deployment of weapons into space? Should the United States deploy systems on the ground or in space to destroy or degrade enemy satellites?

Space continues to be viewed by many in a romantic light, as a place to be kept free of man's quarrels. Central to this idea is that space should be exploited solely for peaceful purposes and the benefit of all mankind without national claim or jurisdiction, a proposition that implies that "peaceful purposes" must exclude military operations.[54] In the eyes of some, combat is one of the more abhorrent activities that could take place in "pristine" space.

The cry against the militarization of space still can be heard loud and clear in debates raging today over space force-application weapons. The images conjured up in the minds of space sanctuary advocates are literally comical and are intended to chalk up a political win in this war of images. For example, Congressman Peter DeFazio (Oregon), argued in 1995 before the House of Representatives, and published in his newsletter, that "the Pentagon has already squandered $36 billion on Star Wars, with nothing to show for our money but a few faked tests and a Star Wars cartoon video that is about as real as George Jetson and his family."[55] Senator Harkin chose to speak about his concerns in more dramatic terms in the debate over SBL: "It's back again, and I am not talking about 'Freddy from Elm Street.' I am talking about star wars, . . . a star wars weapon system called the space-based laser. . . . You can think of it as a giant, deadly flashlight, able to zap up to 100 missiles with the amount of fuel on board, or zap a maximum of 5 to 10 theater-range missiles launched simultaneously, or maybe zap 15 to 20 ICBM's launched simultaneously.

"SBL," he continued, "has been the dream of some Star Wars' enthusiasts for a long time. I think they saw too many of the Star Wars movies." It is like "that snake has not been killed."[56] As the reader can tell from this short excerpt, *Star Wars* and demonic images are favorite symbols.

Steve Aftergood, of the Federation of American Scientists, similarly played hard upon this theme in his criticism of the Pentagon's laser programs. He argued, curiously, that "our political and military leaders may have been seduced by the spectacular efficiency of Hollywood lasers."[57] The Pentagon's scientists also have been accused of engaging in research that could have serious consequences for those on earth. Lord Solly Zuckerman, for example, warned in 1987 that certain kinds of space-based laser weapons would ignite substances on Earth.[58]

According to laser expert John Rather, a country that was the sole possessor of space lasers would have "the longest 'big stick' in history, . . . the capability for unilateral control of outer space and consequent domination of the earth."[59] Other critics have warned of unintended consequences that might accompany the deployment of space weapons:

> The effect of even moderate amounts of space combat on the international banking system . . . would probably bring untold hardship, and even death, to millions of people. There could also be grave effects on the homeostatic systems of near space and the outer atmosphere from the very deployment, let alone the use of such weapons. Thus even if their close military links to weapons for planetary destruction could somehow be checked or neutralized, and even if conventions were observed whereby it was never "done" to attack the vulnerable but essential parts of space systems which are in fact down at ground level, the actual physical effects of space weapons would make it impossible to localize and isolate space warfare . . . by mutual tacit consent.[60]

Many fear that a quest for space weapons signals a quest for space control, which in turn signals a desire to control the planet. Accordingly, it has been stated that "the idea that overall military control of our planet might be achieved from space is now so venerable that references to it as 'the new high ground' are ludicrously inappropriate."[61] These are the images that helped kill President Reagan's Strategic Defense Initiative and which will continue to be applied in campaigns to prevent the "weaponization" of space. There was much criticism along these lines following the Pentagon's decision to fire MIRACL at the MSTI-3 satellite in October 1997 and in the late twentieth century surrounding the development of the *ground-based* National Missile Defense system.[62]

The most frequently heard case against taking weapons into space is that such actions would fuel an arms race in space. Space is, in the eyes of some, "the last place safe from human wars[,] . . . the last refuge of a sacred presence in the universe."[63] Ronald Reagan's ideas for a comprehensive ballistic missile defense, though considered feasible in the long term by many, were nonetheless ridiculed by political opponents, academics, and scientists as utopian.[64] But his real "crime" was that he proposed deployment of a series of weapons that would take mankind's wars into "the heavens." Following Reagan's proposal to deploy a defensive system to make ballistic missiles obsolete, Senator Edward Kennedy (D-Mass.), a critic of SDI, dubbed the proposal "a reckless Star Wars scheme."

As political and campaigning tactics go, the ability of SDI critics to attach the *Star Wars* label to the president's program was a master stroke. Reagan's political opponents viewed him as a grand wizard bent on permanently and detrimentally transforming the space environment along with the very soul of the United States. That label was, in fact, the glue that held opponents of SDI together, and it served as a rallying cry against what they would refer to as a high-priced, dangerous, nonsensical effort to militarize space with antiballistic missile systems and antisatellite weapons.

But would the introduction of weapons into the space environment be as bad as the critics portrayed? The question cannot yet be answered to satisfaction because we lack the necessary experience to make the evaluation. It also cannot be answered adequately because we are not posing it within a strategic context; there are no stakes, no political-military ends, and no moral or policy backdrop against which a judgment may be made. Yet even though we lack the necessary observations to thoroughly understand the impact of combat in space, by drawing upon our experiences on Earth we can come to some preliminary conclusions in the abstract as to whether the consequences of combat in space would be as apocalyptic as space sanctuary advocates believe.

Even if a spacecraft were shot down by accident, the implications would not be so grave when compared, for example, to an accidental launch of a nuclear-tipped missile. So, asked Walter McDougall, "why is it more important to protect pristine space, where nothing lives, than the crowded earth?"[65] This is a good question. It is a question that has never been addressed fully by space sanctuary enthusiasts (apart from making the apolitical, astrategic point that attacking unmanned targets in space makes wars "more likely" or is escalatory), who tend to use only one "strategic" framework for assessing the implications of space weapons: the framework of apocalypse. Any use of weapons in or from space will bring doom to "our planet." Arms races will drain the life blood out of the national budget. Interception of nuclear-tipped ICBMs in space is a precursor to a treacherous instability, the kind that could lead to nuclear holocaust. A decision by the United

States to use the space environment for protection will bring the acrimony of the entire world against Washington, asphyxiating U.S. national and economic security. This is not strategic thought—this is the worst-case, even unimaginable-case scenario played to the hilt.

Bearing in mind that final judgments about the implications of any military action are heavily dependent upon the political and strategic circumstances, one can imagine that in most situations, destruction in space will be preferable to death and destruction on the ground. To confine destruction to the environments where people live is a "moral" solution that is as rich in irony as it is puzzling.

Star Wars II

Former Senator Charles Robb published an article entitled "Star Wars II" in the December 1998 issue of the *Washington Quarterly.* This article by a U.S. official on the subject of space sanctuary, a rarity, may be said to represent best the controlling head of Janus, or that body of opinion that holds that there is significant moral and strategic justification for limiting activities in space of a military nature. For this reason, Robb's arguments merit detailed exposition and evaluation.

In answering Senator Robb and those who sympathize with his position, I do not make a case that the United States ought to launch immediately into programs to assemble an array of space-control and force-application weapons. This would constrain us to a serious and careful consideration of politics, strategic and operational efficiencies, technology, and resources.

The Abuse of History. In brief, two-term Virginia senator Robb expressed relief that, to date, the United States has refrained from placing arms in orbit. He warned that calls by some senior service officials to develop antisatellite and space-to-ground weapon technologies to ensure U.S. space control and future space dominance were misguided. To write this next chapter in our military annals would be "a mistake of historic proportions." Moreover, to pursue these costly programs would be futile, for as soon as this innovation in warfare were established, neutralizing counter developments would be inaugurated by our enemies, depriving the United States of the "lasting advantage" it was seeking.

Moreover, to embark on this path would recklessly disregard U.S. history. Senator Robb referred the reader back to the debate over the deployment of multiple, independently targetable reentry vehicles, which allow many nuclear warheads to be placed on a single ballistic missile and guided independently to different targets. This 1970s bid to get a leg up on the Soviet Union backfired, he contended, with the real consequence having been a robust Soviet campaign to match the United States and eventually surpass it in MIRVed capabilities. MIRVs "created a whole new source of instability," since each side had a stronger incentive to strike first in order to knock out missiles on the other side.

Senator Robb's lesson is that once the United States takes that fateful step to deposit weapons in space, other governments will not sit by idly. They will deploy weapons to attack the satellites upon which the country so greatly depends and strike targets on U.S. soil. In an arms race such as this, "America would lose the most." "Military history," he wrote correctly, "can be depicted as a progression of frontiers from land to sea to air. . . . Over the last few decades the military frontier has moved to outer space." He rightly noted that the United States today depends mightily on world-circling spacecraft to execute such crucial military functions as communications, intelligence, navigation, and weather monitoring.

However, he wrote, we must now put a stop to this advancement in arms. We must refuse to move to that next level where the United States deploys systems to destroy a foreign satellite or exploit space weapons platforms to "strike an opponent anywhere in the world—on the ground, in the air, or in space—in a matter of seconds." The distinguished member of the Armed Services Committee reminded us that space weapons include space- and ground-based directed energy and kinetic energy weapons (which work through explosion or high-velocity impact), electromagnetic pulse and high-power microwave weapons, space mines, and munitions launched by space vessels.

The senator's apparent attention to military history is ironic, for elsewhere he implied that the relentless forces of military history no longer apply to our age, that the evolution in arms can and *must* be arrested at the edge of space. "Why pursue this option [of deploying weapons in space]," he wrote, "when there is no compelling reason to do so *at this time?*" This allowance, of course, is tantamount to admitting there is an uncontrollable element in the progression of weaponry, and that innovations in technology, tactics, and operations will proceed regardless of what the United States does in this area today. And in this he is correct, for (in the words of Robb), "in the history of warfare not a single breakthrough in weapons technology has gone unchallenged for a significant period of time by a neutralizing counter development." Tomorrow's chroniclers of military space, in other words, may be expected to include chapters on the proliferation of foreign satellite and weapon systems to neutralize and match the United States' present-day orbital advantages.

Others have argued that the United States has maintained a rather consistent "sanctuary strategy" over the past forty years in order to avert political challenges to U.S. reconnaissance satellites and other information-handling spacecraft. The United States has done this, so the argument goes, owing to the "inherent, destabilizing nature of weaponization" and, conversely, to "the inherent, stabilizing effects" of reconnaissance satellites.[66] Therefore, to pursue force-application options or develop space-control combat capabilities would represent a major "discontinuity" in U.S. national strategy.

The proponents of this notion that the United States has a "tradition" of space sanctuary are correct, but only insofar as the country surely has not committed itself one way or the other and has taken no definitive action to the contrary. The United States has a sanctuary "strategy," in other words, by default. The absence of a clear policy vision for space has left the country vacillating between periods of intense commitment to preserving space as a sanctuary from combat (led by Presidents Carter and Clinton, for example) and periods of rather intense commitment to various force-application and antisatellite programs (as occurred under the leaderships of Presidents Kennedy, Johnson, Ford, Bush, and Reagan).

Some may view President Eisenhower as the father of the space sanctuary idea, but in truth he not only authorized space weapon research, he also understood that "if space is to be used for military purposes, we must be prepared to use space to defend ourselves." Incoherence in the executive branch of government with respect to this issue, however, has been a more common trait over the years. President Johnson, who oversaw a major expansion in U.S. space weapon research and deployment, also made some startlingly incongruent statements. In an essay published in 1965 he wrote that "the technical accomplishments of the Age of Space will be enemies of war," and "it is my conviction that space can be mankind's first real avenue to peace by giving all men and all nations a common self-interest and joint venture."[67]

Hardly the result of a clear vision and deliberate strategy, competing national visions and diametrically opposed understandings of "strategic stability" have prevented the requisite political consensus from moving forward to deploy the F-15 miniature homing vehicle ASAT, the army's KEASAT, and elements of the Strategic Defense Initiative. Not only has the United States had numerous ASAT research programs, it also deployed for eight years, on twenty-four-hour alert, two Thor ASAT launchers at Johnston Island and two training ASAT launchers at Vandenberg Air Force Base. In 1970, the launch personnel were moved to Vandenberg and the alert changed to thirty days. Events precipitating the dismantling of the warheads from the ASAT launchers in 1975 included shortages of spare parts for the system, a decline in training, and Hurricane Celeste, which caused heavy damage to the launch site.

Local political opposition to the deployment of nuclear Sentinel ABM systems in major residential areas of the United States (Seattle, Detroit, Chicago) contributed heavily to the decision not to deploy these ground-based space weapons—not the dedication of the American people to keep space "safe."[68] The United States deploys no space combat arms because policy has left a void with respect to defining national defense actions in that environment and, at least in the earlier years, nonnuclear space and ABM weapon technologies (especially technologies for the SDI) were too immature to turn space combat policy visions into a reality.

Space may be "kept safe" only if the rest of the world agrees to play by U.S. rules and adopt U.S. concepts of strategy. Moreover, if it is to keep space safe, the United States must have it within its power to control the proliferation of relevant space weapon technologies and shape critical political-military developments abroad. It is my contention that the rest of the world will not "tow the line." Neither Washington nor any other government is sufficiently influential to effect the development of foreign strategies and military forces for exploiting space or denying space to the United States.

Salvation Lies in the Labor of Diplomats. Senator Robb maintained the United States must arrest the development in arms that will lead to space weapons by boldly taking two steps. First, Washington must unilaterally forgo advanced prototyping and testing of space weapons. Second, the United States must endeavor to find a political solution. The senator did not cite one prior case where a political artifice, such as an international treaty, significantly impeded improvements in weaponry. He nevertheless argued that we must expand the 1967 Outer Space Treaty "to prohibit not just weapons of mass destruction in space, but all space-based weapons capable of destroying space, ground, air, or sea targets."[69]

Although there are serious implementation challenges associated with a treaty that seeks primary enforcement through "international consultations" (and the enforcement challenges necessarily would rise exponentially under an expanded regime), Robb insisted that "we should also *explore* a verification regime to allow inspection of space-bound payloads." Commit now, in other words, explore later ways to enforce the provisions.

But the senator wrote earlier in "Star Wars II" that "attempting to monitor weapons in this vast volume of space would be daunting." In fact, he observed that verifying the existence of satellites is not nearly as easy as counting "concrete silos at missile wings or submarine missile tubes at piers or bombers on airfields." Robb also recognized that most space payloads "are built and launched with great secrecy and can operate at any distance from earth." So while making a case that we should avoid an arms race in space because we cannot know with any confidence what weapons other countries might deploy, Robb effectively lays out a counterargument that to contrive a robust, multilateral space arms treaty and inspection regime would be a very difficult undertaking.

With the bizarre cast of characters comprising our present-day international system, it is curious that the senator and other supporters of arms control would propose to base fundamental security interests in space on a multilateral treaty. Indeed, Robb raised the prospect in order to further his own argument that the

United States ought to *anticipate* that other governments will violate their solemn treaty obligations.

Case in point. The senator devised two scenarios wherein the United States deploys a space-based weapon force powerful enough to defeat enemy missile strikes, destroy foreign space capabilities, and defend U.S. and allied satellites. He outlined a scenario wherein the United States possessed a capability to "wreak havoc" in and from space that prompted vigorous foreign responses. In response, China placed nuclear weapons in space (to attack U.S. satellites and deliver warheads to earth "in minutes") in direct violation of the Outer Space Treaty—the very treaty he proposed to repair and repackage to fortify the United States' security. The senator also stipulated that Russia, besides fielding lasers and kinetic-kill munitions to destroy satellites and ground targets, reneged on the START treaties. Lawless and dangerous states scrambled after "commercially available cruise missiles."

A couple of points may be taken from the senator's scenario. First, treaties, which lack a centralized judicature and a policing force having requisite enforcement powers, are never reliable guarantors of security, leading one to question Senator Robb's enthusiasm for a souped-up Outer Space Treaty. Second, one may ask, what would block China or Russia from pursuing such weapons programs in any case? Why do we assume that, by suppressing the development of space weapons in the United States, Iran or North Korea will not pursue cruise missiles, that China will not adopt a concept of operations that begins with nuclear explosions in outer space to knock out U.S. satellites, or that some future Russian government will not at some point revive its antisatellite programs?

Initiatives to ban, reduce, or control arms using international treaties and conventions have a long and discouraging history. There is no hierarchy of power, no supreme court, no monopoly of strength to enforce and provide the sustained discipline required to meet the objectives of arms control. Critics of space arms desire to prevent the introduction of all armaments into a new environment. To the degree space is already militarized, many critics of space arms seek to reverse this situation to bring about disarmament and attempt to achieve space control without weapons.[70] Yet there is no practical, reliable way to monitor and verify activities in space, and the promises of enforcement using multilateral means are dubious.[71]

The Second Lateran Council gave visibility to the Catholic Church's twelfth-century campaign to enforce peace and block social challenges to the aristocracy by banning the use of the crossbow, bow and arrow, and ballista (a machine for throwing heavy stones). Diplomats at the First Hague Conference (1899) declared their support for the prohibition of throwing projectiles and explosives from aerial platforms—a clear attempt to prevent the "weaponization" of the air. The United States, United Kingdom, and Japan agreed during the Washington and London conferences of 1921–22 and 1930 to far-reaching proposals to reduce naval arms (leading the British and U.S. governments actually to sink perfectly good battleships and break up their respective naval powers). Today, there are those who would look for peace and stability in international treaties that ban weapon systems from the space environment.

Efforts to stop or mollify hostilities in the "next" war by banning or reducing weapons on the land, in the air, and at sea all failed. Despite the fact that the ban was reaffirmed several times, the crossbow and related implements of war proliferated and evolved into more effective forms of weaponry. Air bombardment operations were perfected during the first half of the 1900s, used in World War I, and came to dominate strategy-making and campaign planning in World War II. Indeed, in order to effect a ban on military aviation, it would not be possible to support a commercial aviation industry—so the proposal to prevent aircraft from being used as bombardment platforms never stood a chance. (We will encounter the same type of problem with respect to spacecraft.) The naval arms treaties did not alter the course of events that led to aggression by Japan and Germany in World War II. Indeed, the acceptance of weaker positions and arbitrary limitations on major weapon systems by Washington and London even may have encouraged Tokyo and Berlin to act more boldly. The conditions were created at these conferences, which, "in the name of peace, cleared the way for the renewal of war."[72]

A primary tool in the battle against the arms race in space is the 1972 ABM Treaty. The domestic controversy surrounding the treaty centers on the U.S.-Soviet agreement to prohibit deployment of a nationwide missile defense. Critics of space weapons perceive this treaty, which bans space interceptors that perform an ABM function, as a hook for restricting nonnuclear space weapons. It is, in fact, the only robust legal device available to achieve limits on conventional weapon deployments in space. As a life force in the world of space arms control, the death of "the treaty" would produce profound anguish and bitter disappointment within the space sanctuary community.[73]

Racing with Arms (Again). Senator Robb maintained in "Star Wars II" that the primary reason our potential adversaries might break treaties would be to respond to U.S. provocation in space. In fact, history supports a different conclusion: a government violates a treaty when it perceives it is in its interest to do so. Japan's early departure from the London and Washington Conferences on naval arms control, Hitler's aggression against Prague, London, and Moscow (to name a few), the calculated Soviet Anti-Ballistic Missile Treaty infractions, and Iraq's 1990 contempt for Article 2 of the UN Charter all scream out against resting security in space on Robb's proposal to bolster the Outer Space Treaty.

The logic used by the senator is simple: by deploying space arms we encourage potential adversaries "to deploy weapons into space that could quickly destroy many [U.S. satellite] systems." Indeed, "a space-based arms race would render many of these more vulnerable to attack than they are today." Yet is it not really the other way around? Is it not the United States' military communications, intelligence, and geographic positioning satellites that pose the real challenge to the strategic interests of Baghdad, Pyongyang, Beijing, and Moscow? We ought to recognize, as was impressed upon the reader in chapters 4 and 6, that U.S. satellite constellations will be the primary motivation driving other countries back to the drawing board to devise counterweapons to present-day information-gathering and data-handling advantages in space. The United States will not have to develop space weapons to encourage foreign ASAT developments.

This presents us with an even larger, more perplexing question. Just because the United States unilaterally refrains from developing technologies and systems to apply force in and from space, why do we assume that other countries will pause right along side it? Senator Robb writes that "it defies reason to assume that nations would sit idle while the United States invests billions of dollars in weaponizing space, leaving them at an unprecedented disadvantage." Others assert that "unless provoked by extensive US space weaponization, America's adversaries will not be inclined to pursue space weapons."[74] But not all innovations in the implements of war stem from provocation. Indeed, there are many other motivations that can and have led to arms acquisitions. It defies reason, in other words, to assume that a state will sit by willfully and pay no attention to improvements in its own military tactics or strategic advantage and military developments in neighboring lands, and only watch the moves made by a far-off power, the United States. Conversely, why should the United States sit by as foreign imagery satellites proliferate and its military opponents acquire capabilities to locate and target American men and women on the battlefield?

All major breakthroughs in weapon technologies are not somehow channeled through the United States. The country does not have a monopoly on knowledge, nor are foreign capitals ignorant to possible improvements in their own military arsenals and tactics. Which brings us to the senator's reference to America's MIRV experience. Washington conceived of MIRVs in part as a cost-effective way to overwhelm a very limited Soviet missile defense, and it deployed MIRV weapons on the assumption that Soviet MIRV deployment was imminent. The result, according to Senator Robb, is that "the Soviets did not sit idle." The cold war thaw and subsequent reduction in strategic missile forces in Russia and the United States, begun in earnest by Presidents Bush and Gorbachev, have brought us to the point in history where MIRVs may now be viewed by Robb as a mistake. To suggest that somehow the Soviets might not have pursued this technology had it not been for the United States' decision to deploy MIRV technology certainly helps the senator to frame his case. But to do so, he must strip the MIRV experience of its full strategic context.

With the sizable Soviet commitment in the mid-1970s to vigorous modernization of its strategic offensive forces (a decade when Soviet missiles eventually surpassed U.S. missiles in number—despite the SALT and ABM treaties), the strategic efficiency made possible by MIRVs made sense to U.S. defense planners. It also was generally expected that Soviet strategic planners (who demonstrated true MIRV technology as early as 1973) would fully exploit this new innovation.

Robb used the MIRV discussion to impress upon the reader that by orbiting arms we will repeat our MIRV experience, only "this time with far graver consequences." In his emotive conclusion the senator warned, "If we weaponize space, we will face . . . the image of hundreds of weapons-laden satellites orbiting directly over our homes and our families 24 hours a day, ready to fire within seconds. . . . This would be a dark future, a future we should avoid at all costs."

It is difficult to fathom that even Senator Robb believes that there is a weapon system in existence, or even on the drawing board, more horrifying and strategically stressing than MIRVed nuclear warheads deployed on land-based and sea-based ballistic missiles pointing our way! It would be hard to argue, in other words, that the consequences of losing satellites in space (where direct collateral damage leading to loss of human life is highly improbable) and of suffering a conventional strike from space (which would have limited destructive force) would be graver than nuclear annihilation. The space warfare schemes envisioned today do not nearly equate to the MAD environment of the cold war.

Another point often overlooked in this discussion is that Americans already have weapons "hanging over their heads." The Sword of Damocles is still there, hanging by a thread. The nuclear forces deployed by Moscow and Beijing, and possibly soon North Korea, may be launched on a moment's notice, and they would arrive at their targets in North America in minutes—causing far greater damage, than the space strike weapons referred to by the senator. We will see more countries with long-range missile capabilities in the next century. North Korea, Iran, Iraq, Pakistan, and India today are adding range to their missile forces, and they all have programs to manufacture weapons of mass destruction.[75]

The United States has no active homeland defenses against ballistic missile threats short of preemptive strikes on enemy territory. Our most hopeful option for protection against missiles carrying nuclear, chemical, or biological warheads may well lay in a decision to deploy active defensive weapons—in space. Ironically, space weapons may be our salvation.

Walter McDougall once explained that "critics have applied to space-based lasers the fallacy of the last move—that every technological fix inevitably triggers a counterfix. True enough, but technological advance can be stopped, if at all, only by universal political will."[76] There is, of course, no universal political will, hence the need to consider first the political-strategic requirements of one's own country. International concerns and power realities, while not trivial, represent only one defense planning factor. Moreover, we ought not to ignore the rather blatant double standards frequently applied in the world press and by foreign leaderships. When in 1999 Moscow tested an ABM interceptor, issued a war-fighting doctrine that trumpeted Russia's rights to use nuclear weapons first, rather than in response to a nuclear attack, and began a renewed military campaign (involving the use of ballistic missiles) to rein in the breakaway republic of Chechnya, we did not hear about how Moscow had behaved in a provocative, destabilizing manner. Such cries of indignation seem to be reserved, instead, for the United States.

The Robb proposal is, in the end, suspect, if only because purely political solutions often are inadequate. Tough strategic and military problems sometimes deserve bold strategic and military responses.

What Is the Bottom Line?

The divided nature of the U.S. space policy regime has permitted other acts of policy making other than the official declared policy to dominate decisions in this area. Ardent defenders of declared policy will say that the written policy's virtue is its flexibility—with it, all things are possible. But this is just another way of saying that the United States does not have a firm policy, that there is no meaningful guidance in what is declared and "endorsed" by the president. Policy, after all, forecloses some options and possible courses of action and selects others.

Can it be healthy for the country to abstain from policy making in a "geographic" environment that influences national security so profoundly? To abstain from decisions in the area of policy, to remain inactive or noncommittal, is to provide opportunities to external forces to shape activities in space. Moreover, we would argue that to abstain is to leave national policy making activities to other internal forces, only under these conditions policy may rise up in incongruous and grotesque forms, because it will be left to develop without the benefit of a vision. With myriad unguided internal forces at play, public opinion on defense space matters will remain unstable. The strength and practical influence of public opinion on a subject, which consists of a collection of individual opinions, can only be realized when there is a growth in the number of people who share it.[77] To leave this stage, in other words, to forego meaningful policy making and public education in the face of all that is happening in space, is to abandon the one great force in the United States that must be present if the country is to be able to secure its interests in that arena.

Domestic political problems lie at the heart of current national difficulties in space. Opponents of a more robust U.S. space power will raise budgetary concerns, explaining that new weapon systems will drain the country's coffers, crushing domestic and other defense programs. Others will point out the daunting technological challenges of designing highly sophisticated weapon systems. And still others will argue that there is no threat to warrant the expansion of U.S. space power. And while these later concerns are important considerations, the moving force in the emerging debate resides in the answer to this one question: to what extent should the country use space for peaceful, military purposes?

When space defense subjects are raised in national forums, Janus's other head always is there to remind us of our self-torn state. The two-faced space policy regime constructed by the U.S. government continues to baffle those having budgeting powers and confound many in DoD who, in the very midst of a yawning policy vacuum, must carry on with mission planning, generate requirements, undertake research, development, testing, and evaluation activities, acquire weapon systems, and develop doctrine. The real world actions of the United States in the area of military space are generally more responsive to the guidance and points of philosophy of the ABM Treaty and theories of strategic stability, not the generally tough rhetoric in U.S. National Space Policy.

The question may be raised whether a written policy is even necessary for the national engagement and growth in space. We have seen that policy may come in many forms, and that what we say and do is often more consequential than what we declare in official directives. Today, we may have a written policy, but it is not a controlling policy. Current policy is the result of political compromise, watered down to meet the needs of each agency. This is "vision" by consensus, rather than by foresight or a conscious expression of political will. Declaratory policy can have a profound role to play, but it has become habitual, and therefore thoughtless. Like religious creeds we may sometimes recite by rote in our places of worship, the ritual issuance of presidential space policies may not carry much meaning for our day-to-day living. That space is a war-fighting medium similar to the land, air, and sea media is ultimately meaningless unless it is backed up by appropriate organization, programs, doctrine, and strategy.

Combat operations in space dominate nearly every politically charged space policy dispute, including disputes about space debris, strategic stability, international provocation, and the "militarization" of space. There is nothing in the written National Space Policy to prohibit the investigation and development of space combat systems, although there is a notable void in the overall policy regime as to whether the United States ever will or should deploy ground or space-based weapons to influence the course of a crisis or war in space. Hence the arguable judgment of some is that to deploy space weaponry would be to break one of the great taboos of the last forty years.[78]

There are powerful waves in the U.S. political system which, in the pitch and roll of the traditional debate over whether space should be a true war-fighting environment or a combat sanctuary, unsettle the overall U.S. defense space policy regime. Such disturbance in governing has led to paralysis when it comes to deciding basic military questions. Space force-application tactics might be the most effective way to achieve certain national security requirements, but there is today no political consensus to consider through careful analysis expanding development in this direction.

National security requires a judicious balance of defense, foreign policy, and economic considerations. I recognize, and forty-plus years of U.S. space policy tradition recognizes, that it is important that all nations develop their infrastructures for space in a stable environment. There is today relative harmony in space. There have been few instances of jamming and, thus far, no published attacks on U.S. satellites by foreign laser technologies. However, political and strategic environments change and technologies evolve. With technology comes opportunity, the opportunity to do something a little differently. The emergence and proliferation of new satellite and weapon technologies will ensure that the conflicts raging on Earth will find their way to space. In transitional times, the only reliable compass point for the president as he looks for a way forward is his obligation to provide for the "common defense." Theodore Roosevelt once observed that "as yet no nation can hold its place in the world, or can do any work really worth doing, unless it stands ready to guard its rights with an armed hand."[79] The quest for strategic stability and the compelling desire to maintain a composed environment for international diplomacy, though important in the overarching national security picture, must not take precedence over the steps we must take to guard against old and new menaces.

Grand assumptions about how the "international community" will react to U.S. deployment of weapons are always suspect in the absence of a given political-strategic-diplomatic context. One would have difficulty saying, for example, that had the United States deployed a space-based weapon to defeat ballistic missiles that the resulting "bitter feeling" in the world would have prevented the United States from forming the successful coalition to defeat Saddam Hussein. I suggest, hypothetically, that had it been generally known that Hussein had nuclear weapons, this factor alone might well have amounted to a far more powerful disincentive to forming a multinational war-fighting coalition than any unrelated weapons deployment decision by the United States.

Although stability in international relationships, solid allied relationships, and a gentle atmosphere in the world are eminently desirable and to be strived for, they are not sufficient conditions from a national security perspective. As long as the United States remains at the pinnacle of power in the world, and as long as it continues to be receptive to public and foreign opinion on defense matters (a characteristic of its republican form of government), the many boisterous protests against U.S. military programs will take place. There will always be attempts by leaderships in foreign lands to influence U.S. arms procurement decisions through arms control and public rhetoric—it costs so little and the potential gains are so great.

The current peaceful international environment gives the American people the luxury to be ambivalent about the conditions of freedom in space. Undeniably, Americans are beneficiaries of stability above Earth's atmosphere, but we have stability not because we have refrained from using space as a combat medium or because there is universal acceptance of a principle that space must remain a sanctuary from war. Space is relatively safe today because capabilities for cheaply and routinely exploiting space have not yet arrived. We have every reason to expect that this will change.[80] This leads us to consider finally this question: What must be done to ensure the defense of the United States in the age of satellites?

9
PUTTING ON A NEW FACE

Maturing the United States' Policy Vision

Both of the United States' policy faces have blemishes. The defect of the one is the conscious decision to remain disinterested or unconvinced of the need to shape national involvement in space as it concerns engagement with potential enemies. The defect of the other is its conviction that space is an alien environment that can and must be kept free of the more discordant activities that take place on Earth.

The deficiencies of both are set in even greater relief when one considers the recent profound changes in the world security environment, bustling international commercial space industry, relentless integration of satellites into military activities, and increasing importance of space activities in everyday life. A comprehensive reexamination of the United States' role in space is long overdue.

To be sure, introspection is not easy. We are not inclined to delve into the deeper meaning of things, but rather readily accept what has been handed down by previous generations. Stability and conservatism in these matters are generally good. Yet stability in policy is not necessarily good; indeed, it is potentially *harmful,* if the environment within which the country must act is shifting and changing.

A genuine effort now must be made to understand fully the relationship between space and security in order to craft a policy befitting national goals in the age of satellites. In the end, without adequate institutional backing, current National Space Policy is nothing more than a parchment, a piece of paper, which per se appears to be an authoritative and meaningful backdrop for the evolution of U.S. space power. As we have seen, however, this is hardly the case.

From an analyst's point of view, it is intellectually gratifying and convenient to be able to turn to one source and read "U.S. space policy." But what is written down is neither the inspiration behind the country's activities in space nor the most accurate description of national purposes recognized and practiced by the leadership. Why declare that things shall be done when there is no power or will to do them? National actions for exercising in and exploiting the space environment define the country's real policy.[1] Although documented declarations consider an attack upon a U.S. satellite to be an attack upon U.S. sovereignty, and possibly even an act of war, we have yet to see what the true policy is in this regard, primarily because the country lacks experience in satellite combat. The leadership, in fact, may not respond in the manner declared in official policy.

Prudence demands a fresh look at this subject. The real controversy here is not limited to means (resources, weapons, organizations, or even strategies), but concerns more fundamentally the ends (national vision and purpose). Policy makers have yet to define clearly what we, as Americans, want to do in space when it comes to matters of national security. *What is it we want to do?* What do the country's leaders believe needs to be done to improve defenses over the long term in light of the challenges posed by the strategic environment early in the twenty-first century? Is space an arena available for strategic and tactical exploitation, for offensive and defensive military purposes? Or is it, can it be, a haven from earthly hostilities? Where is policy vis-à-vis the interests of the United States?

These are first-order questions, which, left unanswered and unresolved, will return to trouble future administrations, thwarting the ability of presidents to plan appropriately for defense, devise suitable strategies, speak with one voice to the American people and foreign governments, organize efficiently, develop the most effective tools for the nation's commanders, and budget effectively.

Certain assumptions underlie this book. Perfect objectivity is not possible. But who ever begins with a perfectly clean slate, with no experiences to shape one's judgments, no prior learning to inform one's reasoning, no acquaintance with history to remind one of all the possibilities of human nature—good and bad? I have attempted, nevertheless, to be objective about the facts. And in this vein, it is only right that basic assumptions are clarified, although they manifest themselves well enough in the preceding chapters. The same should be expected of anyone who sets out opposing arguments.

The basic fact from which we must proceed is that humanity is divided. Man has, and will forever have, the capacity to behave in both good and evil ways, maliciously or benevolently toward his fellow man, for or against the common good. Spiritual conversions, political revolutions, social reconstruction, international agreements, altered environments, or fundamentally new economic situations cannot change this basic fact: man has a dark side.[2]

A second assumption is that there can be no universal political will. Consequently, there can be no consistently reliable political solution for banishing wars from future history. Although some political architects may recognize man's divided nature and his conflicting interests, they nevertheless believe that there exists an untapped political will among men to live right among one another. What fails us, according to this view, is our ingenuity to devise the proper institutional arrangements to encourage man to act lawfully and to dissuade him from acting in a way contrary to the interests of his fellow man. Yet not all governments and peoples desire peace (at least not the kind of peace most Americans would endeavor to achieve), justice, controlled relations among states, and lawful behavior. There will always be "outsiders" who desire to play by different, sometimes dangerous, rules. Political solutions for international problems are sometimes possible, but we should never place our full trust in them.

As a consequence, there is a third assumption: wars are ever a part of history. Wars are an attempt by political leaders to settle by force differences of interest relating to sovereignty, morality, wealth, and prestige, among other things. Even if all governments and peoples were basically good, we would still have conflicting interests and disagreements among states. Conflicts of the future will look different from past wars, which featured massive opposing armed forces. Future wars will feature such hostile acts as cyberwar and terrorism. Our last war, in other words, was not the "last" war. And we can state with great confidence that we are not now at the end of history but in an interwar period.

So what are we to do in the face of this bleakness? What can be done? Although we are condemned by the nature of things to suffer wars, we are not condemned to lead predetermined lives. Although wars are to be expected and sometimes even predicted, they are not inevitable. We can improve our circumstances and sometimes even avert wars (but we can never alter our circumstances enough to ensure that wars will never again disfigure our lives). *Political prudence, foresight, and a predisposition not to wait for events to determine national actions are qualities we must look for in our statesmen. We do not have to await some catastrophe before we can know how to improve national security.*

With these assumptions in mind, the U.S. leadership should look anew at the relationship between space and security and consider the following courses of action.

Vision

The national vision for space will decide everything else. The country's leadership cannot undertake a comprehensive review of policy or make bold decisions that will affect the future unless there is general agreement on a vision, which may be expressed in basic principles that set out the relationship of space to security. I believe that Americans share a common vision for space and that it centers on the idea of freedom. Freedom for all governments and private interests to come and go as they please, to function in space as they please, to stay as long as they please—provided that these activities pose no harm to U.S. interests or security—fairly characterizes the United States' vision for space as expressed in more than forty years of declared policy. That said, there has not been clear agreement on how to realize that vision.

Expanding upon this basic vision slightly, how can the country ensure that space remains free (for surely this is not something officials want to leave in the hands of the country's enemies or to fortune)? The past teaches us that power, strength, endurance, versatility, and flexibility are all good qualities within a national security strategy, qualities that befit the global responsibilities of the United States and help ensure its freedom to act in the other environments. *It is a desirable goal, therefore, to propel the United States generations ahead in military technology, and to use that technology within sound strategic and institutional frameworks to achieve vision-inspired ends.*

Consonant with this understanding, *the United States should strive to remain the preeminent military power in space.* This is a laudable and noble end. To be powerful in space does not mean to act imperialistically.[3] It does mean being mentally and physically capable of acting freely in space, if need be in order to control specific orbits. Being in a position of dominance will be important. The United States has not had to suffer wars imposed upon it precisely because it has been strong. Indeed, it has been the strongest power on the face of the earth for at least half a century. Strength and defensive powers are good things in space as well.[4]

There is at least one possibility for exhibiting U.S. strength in space at the levels of vision and policy in a new, more invigorating sense. A "Monroe Doctrine" in space, for example, may be a constructive way to view how Americans should conceive their role. This doctrine, or more accurately, this U.S. policy *declaration,* did not reflect an attempt by the United States to colonize or even control the Western Hemisphere but amounted to an unambiguous statement about the importance of this region of the world to the United States—simply put, Washington would not tolerate attempts by foreign powers to control or subjugate parts of Latin America.[5] To paraphrase Monroe's declaration to fit the current national situation, "We owe it . . . to candor, and to the amicable relations existing between the United States and those powers, to declare that we should consider any attempt on their part to extend their system to any portion *of outer space* as dangerous to our peace and safety." It is impossible, Monroe stressed, and so it may be stressed today, "that we should behold such interposition, in any form, with indifference." This statement does not imply a desire to be a global constabulary or achieve an all-encompassing domination in the space environment. Such a declaration does reinforce the idea that *Washington should oppose all attempts by foreign powers to establish permanent, or even situational, control over any of Earth's orbits.* Strategic thinking, military concepts, and diplomatic measures, for deterrence and enforcement purposes, naturally will fall out of this vision.

Policy

One glaring weakness of current declared policy is the continued inattention to details and conflicting messages regarding the requirement for military measures to enhance deterrence and defensive possibilities in space. If freedom of space is our guidestar, what is being done to nurture and protect it? Are not U.S. policy makers setting a bad precedent by unilaterally restricting national activities in the force-application and space-control areas, limiting in effect the country's freedom to exploit space?

Since President Carter's precedent-setting comprehensive review of U.S. space policy, each subsequent administration has issued its own space directive. I question whether that tradition has any real value, and even whether there may be harm in maintaining a perfunctory, written space policy that leaves a false impression that the country has a coherent policy to face evolving military challenges. *The National Space Policy parchment does not secure the United States' rights in space—U.S. power does.* U.S. policy will be affirmed and recorded when decision makers act. The American people and foreign governments will pay attention to actions, including serious procurement programs that require money and direction. Perhaps we should mature ourselves out of the written space policy business. Policy matters, by default or by conscious decision, *must* be dealt with in the budgetary process. Policy makers do not routinely publish formal written policies, after all, providing specific guidance for the functional development of U.S. sea, land, or air power.[6] Can we proceed similarly in the space area?

Currently, policy is a product of bureaucratic "coordination." In government, this means that every bureaucrat, or "stakeholder," who has an interest in space policy subjects has an opportunity to contribute to and spin the written policy. Each has a chance to approve, modify, or disagree with what is proposed. The bureaucracy, a body of unelected government officials, weighs in heavily in this process. The bureaucrat is a conservative animal, after all, who is neither expected nor encouraged to have vision (which is not to state that there are no visionaries in the public work force). President Reagan did not "staff" his vision for the Strategic Defense Initiative before his March 1983 speech, and yet he initiated fundamental changes in approaches to national security. His executive decisions (establishing the SDIO, fighting for appropriations, supporting his vision politically and diplomatically), not his writing it down, made his vision real.

Rhetoric, capabilities, and deeds reflect real policy. At present, based on national ABM Treaty commitments and fragmentary funding for limited and constrained space combat capabilities, U.S. defense space policy hedges more toward the belief that space can and should be a sanctuary from combat. The president, in consultation with Congress, needs to reassess basic space policy to resolve its split vision and readjust current goals and principles. Otherwise, the country cannot identify reliable guidestars for strategy, deliver a consistent message to the American people or foreign governments, organize properly to implement strategy, or come to any resolution on long-term program goals. Policy incoherence will not allow officials to budget, prioritize, adequately advocate for resources for future space forces or even render judgments about the relevance of existing treaties to twenty-first-century national security.

Current policies do not consider fully the hostile possibilities behind proliferating space systems and technologies. Readers of the National Space Policy are left in the dark as to the nature of possible threats facing U.S. satellites and the more specific rationales behind the space-control and force-application missions. Foreign capabilities and technologies for challenging the United States in space have evolved over the past two decades—but there is no background discussion in the recent policy documents demonstrating today's need for assured space-control capabilities.

A twenty-first-century defense space policy must align with the realities of increased military dependence on space. *A thoroughgoing assessment of trends and threats should bring to light how unprepared the United States is to deal with contests for control of space, and that Washington could perhaps do more in the decades ahead to enhance the speed, precision, and overall effectiveness of its military power by exploiting space in ways heretofore considered politically unacceptable.*

As an environment that does not in any perceptible way preclude warfare, we may expect space increasingly to take on combat importance and host combat activities. The following are not unreasonable phases in that evolution. *Phase 1:* satellites are used to support warfare, and the advantages of space are exploited by more countries. *Phase 2:* deliberate interference with satellites involving signal jamming and lasers increases over time in intensity and sophistication. *Phase 3:* countries ratchet up the levels of destruction. In this phase, countries attack satellite ground stations and terrestrial communications links. In the interest of increasing military efficiency and effectiveness, and reducing direct collateral damage, countries take destruction to space by attacking satellites from the terrestrial environments. *Phase 4:* technical advances allow space to be exploited primarily by the United States and later by other countries to improve defenses and deterrence, especially defenses against increasingly long-range and proliferating ballistic missile systems. Other active defenses in space evolve, including defense of space assets. *Phase 5:* possibly concurrently with phase 4, the need arises to gain new efficiencies in offensive warfare and deterrence, and nations place weapons in space to neutralize or degrade targets on Earth.

The successful uses of air power over time helped ensure American public acceptance of air dominance. Today, U.S. military air activities during crises and wars do not usually generate criticism (at home or among the United States' allies) for being too "provocative" or significantly detrimental to international relations. (The United States will always receive criticisms from adversarial states, especially for its air activities, and suffer reactions based on a more fundamental resentment of U.S. power or an existing hostile relationship.) The U.S. Navy dominates the sea and operates under clear rules of engagement. There are today no challenges to U.S. sea power. Sea power is an accepted fact. In fits and starts, we are going to space. Why not strive to accomplish the same balanced expression of power in space that the country has achieved at sea and in the air? A clear case can be made that the armed forces must be able to do in or from space the same things they do in or from any other environment in order to protect and promote U.S. interests.

In the end, a written policy never forces an administration to do what it does not want to do. The most desirable way to proceed in order to rectify and refine current defense space policy is for the country's chief executive to lay out his vision to the American people and set about implementing it. The presidential office is ideally suited for mobilizing, focusing, and instructing public opinion. The office provides a sense of legitimacy, authority, and strength to the policy of the government. The president should unambiguously revitalize the national security establishment to enforce the following observation made in current (unimplemented) policy: *It is the policy of the U.S. government to use space, like it uses the land, sea, and air environments, to improve U.S. national security and warfighting effectiveness.*

Establishing a national consensus requires a delineation of specific national security space goals by the administration and its implementing departments and agencies. Goals that are too broad lack specificity and firm direction. Speaking generally to U.S. commitment to freedom of space and the peaceful uses of space works well at the level of principle. At the level of policy, however, the expression of U.S. vision will require a commitment to building, inter alia, military forces, and pursuing in the near-term, *type x* forces. The administration must cultivate congressional and public support for these goals. Industry, in turn, must be convinced that the pursuit of these goals will serve its interests. The president can clarify existing declared policy documents by sharpening issues and the relationship between space power and national power. These actions, consonant with an agreed strategy, will provide direction required for developing appropriate doctrine and developing new concepts of operation.

Failure of a future president to accomplish this will leave the action to Congress. Although Congress cannot consistently push a policy (it is not an executive body), it can provide the forum for beginning a national debate on the relationship of space to security. Beginning from the premise that there are competing policy visions, Congress could request that the administration prepare a report on this subject and appoint an independent commission to evaluate what space means to U.S. national security. The report and commission's findings would be the basis for public hearings on the subject. The hearings would be a prelude to a congressional debate to answer this question: Is the space environment like the land, sea, and air environments from a military perspective? Congress would then pass a resolution recommending a course of action to the president.

All should welcome a debate about threats, technical facts, military needs, defense options, and policy problems—a decision to continue to operate with the same ambivalence for the next ten years would be insufferable. I believe that a debate on this subject likely will ratify the statement that, for the purposes of national security, space is like the land, sea, and air. An opposite finding, however, would lead to erasing language from existing policy documents that suggest space is in any way a war-fighting medium.

A well thought out policy, even if it resulted in a negative decision on the question posed above, will enable costs to be controlled, if only because the issues at hand will have been studied thoroughly before a major financial commitment is made to programs. In other words, there will be a better agreement on which programs to fund. Many systems have been canceled or delayed because of the ineptness of current policy. As taxpayers, we should insist that the leadership examine the basis for decision making in acquisition. *Cost savings may be expected in programs that are not pursued or that are pursued* and *completed.* While divisions almost certainly will continue to exist after a period of debate, a line will have been drawn, and a more definite policy path established.

Strategy

The policy decisions made will be momentous for the makers of strategy. Not only will it be important to consider broader foreign policy objectives concerning the United States' position and involvement in the world, once the country knows its defense requirements and recognizes what it must do in space, the defense leadership can begin to consider seriously developing a strategy for U.S. space power to support those higher ends. The ends of high policy, after all, cannot be reliably orchestrated unless a series of architectonically lesser objectives is identified and implemented by identifiable means. A new vision will lead to not only new policy options but also development of new strategic concepts, military doctrine, acquisition strategy, operational architectures, and tactical choices.

A policy decision to treat space as a sanctuary would lead to a strategy that focused less, if at all, on military operations in space for space control, and more on diplomatic means (treaties, arms-control agreements, sanctions). U.S. policy makers might turn to the United Nations to plead their case and further the country's policy vision that no nation ought to commit hostile military acts in space (although just what constituted a "hostile act" and an improper military use of space would have to be defined precisely). Deterrence strategy would have to rely on retaliatory measures that did not stray into the exoatmosphere. Likewise, current concepts for ballistic missile defense, which call for missile intercepts in space, would become problematic. Military planners would have to consider the implications of this general policy decision on current doctrine (*Joint Vision 2020*) and key intermediate goals such as information dominance.

On the assumption that U.S. leaders will recognize and embrace a policy that requires a more active exploitation of space for national security, we may consider the following. The decision to make more active military use of space would support the general U.S. foreign policy objectives assumed in the later half of the twentieth century, which reflected an understanding that U.S. preeminence in the world *enhances* security and fosters an international climate more receptive to democracy, the rule of law, and free markets. *A strategy to support these ends demands both economic and military strength in all of those environments where the United States has significant activities and interests.*

The United States can undertake military activities in space, as it can in the other strategic environments, for negative or positive reasons—for purposes of denial or active encouragement. Washington may undertake a long-term campaign to prevent space dominance by foreign powers (the Monroe Doctrine applied to space). Or it may strive to ensure U.S. freedom of action in space in order to ensure more actively its access to that environment so as to further its national security and foreign policy objectives. Washington could strive to implement its strategy collaboratively with its NATO or Commonwealth allies, or it could choose to guarantee national security unilaterally, or, more likely, use some combination of these two approaches.

Whether it pursues positive or negative ends at the level of strategy, in concert with allies or alone, deterrence will remain the nation's number one objective. A twenty-first-century national security strategy may be expected to feature space control, to ensure that U.S. forces have the space superiority, "forward presence," and the information they require to succeed. The U.S. way of war requires superiority in all of the war-fighting environments.[7] *The ability to control space, therefore, will be a critical component of a future deterrence strategy to prevent hostile acts in space and on Earth.*

U.S. and allied navies have long kept the sea lanes free, primarily in a deterrent mode. After all, there have not been many engagements at sea over the past fifty years while freedom of the seas has been enforced by U.S. naval forces. The U.S. Navy's occasional visits to the Gulf of Sidra, which Libya views as its national waters (a legal position with which the United States has not agreed), are bold demonstrations of the U.S. commitment to the freedom of seas principle. This kind of activity also may be applied to space.

Deterrence will demand reliable defensive countermeasures that allow U.S. and allied systems to detect and survive attacks against space systems and identify attacking systems. For deterrence to succeed, operators need to know when orbiting systems are being interfered with and who the culprit is. The United States will need to deploy warning and verification sensors on satellites to supplement existing ground systems.

Offensive nonnuclear combat capabilities to affect the course of war in space also could help deter hostile acts in peacetime and crises. What is radically different, after all, from placing something into space to be called down on demand (or launched on demand in a crisis situation or during negotiations) and threatening projection of force using the current array of highly responsive land, sea, or air strategic forces?

What about war fighting? *The U.S. defense leadership needs to think long and hard about the implications of access to space in wartime as well as the possible hostile uses of space.* There are, broadly speaking, two aspects of space warfare: controlling space in the classic sense (from Earth or in space) and striking targets on earth from space. Americans will have to live with the fact that other countries will be operating in space, so that even if it were a desirable objective, the idea of completely "controlling" space is not possible.[8] The key space-control questions are: What do we want to control, for how long, and for what purposes? Policy and strategy requirements will determine whether capabilities will be required to achieve a selective or more globally encompassing space-control objective. Moreover, for reasons of strategy and geography, the United States may well be driven to develop space strike warfare assets. The urgency to react to crises halfway around the globe may compel future commanders to look to space for the timely arrival of forces (without having to transport tanks with C-5s or deploying bombers).

On a slightly different plane, the United States also may consider a political strategy for securing its long-term space technology advantage. Although it still must guard against the transfer of critical military technologies, *capitalism ought to be set loose to advance the development of satellite technologies and services (including imaging services), which would allow U.S. industry to play to its strength—technological innovation and application—which in turn would provide the United States significant technological advantages in the years ahead.*

Current U.S. policy is to bludgeon even its closest allies to tighten restrictions on, for example, the distribution of space imagery. In fact, this heavy-handed approach has had the opposite effect. It has pushed foreign governments to reject cooperation with the United States and pursue their own military space programs. Current policy restrictions on the distribution of satellite imagery ("shutter control" in times of crisis) does not apply to foreign satellite owners, which has had the unintended effect of spurring foreign competition. The European Space Agency's Ariane rockets, French Helios satellites, and ESA's Galileo navigation system were begun, in large part, to avoid overdependence on American generosity and the availability of U.S. launch capacity, reconnaissance, and the GPS systems.[9] An enhanced international cooperation strategy and policies that ensure U.S. industry dominance would be very effective in assuaging long-term security concerns about the proliferation of space capabilities. We must learn otherwise to cope at the levels of policy, military strategy, and doctrine with the consequences of space technology proliferation and the more active involvement of foreign states in space during crises and conflicts.

We have policy mugging today because we have inadequate policy making. Yet is it not clear that the United States will not stay ahead technologically by holding the rest of the world back? Not only must we expect those attempts to fail, they also have the potential to do tremendous political harm as well as stymie business opportunities for U.S. companies. The way to maintain the country's lead is to press ahead technologically and maintain a stable policy regime appropriate for the age. Surely Washington will want to influence what people image from space, and U.S. cooperative relationships with allied and friendly governments will help the country to do so. But one thing is certain—we will not be able to bludgeon our enemies into cooperation. For those times, the United States needs to have in place more assertive means and doctrines to counter hostile activities in space.

A number of studies are in order to evaluate the political and military utility of space systems and weapons, involving considerations ranging from diplomacy and foreign policy to such military matters as force projection and interdiction in and from space. Policy makers must also study tradeoffs between the maintenance of more conventional missions and weapons and introduction of more missions into space. If a determination is made that space weapons would improve national security, further analysis would be required to map out a path to take to introduce these tools in the arsenal and military strategy and a time line from which to plan.

Rhetoric

Ideas have consequences, and what we call things matters. Not only is there self-evident benefit in using language that relates ideas and issues clearly to the American public and those who implement policy (i.e., effective communication), but the effectual public presentation of opinions also is important. Political speech shapes and feeds into public opinion by identifying with those principles and sentiments that define U.S. political life. Hence, Abraham Lincoln could observe that "in this and like communities, public sentiment is everything. With public sentiment, nothing can fail; without it, nothing can succeed. Consequently, he who moulds public sentiment goes deeper than he who enacts statutes or pronounces decisions. He makes statutes and decisions possible or impossible to be executed."[10]

Ill-considered rhetoric in the highest offices can contaminate and debauch public opinion, which in turn can impair the ability of the United States' leaders to govern effectively through discussion and argument. *Politically effective speech about the military uses of space must refrain from employing statements that are counterintuitive, against reason, and cynical.* The burden of proof is on the theorists to show (contrary to what experience demonstrates) that "1 + 1 = 0" or that "the Sun rises in the west"—that the enemy's strength enhances national security, that U.S. vulnerability equals U.S. security, that stability in the international system is achieved by a reduction in the country's ability to defend itself, that instability automatically follows the deployment of a weapon system in space. The susceptibility of U.S. leaders to subscribe to this twentieth-century, nuclear-age line of argument unduly burdens and hopelessly confuses public discussion about the proper relationship of space to national security.

There is such a thing as being too clever, or eschewing simple insight and a manifest train of reason for academic ingenuity or pseudoscientific rigor. It remains to be seen whether the cleverness relied on by the strategic theorists of the previous century, as it was applied to the deployment of robust national missile defenses, will result in future tragedy. The United States still stands vulnerable to small and large missile attacks because for decades the ideology that has grown up around theories of strategic stability has permitted the threat to outpace policy and development of military countermeasure capabilities. In another example, the United States and Great Britain were fortunate to escape destruction at sea at the hands of the Japanese navy during World War II after years of following the counterintuitive advice of their own diplomats to allow for a degree of proportionality and equality in the balance of international naval arms. Japan's withdrawal from the Washington-London system of naval arms control allowed some time, at least, for the two Allies to provide by late 1943 stronger navies to deal with the challenges posed by the Axis powers at sea, especially in the Pacific. Tragically, the Allies needed stronger navies in 1941 and 1942 as well.

Public rhetoric must reflect the perils of vulnerability and the folly of relaxing the country's guard when an enemy's offensive capabilities increase to match its own. "Stability" is generally a desirable condition in the international system, but the shapers of public opinion must clearly convey that it cannot be placed on a pedestal higher than national defense.[11] When considering national security, one must look hard at defense objectives, which necessarily involve choices concerning the procurement and employment of the implements of force. It is not all that obvious that the deployment of a space weapon is fundamentally more destabilizing than deployment of a long-range weapons platform in the air, sea, or land environments. To the contrary, in this strategic environment, *whatever augments U.S. power should be communicated as being stabilizing for the United States as well as its friends and allies.* National strength begets true stability.

"Moulding public sentiment" with respect to defensive uses of space and defining the terms of the debate make it possible for associated policy decisions and statutes to be executed. A decision to speak about the space environment in the same manner that we speak about the land, sea, and air as it pertains to national security will mean a conscious decision among policy makers to refrain from referring to space as a quiet haven from conflict, or a "heavenly" place unaffected by earthly activities. "Peaceful uses" of space does not mean "nonmilitary," because we know from experience that the military arm of the state is a major contributor to the conditions of peace. The consideration of the "space weapon," moreover, in isolation of all else, is not the basis for rendering a judgment about its desirability; rather, evaluation based on a weapon system's purpose, cost, and requirement for its use should be the ultimate arbiter as to whether it should be included in the arsenal. *U.S. leaders must speak smartly to strategy, war fighting, and deterrence matters and convey to the American people (and foreign observers) the importance of national power, to include military strength, for basic security and foreign policy ends.*

The national leadership also must be clear in its rhetoric for the sake of deterrence. The U.S. unwillingness to protest seriously the Russian use of a laser in 1997 to dazzle and nearly blind an airborne U.S. Navy officer speaks volumes.[12] If the president will not defend U.S. interests in a case in which a U.S. citizen is injured, why should we believe he will react with resolve to retaliate for or resist hostile attempts to dazzle and blind U.S. satellites? Official U.S. documents convey that space is of high national importance, thereby implying that space control, or control of the "high ground," is its prerogative and unbounded by international law. What remains to be seen, however, is if, when, and how the United States will choose to exercise this prerogative.

Similarly, an unambiguous public statement by U.S. officials at the outset of a crisis or war that puts all commercial operations on notice that the United States is watching could help immensely to reduce the degree of leverage U.S. opponents can get out of commercially available space services. Washington could state the following: "The United States supports peaceful commercial satellite operations.

However, evidence that a company has utilized satellite services in a hostile manner will be used to justify military action to neutralize or degrade the offending satellite systems in order to protect American interests or troops."

In other words, U.S. authorities can behave in much the same way as they would to prevent a violation of neutrality laws on the high seas. The effect would be to put all commercial providers on edge during critical times. In many cases actions will talk unambiguously. Consistent practice along these lines will make clear U,S. intentions in future crises and conflicts.

Diplomacy

The United States will need the political support of its allies and friends as well as their involvement in military space activities, to include economic contribution through collaboration in system development and participation in operations. *Washington should consult early on with its closest allies, which include the Commonwealth countries of Canada, the United Kingdom and Australia, the other countries of the North Atlantic Treaty Organization, and possibly Japan about its future plans for space.* Not least important, the United States will need to make sure it can provide firm, credible security guarantees to its allies to support its partnerships.

Indeed, where are our modern alliances with respect to military activities in space? There is no obvious reason not to have enhanced collaboration with friendly states in this area—in fact, it may not make sense to go it alone. *Allied involvement in the protection and prevention aspects of space control are especially plausible.* Satellites are only a part of the system. There are ground elements, and surveillance activities that are critical to all military space missions, and there are undoubtedly several contributions U.S. allies can make in these areas.

The worldwide capability of the United States is not easily duplicated, and it should use this fact to improve relations with other nations and shore up alliances against those countries and entities that would do it harm. A technology gap with U.S. coalition partners, including in the area of "command, control, communications, computers, intelligence, surveillance and reconnaissance systems (C4ISR)," remains a serious concern expressed by U.S. defense planners who are struggling to ensure military interoperability among all friendly forces.[13] The United States should work to ameliorate this concern through greater technical and operational collaboration. A collective, cooperative approach may ameliorate some of the policy difficulties that may attend decisions to acquire specific weapon systems.

Alliances make sense from a strategic and military perspective, where system interoperability will be important, but also from the perspective of industry, because the industrial alliances the United States forges to get to space and exploit space will be fundamentally important in the future. *U.S. policy makers will need to put together a compelling argument for space from an industry perspective that will allow for possible international partnering.*

It is commonly held that there is a tradeoff between the capabilities we can exercise in space and the standing of the United States in "the court of world opinion." Should the United States deploy space weapons, it is often assumed that there will be "a wave of disquiet" throughout the world. Space has become more internationalized, after all, and there is now greater interdependence, which will make the deployment of space weapons all the more difficult. Let us not forget, however, the beneficial and stabilizing effects of the powerful U.S. Navy on the high seas and of the not-to-be-outclassed U.S. Air Force within domestic and international air spaces. It is a comfort to know, after all, that hostile states and nonstate thugs have to consider the consequences of attacking vessels and planes peaceably traversing the seas and the air.

Washington must learn how to relate defense space subjects with confidence and conviction to the international public and foreign governments. The "shock" of U.S. deployment of weapons in space eventually, like the introduction of many other novel or revolutionary weapon systems(such as aircraft carriers and Stealth bombers), may be expected to wear off in time. After all, we must square this concern with the fact that we already deploy some pretty "ugly" weapons (including nuclear-tipped ballistic missiles). World reaction will depend on how U.S. authorities understand their new capabilities and how they choose to talk about them with allies and other countries. Perceptions are realities for other nations, and in some cases foreign perceptions will not match reality. Let us make sure that the perceptions of other government leaders are accurate.

In the end, it may be difficult enough to work with other countries on one program or in a particular mission area. Indeed, officials should study how the United States can cooperate in space with allies and friends. It may well be that the United States can only hope for cooperation in the political and economic arenas, and that truly effective space defenses will be achieved primarily through self-reliance.

This is not the place to review the pros and cons of arms control. A great deal of ink has been spilled over the years on both sides of the debate over the utility of arms control to national security, especially as that debate involves space arms.[14] Not one to discount entirely the possibilities of arms control, I am nevertheless obliged to explain this skepticism and to emphasize the need to consider alternative measures to a primary reliance on international agreements for security. The possibility of avoiding a "spiraling arms race" posing threats to satellites, which support strategic deterrence and stability in space, contributes strongly to arguments in favor of refraining from deploying weapons in space.

Yet this caution against succumbing to an "action-reaction" cycle hypothesized by arms controllers does not reflect reality. Arms-control ideology makes a troubling underlying assumption that should give any political leader pause. That assumption is that governments behave as would Pavlov's dog. Ivan Pavlov, noted Russian physiologist, conditioned his dog to salivate at the ring of a bell, much as it would were a plate of food presented. Pavlov believed this "conditioned reflex" operated in all acquired habits and even in the higher mental activity of humans; stimulus response, stimulus response, action-reaction. Thus, from the perspective of those who believe that human activities are really no more than reactions to stimuluses, we can derive a theory of behavior leaving no room for rational thought or choice.

The parallels here with arms control are obvious, as a basic presumption of arms control is that when State A is exposed to the stimulus of a development in arms by State B, State A is expected to react in the form of a counterdevelopment (regardless of national security requirements). Hence, the "spiraling arms race." Take this theory on its face, and one would have to assume that the United States' neighbor to the north would fortify its borders and spend itself silly producing arms. But Canada, sitting so close to the world's strongest military power, has not reacted thusly.

If arms control is to work, mutual trust must be worked into the relationship. But this is where the objective of arms control starts to look strange. Arms controllers are burdened by a fundamental paradox. In times of peace among rival states, when international trust receives high acclaim and few reasons arise for calling that trust into question, arms control is most active and apparently most successful. Yet during such placid times, arms control is also least required to the stabilization of the international relationship, as other diplomatic-strategic factors tend to dominate the course of relations. When tensions are high among those same rival states, international agreements to control arms, which by definition are based on trust, tend to be an impossible undertaking and a most suspect approach for ensuring security.

The East-West arms agreements successfully negotiated during the cold war may be said to have resulted from a combination of factors, including misplaced trust on the part of Washington (which left U.S. officials in the position of having to deal with repeated Soviet arms-control violations) and the Soviet strategy of using arms control to stall progress in U.S. weapons programs. The success of the START treaties had more to do with the change in the political regime in Moscow in 1991 and the general impoverished state of the Russian Federation.

There are also politico-cultural perspectives to consider. Diplomats, who by their very nature assuage the adversary and attempt to soften the sharp edges of controversial issues, negotiate arms-control agreements. The struggle to reach a consensus with the other party (or parties) inclines negotiators to strive for ambiguity in order to downplay or bridge fundamental or technical differences. The problem is that the ambiguities that are invariably written into agreements may be interpreted differently by each party. The United States chooses to interpret arms-control agreements as legal international agreements (as opposed to a political agreement), despite the absence of a judicature to which parties may appeal. The United States can litigate with Canada and the United Kingdom, but can it do the same with North Korea or Iraq?

Given that governments, not weapons, make war, and that the different objectives pursued by governments make wars more or less likely, it behooves a government leadership not to watch the arms-control "birdy" but to put that distraction aside and pay close attention to the play and essence of the political decisions being made and to the insights provided to us by history concerning the reality of interstate relations. In the end, constraints on weapons ought to be sought only if such reductions align with national security objectives.

Arms control is serious business. Arms treaties are to defense planners what constitutions are to lawmakers. The negotiators and drafters of these fundamental documents define the parameters for those who implement them. Indeed, arms-control agreements establish supervisory institutions that are intended to ensure peace automatically, without having to rely on decisions by government leaders and citizens, who, when these agreements are in play, ultimately are made to be inconsequential in certain war-and-peace decisions facing the country. These documents, not easily modified or abandoned, remove important decision-making powers from elected and appointed officials by defining what is and is not subject to change through deliberation or policy making. Because they decide fundamental questions at these outset, in other words, those who draft arms-control treaties have the highest obligation to make sure they "get it right."

Organization

Establishing a national consensus requires a delineation of specific national security space goals by the administration and its implementing departments and agencies. *The administration must cultivate congressional and public support for these goals, but it must first sell the new ideas to itself by organizing to succeed.*

Do we still require the National Reconnaissance Office? Today we need imagery to support troops, but does the country require a separate, hybrid agency that also has allegiance to the CIA?[15] Perhaps the NRO is the best way to handle the defense intelligence business, but we need to know why that is. Another point of consideration is the value of a National Space Council, which in the past has focused space and security matters for the White House and helped govern the interagency process. The council could ensure that the views of commanders responsible for space defense are hooked into the Cabinet-level development of space policy. The council, or a national space executive, also could serve as the president's "brain trust" for high-level space policy matters and perform an important national education function. What other organizational mechanisms may be used to ensure that defense space matters retain a high profile in the nation's seat of decision? Are the space commands appropriately structured and do they have the right missions? Given the armed forces' urgent requirement for reliable access to space, should NASA have the lead in the development of new launcher technologies?

Should the country inaugurate a space corps within the Department of the Air Force? Shifts in the balance of power inevitably bring about changes in the nature of threats to security. Modification of the threat picture may compel a change in national policy and a reevaluation of the role the armed services must play in the new security environment. The strategic concept of a service is fundamentally a description of how, when, and where it expects to protect the nation against some threat. It may well be the case that a service's reason for being will depend upon its ability to develop a new strategic concept related to the newly perceived threats and the changed international structure. As a service's strategic role changes, it may be necessary, according to Samuel P. Huntington, for the service to "expand, contract, or alter its sources of public support and also to revamp its organizational structure in the light of this changing mission."[16] One might add that when a whole new spectrum of threats comes to be recognized, consideration of a new service structure may be justified.

The unavoidable question posed to U.S. policy makers and defense planners will center on this issue of whether new organizations may be required to exploit advances in space, to develop and employ space war-fighting capabilities with strategic effectiveness. A future military structure will require incentives and a reward system for servicemen who dedicate themselves to space defense. The easiest thing to do organizationally, though not culturally, would be for the air force to become an air and space force. Today's air force has not manifested a willingness to give space a higher priority in its budget. There has been a great deal of rhetoric, and fragmentary, capricious funding for space planes and space-control technologies. Continued air force objections to space would compel a move to reorganize it out of the space management and operations business. *However, we have not had circumstances that allow a fair test of the air force's commitment to the space mission. To date, the air force has not been given an adequate opportunity to lead because it has had to work within a poor policy framework.* The answer to the question of whether the country requires a separate space force is not so clear.

Military space activities today also may be considered too organizationally splintered, having to deal in both white and black worlds. A case may be made, after all, that the Soviet Union knew what the United States was doing in space during the cold war but the American people did not. At some point one has to ask whether such an arrangement, carried to the extreme (as it has been in this country), is politically healthy. Undue classification of programs may well have hindered weapons development by obstructing the general education of the American public on national defense requirements, and may be partially responsible for the lack of a space combat capability today.[17] The task of future administrations will be to strive to normalize space operations, which may mean removing the "black magic" aspect, certainly some of the "blackness," at the least.

The organization of Congress also deserves extensive consideration. Not only are defense space matters largely ignored by those responsible for authorizing and appropriating funds, but it is not at all clear that the armed services and intelligence committees are structured appropriately to raise and handle efficiently space security matters, especially as they may impact support to the war fighters. One way to ensure that this subject can be addressed appropriately is to establish subcommittees of the Armed Services Committees for space and information warfare, although care must be given to ensure a proper understanding of space defense as opposed to information warfare. To what degree should the intelligence committees continue to influence satellite programs that impact the effectiveness of the armed forces?

Studies to examine effective and efficient organizations to carry U.S. space power into the future are required. *However, ideas for reorganization are no substitute for policy and strategy. Organizational plans only make sense if there is sound policy and a dedication in the federal legislature to support and fund space defense programs as part of a larger package of policy and resource commitment.* The task before a future president will be to institutionalize space, probably most visibly in a service, but the vision and policy commitment must precede this activity.

Programs

The purpose of this section is not to advocate a specific space weapon force structure. Indeed, the strategic, military, and economic value of many such weapons is not clear. What is clear is that commonly heard arguments postulating that tools of combat for affecting space activities should *not* be pursued for fear of upsetting strategic stability and international relations are politically and strategically shallow. Should such arguments continue to master our politico-strategic dialogue and defense planning activities, they may do great harm to future security.

Defense programs providing technical capabilities that deter and defeat hostile actions against U.S. satellite systems will be a fundamental part of U.S. national security strategy. In this vein, *the country should strive to ensure that it has the situational awareness (adequate sensor capabilities in space and the other environments) so that it may act promptly and decisively.*

Washington also needs to ensure that it has adequate, responsive, and reliable access to space. The United States does not yet treat space operations and launch as it treats commercial air, for this should be the goal to which it aspires. All too often doubts are raised about the adequacy of the country's launch infrastructure, the reliability of its operations, and the vulnerability of its coastal launch sites.

In a future strategy to control space, smashing enemy ground stations may be politically counterproductive and militarily ineffective. Given that increasingly routine space operations may be used for hostile purposes, combat measures may be required in the future. The U.S. Navy employs vessels to ram and disable ships violating international law on the high seas. The United States may well need a similar capability in space. *Within the policy framework proposed above, strong consideration should be given to accelerating development and space testing of the U.S. Army's ground-based KEASAT weapon system.* It is the only weapon system of its kind under development that may be fielded in the near future. A capability for responding in a forceful manner in space will give policy makers one more tool to manage national security effectively. Subsequent studies and consideration should be given to development of ASATs using different approaches and technologies, including nondestructive means, in order to ensure that the United States has the most militarily and politically effective tools for space control. Planners also should consider the operational advantages of a military space plane or operations vehicle. One may legitimately ask, without a military space plane, can the country be assured of its preeminence there?

Over time, the political acceptability of defensive and offensive combat operations in space probably will change. Novel systems may be required to protect and repair objects in space and augment land-based national missile defenses to provide more comprehensive defense or protection of terrestrial assets. *Washington should consider reviving a program to develop space-based kinetic-kill weapons to intercept and destroy ballistic missiles in space.* Using a concept of operations that leverages the space environment is the only real serious response to the proliferation of long-range ballistic missiles, the increase in the number of launch points, and the likely future use of multiple (possibly nuclear) warheads, countermeasures, and submunitions, which may carry biological and chemical weapons. A constellation of interceptors could intercept missiles in the early phases of missile operation and provide multiple intercept opportunities. A space-based defense system also could provide global coverage—an important feature when one considers that the source of the threat can shift from region to region, or emanate from the sea.

Directed energy may provide a means in the more distant future for intercepting missiles rapidly and more efficiently. Research and study should continue on the possible military and strategic utility of laser weapons. *Within the policy framework proposed above, work should continue on the airborne laser and space-based laser projects.* Not only might an operational ABL system help to "thin out" enemy theater ballistic missiles in order to reduce the burden on other land- and sea-based theater missile defenses, it could provide a mobile capability for degrading hostile satellites. LEO satellites would be particularly vulnerable to assault by ABL. ABL will help advance directed energy technology so that one day it might become feasible to deploy space-based lasers to perform a variety of missions.

Defenses are good because they provide options. Said differently, refusal to develop and acquire appropriate, effective defensive systems may be very unwise. There is not likely to be a good practical case, one that made military and economic sense, for a massive movement of missions to space. When it comes to forging military forces (both size and type), clarity of purpose and identifiable service missions are vital. In the end, space weapons, like any other weapon system, will have to be evaluated on a case-by-case basis, with military effectiveness and affordability among the primary criteria. Of course, political and strategic criteria always are at play when it comes to arming the state and will have to be taken into account.

In an age when it takes ten to twenty years to bring a weapon system from concept phase to the field, attempts to put off finding answers to fundamental questions—refusing to go as far as one must to converse with the American people, their representatives, and elected officials on these subjects for fear of hearing "the wrong" answer—risk forfeiting at least a part of the United States' military-technological edge to foreign powers. Other states will not be otherwise inhibited in the application of technologies to new military endeavors.

Policy makers also need to examine the wisdom of keeping so many of the country's space activities in the classified "black" world, especially as they relate to the development of weapon systems. By having a secret space policy with respect to the development of important military tools, do we not exacerbate our political problems? How long do we "talk softly" about ASATs in order to avoid "impolitic" statements? To what extent do we impede prudent government practices by refusing to address many technical and even high-level operational questions in a public manner? Programs in the black world are not supported by a framework of public argument and justification, which means they may be terminated without due consideration in open forums.

It is possible to talk generally regarding space weapons and concepts of operation without getting into compromising details and specifications. We can talk about "stealth bombers" in general terms, for example, without giving up the secrets of low-observable technology. There is no reason we cannot apply the same circumspection to space activities and systems. Clearly, there is a need to think intelligently about space systems whose existence and/or function should be kept as secret as possible, and this will require some thought as to what constitutes real security needs.

The armed forces will need to take advantage of commercial investments in overall operation, launch, and satellite development. It is in the interest of the United States, therefore, to promote private sector investment, ownership, and operation of space assets and develop a minimalist and predictable regulatory environment to reduce impediments to space entrepreneurs and capital investing in the space industry. The country's engines for wealth creation—its industrial, technological, and service bases—must be a part of its defense space future.

Resources

Resources are a constant bugaboo of defense planners and program managers. Most aspects of the space business are very expensive, so a strong case must be made that the public's money should be spent on a particular program. It is, however, not the case that the country does not have money. The United States is very wealthy, in absolute and comparative terms. So the question that matters is *how* it spends its money.

Current U.S. spending on defense hovers around 3 percent of GDP. To have force levels to meet the threats of the future, including space threats, a defense budget that is roughly 4 percent of GDP, or the same percentage of GDP before the United States entered World War II (in other words, before it adopted a world leadership role), may be more appropriate given current national global responsibilities. In any case, there must be a strategy for affording required forces. The leadership will need to divert money from less productive, less relevant activities. Some missions and forces will fade away for lack of relevance, freeing up money for new missions and weapons.

Policy makers and armed services leaders tend to favor conventional, or traditional, weapon systems. Washington continues to spend heavily on cold war–era systems. The status quo is a powerful agent against change. *The country's defense may well benefit from a dramatic shift in funds to deal with new and evolving threats and invest more in revolutionary science and technology.* Currently, the armed services devote less than 2 percent of their budgets to basic science and technology.[18] But few will deny that advanced technology is important to dealing effectively with tomorrow's threats. U.S. acquisition leaders have identified commercially available satellite navigation, communications, and imagery as among the primary emerging threats.[19] Currently, programs for addressing these vulnerabilities do not match this rhetoric. There is a need, therefore, to introduce key space initiatives into the long-term planning process, and turn research and acquisition priorities into programs.

It all comes back to policy. Without a clearly articulated policy, government and industry officials will be reluctant to commit resources to new space ventures, especially if those ventures do not have manifest congressional support.

Intelligence

There is a rapidly growing requirement for regular, focused intelligence assessments and analysis of threats to U.S. space systems and threats from adversarial spacecraft. An immediate, short-term counter to such threats is to improve intelligence capabilities worldwide, both with national technical means and human intelligence assets. There is a fundamental need for ongoing assessments of the threat and military balance of power that include the space dimension.[20]

The U.S. intelligence community used to keep a close eye on the development of antisatellite technologies by the Soviet Union. Since the end of the cold war, intelligence analysts have not had the luxury of being able to concentrate on a single threat or having access to adequate funding to assess the threat to space systems. A gap in tracking threats to U.S. satellites has grown.[21] More needs to be done by way of alerting satellite controllers to threats and monitoring space activities.

The performance of the intelligence community over the last few years has been disappointing at times. There are numerous and recent examples of U.S. intelligence analysts having been caught off guard. China's interest in directed energy and high-power microwave weapons, first publicly recognized in 1998, was a shock to many. China reportedly has tested lasers to dazzle infrared sensors on satellites and was believed as of mid-1998 to have an operational system in central China.[22] North Korea caught U.S. intelligence analysts flat-footed when it launched in 1998 a variant of its Taepo-Dong missile that included an unanticipated third-stage, which, also unexpectedly, was used to attempt to place a satellite in orbit (and it apparently managed to do so for about 27 seconds).[23] U.S. intelligence experts also failed to predict the 1998 embassy bombings in Kenya and Tanzania by terrorist Osama bin Laden, much as it missed the 1996 bombing of the Khobar Towers in Saudi Arabia and Iraq's plants to manufacture biological and nuclear weapons. The growing list of threats, including space threats, will continue to frustrate an intelligence community that longs for strategic guidestars and which apparently is as underresourced as it is misorganized to deal with the challenges of the twenty-first century.[24]

International Law

The United States must have the freedom to provide for its defense, and *its commitment to international treaties should not contradict this requirement.* The country needs to have a coherent policy first, however, and this will drive the national position on international law. A country's perception of its own defense is a dominant concern as leaders consider this subject. Self-defense is an inherent natural right of all nations, recognized in the UN Charter. The government must have the capacity to determine whether a threat is imminent or aggressive, and then to act either to prevent it or defend against it.

Should the United States modify or abrogate the 1972 ABM Treaty in order to remove the only legal hurdle to deploying space-based ballistic missile defenses and other space-based nonnuclear weapons capabilities that, though not explicitly prohibited, are caught up in differing political interpretations of this treaty? The Clinton administration attempted to adopt just such an expanded interpretation of the ABM Treaty when it reached agreement with Moscow to extend the ban on space-based weapons to theater missile defense capabilities—an interpretation that has never been approved by the U.S. Senate. There is growing controversy surrounding this arms-control treaty, which itself is a sign that the treaty may have outlived whatever usefulness it once had. We must ask, ultimately, how this treaty serves national interests in the current international strategic environment. If a defensive capability that is obstructed by the ABM Treaty becomes compelling, the treaty itself must fade in significance.

There are other questions to consider. International law currently does not specifically address private space operations. Yet private entities can do mischievous and harmful things. How shall we evolve space law to deal with this gaping loophole? Do we want to perpetuate the idea that no part of space or celestial bodies can be owned? Do we want to continue to uphold provisions of the 1967 Outer Space Treaty that hold that states cannot establish sovereignty over space resources? Failure to allow for sovereignty will discourage private companies from investing in the extraction and exploitation of space resources. As legal scholars will explain, legally protected private property is a critical aspect of space commercialization. Indeed, *unleashing venture capitalism (letting private risk do the work) to find and use resources on the Moon and asteroids may be the impulse the United States' otherwise Earth-bound enterprises need to develop reusable launch technologies and other capabilities that will make access to space cheaper and easier.*[25] The country's legal positions, in other words, will have many foreseeable and unforeseeable consequences.

Education

The education of citizens on defense matters is important in an era when the introduction of new technologies over a short period of time may turn the country's weaknesses and dependencies into vulnerabilities. *There is a critical need to establish the right frame of mind in the United States for dealing with emerging threats, for the time is right for serious, deliberate, long-term thinking about space and security matters.* To develop and maintain U.S. space forces, the public must have an understanding of the importance of space operations to national security and a basic conceptual understanding of space missions. Public support is required to ensure that human and material resources for the development of space forces are forthcoming.

The molding of public sensibilities and sensitivities will help to shape the terms of the debate in the United States and abroad. Rhetoric must be a part of the public education package for defense space matters. A good public affairs plan will help to explain why specific space forces are required. Such a plan also must develop an education strategy that essentially expands and clarifies basic concepts that should have been expressed already in a well-defined strategic concept for space forces. *There is a continuous need to build up and maintain public confidence in its space power at a level commensurate with the important role space forces will play in national defense.*

The Bottom Line

There is a pressing need for free inquiry into possible military uses of space. It is reasonable to conclude that the capability to control and use space for important defense missions will provide the United States security in the twenty-first century. In matters of security and space, the leadership should avoid the situation in which the country's policy and defense capabilities are chasing the threat. To wait to address possible new security threats until after they appear fully is poor defense planning, if only because the country cannot overnight develop technologies and deploy systems it might need to defend freedom of space. Indeed, to wait in the face of increasingly compelling evidence would unwisely countenance a period of time in the undefined future when the United States would be vulnerable to hostile external forces. It is best to address the many political, foreign policy, and economic issues surrounding this new dimension in military science during a time when the United States still can choose how to fight its wars, rather than waiting for tragedy or a catastrophe borne by misfortune to turn today's choice into tomorrow's urgent necessity.

APPENDIX

ELEMENTARY DESCRIPTION OF ORBITS

Terms

Achieving Orbit

Satellites achieve Earth orbit by being propelled beyond the atmosphere at about 17,000 miles per hour, a velocity that counteracts the force of gravity. At that speed, the earth recedes from the satellite at the same speed that the earth's gravity pulls the satellite toward it. In orbit, the satellite follows a circular or elliptical path around the earth.

The properties of orbits can vary greatly. Altitude, inclination, eccentricity, and orbital period are the major aspects or parameters of satellite orbits.

Altitude. The minimum altitude for an orbiting satellite is about 60 miles; at lower altitudes, the atmosphere induces drag that causes immediate reentry. Due to the presence of rarefied atmosphere, spacecraft whose entire orbits lie in the 60– to 100–mile altitude band require repeated activation of their engines in order to avoid reentry in less than 24 hours. Satellites operating at 200 miles can remain in orbit for almost a year without use of their engines. Although it may take decades, even at 1,200 miles, atmospheric drag eventually will cause a satellite orbit to decay. Beyond about 1,200 miles is the hard vacuum of space; objects orbiting at that altitude will stay in orbit permanently. At about 22,300 miles altitude, satellites orbiting the earth can remain over the same point on the earth's surface (see Geostationary Earth Orbit).

Inclination. Orbital inclination is measured in terms of degrees above or below the equator. The degree of inclination of a satellite's orbital plane determines the area of the earth a satellite will overfly as it orbits. Satellites in 90–degree (polar) orbits can pass over every point on earth. A satellite in an orbit of 45 degrees inclination will only be able to overfly the tropical and temperate territory ranging from 45 degrees north to 45 degrees south of the equator; it will not be able to view the polar regions directly.

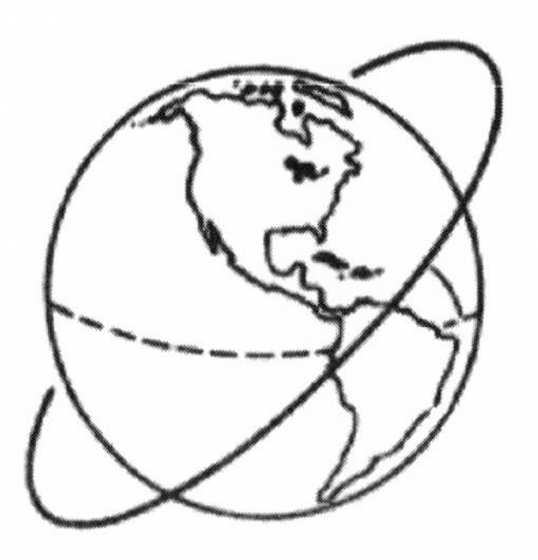

An orbital plane, inclined approximately 45 degrees from the equator. Figures courtesy of Ashton Carter and adapted from his article "Satellites and Anti-Satellites," *International Security*, Spring 1986, 52.

An equatorial orbital plane, inclined zero degrees

Eccentricity. Satellite orbits follow either circular or elliptical paths. Eccentricity describes the shape of an orbital ellipse, in relative terms of "fat" and "wide." A circle is simply an ellipse that is as "fat" as it is "wide."

Orbital Period. The time it takes for a satellite to circumnavigate the body around which it orbits. The higher the altitude of the orbit, the longer the orbital period. In other words, the higher the orbit, the longer it takes to circle the earth. For satellites in low Earth orbit, the orbital period is approximately 90 minutes. For geosynchronous satellites, the orbital period is the same length as the earth's day: just under 24 hours.

Orbital Plane

An imaginary two-dimensional surface (of infinite height and length, but no breadth) that contains all the flight points on an orbit in space. Once launched into space, a satellite will normally maintain the same orbital plane as long as it is in orbit, whatever its altitude and degree of eccentricity. The only way a satellite can change orbital plane is by changing its direction, which requires an (often considerable) expenditure of propellant. Multiple satellites can occupy the same orbital plane.

Apogee

The maximum altitude attained by a spacecraft in elliptical orbit around Earth.

Perigee

The minimum altitude attained by a spacecraft in elliptical orbit around Earth.

Types of orbits include:

Low Earth Orbit (LEO)

Low Earth orbits range from 60 to 300 miles above the surface of the earth. This area of circumterrestrial space is bounded at the lower limit (60 miles) by the presence of atmosphere sufficiently thick to induce reentry and at the higher limit (250–300 miles) by the Van Allen radiation belts. Between these altitudes, the orbital period of satellites ranges from about 90 minutes to two hours. Because of their closeness to Earth, LEO orbits are optimal for reconnaissance satellites that collect imagery of the earth's surface. Low Earth orbits also are used as an intermediate step in the placement of satellites into higher orbits, especially geosynchronous orbit.

Constellation of five U.S. Transit navigation satellites in polar LEO, in five orbital planes

Medium Earth Orbit (MEO)

Medium Earth orbits range in altitude from just above LEO (300 miles) to just below geosynchronous orbit (22,300 miles). MEO orbits are less commonly used than LEO or GEO orbits. One useful MEO orbit is the "semi-synchronous" orbit of about 12,400 miles altitude. This orbital path allows one revolution around the earth every 12 hours or two identical ground tracks over the earth every 24 hours with the proper inclination. U.S. navigation and nuclear-burst detection satellites operate in semi-synchronous orbit.

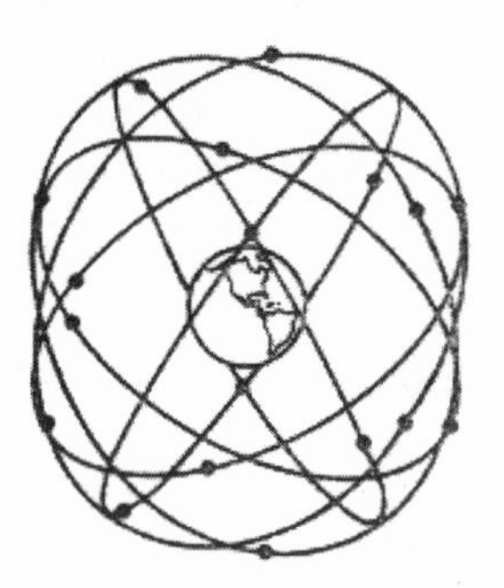

Constellation of 18 U.S. Navstar GPS navigation satellites in MEO, in inclined semi-synchronous orbits in six orbital planes

Geostationary Earth Orbit.

Geostationary Earth orbit is a circular orbit at the altitude of 22,300 miles, with the orbital plane at zero degrees inclination—i.e. directly above the equator. Satellites in this orbit appear to stand still above the same equatorial point, because at that altitude, the satellite's orbital period is exactly the same length as the earth's day. In other words, a satellite in geostationary orbit travels around the earth in exactly the same length of time it takes for the Earth to rotate. Geostationary earth orbit is one kind of geosynchronous Earth orbit (GEO). See below.

Geosynchronous Earth Orbit (GEO)

Geosynchronous Earth orbit is any elliptical orbit centered over the equator that averages 22,300 miles in altitude and makes a single figure eight over the earth daily. Satellites in GEO overfly the same area of the Earth in each rotation. GEO orbits, differing from geostationary orbits in that GEO orbital planes do not remain on the equator, are useful for communications satellites.

Polar Orbit

A polar orbit is one with an orbital plane of 90 degrees inclination. In other words, the satellite orbits the Earth perpendicular to the equator. The earth rotates continuously beneath the orbital plane. Such an orbit (unless it is geosynchronous) allows a satellite eventually to overfly the entire surface of the earth and is useful for reconnaissance satellites. A geosynchronous polar orbit also can be useful if a satellite's purpose is to overfly the same longitudinal swath of the earth's surface.

Polar Orbit: the inclination is 90 degrees from the equator

Sun-Synchronous Orbits

The sun-synchronous orbit is a special type of near-polar orbit. The orbit, approximately 82 degrees in inclination from the equator, or just eight degrees off a true polar orbit, takes advantage of the irregularity in the earth's shape to allow rotation of the satellite's orbital plane at the same rate that the earth's position relative to the sun changes; thus, sun-synchronous orbit remains over the sunlit portion of the earth. Sun-synchronous orbits allow reconnaissance satellites the opportunity to take photographs of the same portion of the earth's surface at the same solar time each day. This allows for better imagery analysis, because the pattern of light and shadow will not vary, except from human activity.

Molniya Orbits

Molniya orbits are highly eccentric ellipses, with their apogees located high above the northern hemisphere. "Molniya," meaning "lightning," is also the name given to the Soviet/Russian communications satellites employing these orbits. The orbits have an apogee of almost 25,000 miles and a perigee of only about 300 miles. The inclination of the orbit, 63 degrees, allows the satellite to spend most of its time over the northern hemisphere, which facilitates communications within Russia.

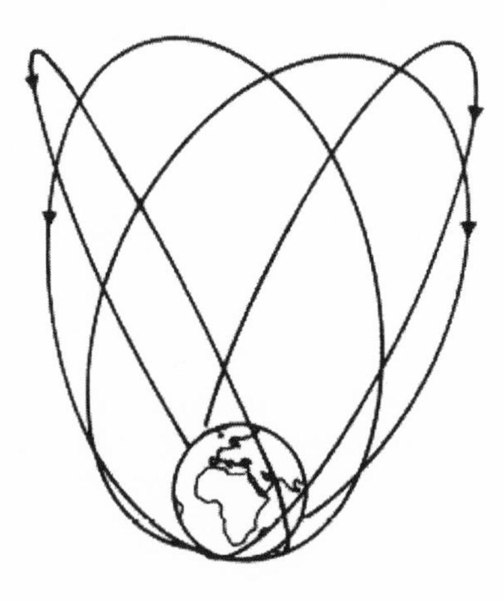

A constellation of four satellites in Molniya orbits, in four orbital planes

High Earth Orbits

High Earth orbits are orbits beyond geosynchronous, or greater than 22,300 miles from Earth. This orbital area contains few satellites, although the United States in the 1960s deployed a constellation of nuclear-burst detection satellites in orbits halfway to the moon. These satellites took eleven days to complete one revolution around the earth. High Earth orbits are sometimes called "super-synchronous" orbits.

NOTES

Chapter 1

1. According to U.S. Deputy Secretary of Commerce Robert L. Mallett, space activities involve some of the most important elements of "our high technology future," including software development, microchip technologies, sophisticated electronics, telecommunications, satellite manufacturing, advanced materials and composites, and launch technologies. Robert L. Mallett, "Next Economic Frontier," *Space News,* April 20–26, 1998, 15. **2.** Institute for National and Strategic Studies, *Strategic Assessment 1999* (Washington, D.C.: National Defense University Press, 1999). **3.** For a layman's look at the history of astronomy, from the speculations of the ancients through the origins of Western astronomy, see Richard Grossinger, *The Night Sky: The Science and Anthropology of the Stars and Planets* (Los Angeles: Jeremy P. Tarcher, 1988). **4.** See Genesis 1:14–18. **5.** For a concise discussion of the principles of orbitology and other characteristics of the space environment, see James E. Oberg, *Space Power Theory* (Washington, D.C.: GPO, March 1999), 23–41.

6. John M. Collins, *Military Geography for Professionals and the Public* (Washington, D.C.: National Defense University Press, 1998), 138, 143. **7.** Walter A. McDougall, . . . *the Heavens and the Earth: A Political History of the 21st Century* (New York: Basic Books, 1985), 20, 21; Alan J. Levine, *The Missile and Space Race* (Westport, Conn.: Praeger, 1994), 1; Frank Winter, *The Rocket Societies: 1924–1940* (Washington, D.C.: Smithsonian Institution, 1983), 22. **8.** Shirley Thomas, *Men of Space* (New York: Chilton, 1960), 24. **9.** Ibid., 28. **10.** Cited in Levine, *Missile and Space Race,* 3, 4.

11. Winter, *Rocket Societies,* 13. **12.** Scientific and Technical Information Division, *Astronautics and Aeronautics, 1985: A Chronology* (Washington, D.C.: NASA, 1988), 152, 53. **13.** William J. Walter, *Space Age* (New York: Random House, 1992), 55. **14.** Cited in William E. Burrows, *This New Ocean: The Story of the First Space Age* (New York: Random House, 1998), 102. In this inspired work, Burrows also relates in colorful detail the pioneering work of Wernher von Braun and the story behind the world's first operational ballistic missile, the V-2 (94–124). **15.** McDougall, . . . *the Heavens and the Earth,* 59.

16. Ibid., 26, 41–65. **17.** Burrows, *This New Ocean,* 134; McDougall, . . . *the Heavens and the Earth,* 120–31. See also Naval Research Laboratory, *Highlights of NRL's First 75 Years: 1923–1998* (Washington, D.C.: Naval Research Laboratory, 1998), 8, 9. **18.** Paul B. Stares, *The Militarization of Space: U.S. Policy, 1945–1984* (Ithaca, N.Y.: Cornell University Press, 1985), 23. **19.** United States Information Agency, Office of Research and Intelligence, "World Opinion and the Soviet Satellite: A Preliminary Evaluation," October 17, 1957, Report no. P-94-57, in *NASA's Origins and the Dawn of the Space Age,* by David S. F. Portree (Washington, D.C.: NASA History Division, Office of Policy and Plans, September 1998), 21–26. **20.** McDougall, . . . *the Heavens and the Earth,* 112–34.

21. F.H. Clauser, in RAND Corporation, *Preliminary Design of an Experimental World-Circling Spaceship,* RAND Report no. SM-11827, Contract W33-038, May 2, 1946, 2. The bracketed comments are mine. **22.** D. Griggs, in RAND Corporation, *Preliminary Design of an Experimental World-Circling Spaceship,* 1. **23.** L.N. Ridenour, "The Significance of the Satellite Vehicle," in RAND Corporation, *Preliminary Design of an Experimental World-Circling Spaceship,* 9–11, 14, 15. **24.** Sir Arthur C. Clarke, "Extra Terrestrial Relays: Can Rocket Stations Give World-Wide Radio Coverage?" *Launchspace,* January/February 1999, 44. This is a reprinted article that originally appeared in *Wireless World,* October 1945. **25.** A. Michael Noll, *Introduction to Telephones and Telephone Systems* (Boston: Artech House, 1991), 48, 49.

26. John L. McLucas, *Space Commerce* (Cambridge: Harvard University Press, 1991), 26–36. **27.** Anthony R. Curtis, ed., *Space Satellite Handbook,* 3d ed. (Houston: Gulf, 1994), 12; McLucas, *Space Commerce,* 31–36. **28.** Michael Fleeman, "Hollywood Looks at the Future through Digital Lens," *Washington Times,* March 13, 1999, A1. **29.** Rachel A. Roemhildt, "Tap-Tap-Tap of Morse Code Is Slowly Going Silent at Sea," *Washington Times,* October 7, 1998, A1. **30.** For a comprehensive and detailed look at services provided by communications satellites, see Curtis, *Space Satellite Handbook,* 11–69.

31. Stares, *Militarization of Space,* 24. **32.** The United States developed high-resolution optics, ultra-fine-grain photographic film, and high-resolution wide-angle lenses during World War II to assist with the strategic bombing campaign, which contributed more than marginally to the defeat of Nazi Germany. The author of a 1947 RAND study that established the basic design concept for reconnaissance satellites suggested that objects illuminated by sunlight may be viewed from space for surveillance or reconnaissance purposes by installing television equipment combined with telescopes in a satellite. According to that report, a "spaceship can be placed upon an oblique or north-south orbit so as to cover the entire surface of the Earth at frequent intervals as the Earth rotates beneath the orbit. . . . If the satellite could accumulate information on film or wire and televise the record rapidly when interrogated by the ground station, a workable system would result." James E. Lipp, cited in Curtis Peebles, *The Corona Project: America's First Spy Satellites* (Annapolis, Md.: Naval Institute Press, 1997), 6. **33.** Information presented in this chapter on the Corona program may be found in Peeble's excellent *Corona Project.* **34.** Jeffrey T. Richelson, *America's Secret Eyes in Space: The U.S. Keyhole Spy Satellite Program* (New York: Harper & Row, 1990). See also George Friedman and Meredith Friedman, *The Future of*

War: Power, Technology, and American World Dominance in the 21st Century (New York: Crown, 1996), 312–30. **35.** Friedman and Friedman, *Future of War,* 312.

36. Curtis, *Space Satellite Handbook,* 93, 94. **37.** Thomas A. Herring, "The Global Positioning System," *Scientific American,* February 1996, 44, 46. **38.** Herring, "Global Positioning System," 46, 47. **39.** Jennifer Harper, "Navy Brass May Scuttle Guidance by the Sextant," *Washington Times,* May 25, 1998, p. A1. **40.** See, for example, Chuck McCutcheon, "Lost in Space: NASA's Quest for a New Direction," *Congressional Quarterly Weekly,* June 6, 1998, 1494–1502; Allen Li, *Managing for Results: Observations on NASA's Fiscal Year 1999 Performance Plan,* GAO/NSIAD-98-181 (Washington, D.C.: GAO, June 1998); Congressional Budget Office, Advisory Committee on the Future of the U.S. Space Program, *Reinventing NASA* [the Augustine Report] (Washington, D.C.: Congressional Budget Office, March 1994); Ruth Larson, "Is the Sky the Limit?" *Washington Times,* October 1, 1998, p. A12.

41. Congressional Budget Office, *Reinventing NASA,* 3, 4. **42.** J.A. Simpson, "Space Science and Exploration: A Historical Perspective," in *A Spacefaring People: Perspectives on Early Spaceflight,* ed. Alex Roland (Washington, D.C.: NASA, 1985), 3–16. **43.** See, for example, Stephen J. Pyne, "A Third Great Age of Discovery," in *The Scientific and Historical Rationales for Solar System Exploration,* ed. Carl Sagan and Stephen J. Pyne (Washington, D.C.: Space Policy Institute, George Washington University, July 1988), 54. **44.** Curtis, *Space Satellite Handbook,* 131–59. **45.** Michael Mecham, "Landsat 7 to Advance Chronicling of Earth," *Aviation Week & Space Technology,* April 12, 1999, 72, 73.

46. U.S. Congress, Office of Technology Assessment, *Civilian Satellite Remote Sensing: A Strategic Approach,* OTA-ISS-607 (Washington, D.C.: GPO, 1994), 40–42. See also Paul S. Harderson, *The Case for Space: Who Benefits from Explorations of the Last Frontier* (Shrewsbury, Mass.: ATL Press, 1997), 56, 57. **47.** Ben Iannotta, "NOAA Will Use Spy Satellites to Track Boats," *Space News,* March 22, 1999, 18. **48.** Curtis, *Space Satellite Handbook,* 70–82. **49.** James R. Asker, "Goes-Next Improves Hurricane Monitoring," *Aviation Week & Space Technology,* August 14, 1995, 62. **50.** Anthony R. Curtis, *Space Almanac,* 2d ed. (Houston: Gulf, 1992), 283–301.

51. Edward H. Phillips, "X-43 Poised to Conduct Hypersonic Test Flights," *Aviation Week & Space Technology,* June 28, 1999, 54–56; Joseph C. Anselmo, "A Mach 8 Vehicle on the RLV Frontier," *Aviation Week & Space Technology,* April 26, 1999, 78–80; Tamar A. Mehuron, ed., "Space Almanac," *Air Force,* August 1998, 28. **52.** Warren E. Leary, "A Rocket Substitute for Shuttles Falls Short," *New York Times,* Science Section, March 2, 1999, p. 1; Michael A. Dornheim, "Engineers Anticipated X-33 Tank Failure," *Aviation Week & Space Technology,* November 15, 1999, 28. **53.** Despite the fact that the cost to build and operate a communications satellite has declined by a factor of twenty over the past ten years, launch costs have remained constant over that same period of time. House, 105th Congress, Second Session, *Statement of Dr. Daniel R. Mulville, Chief Engineer, National Aeronautics and Space Administration before the Subcommittee on Space and Aeronautics Committee on Science, House of Representatives,* copy, September 29, 1998. See also Joseph C. Anselmo, "NASA Chief Rips Industry, Urges New Design Path," *Aviation Week & Space Technology,* May 10, 1999, 32. **54.** There is a high probability that other still-classified procurement projects will be interwoven into the X-33's final design. Bill Sweetman, "Securing Space for the Military: Hypersonic Military Spaceplanes Go Quietly about Their Business," *Jane's Interna-*

tional Defense Review, March 1999, 53. **55.** Industry revenues rose 15 percent from 1997 to 1998, U.S. companies accounting for 46 percent of the total. By early 1998, the spacecraft manufacturing and ground-equipment manufacturing sectors employed 94,000 and 44,100 people, and pulled in $13.5 billion and $11 billion, respectively. The launch industry, at this same time, employed 36,900, who in turn helped generate $7.5 billion in revenues. The satellite service sector is expected to grow well beyond its 18,000 employees at the end of 1997, at which time it was considered to be a $19.2 billion industry. See Joseph C. Anselmo, "Satellite TV Fuels Industry Growth," *Aviation Week & Space Technology,* April 12, 1999, 32.

56. James R. Asker, "Motorola Longs for Future as Major Satellite Builder," *Aviation Week & Space Technology,* June 15, 1998, 63. See also Anthony L. Velocci Jr., "Streamlined Satellite Production on Horizon," *Aviation Week & Space Technology,* November 23, 1998, 49. **57.** Anselmo, "Satellite TV Fuels Industry Growth," 32; Space Publications, *State of the Space Industry 1999* (Bethesda, Md.: Space Publications LLC, 1999), 23–28; Warren Ferster, "Studies Predict Growing Demand for Big Rockets," *Space News,* May 24, 1999, 8. **58.** Some attribute these failures to new business approaches, new cost-cutting measures, adherence to a "better, faster, cheaper" philosophy, and the loss of scientific and technical knowledge as companies have merged and unloaded high-priced (but experienced) technicians, engineers, and program managers. The launch systems in use today are derivatives of systems developed in the 1950s and 1960s—they are basically old, familiar technologies to those who have the experience and know the quirks and patterns of these venerable systems. The young replacements simply do not have the same knowledge base. See also editorial, "U.S. Space Sector Needs Healing on Many Fronts," *Aviation Week & Space Technology,* May 10, 1999, 102. The success rate for expendables remained in 1998 at a respectable, though ultimately unsatisfactory, 95.1 percent. *State of the Space Industry 1999,* 33–35. **59.** Anselmo, "Satellite TV Fuels Industry Growth," 32; "Launches Projected at More than 50 a Year," *Space News,* May 17, 1999, 2. **60.** William B. Scott, "Airline Ops Offer Paradigm for Reducing Spacelift Costs," *Aviation Week & Space Technology,* June 15, 1998, 65; Michael A. Dornheim, "Roton Hops Off Ground," *Aviation Week & Space Technology,* August 2, 1999, 36.

61. Paul Proctor, "Kistler Pursues Funds to Restart Program," *Aviation Week & Space Technology,* December 13, 1999, 84, 85; Julia Malone, "Competition Expected to Cut Costs," *Washington Times,* October 31, 1998, p. C1. **62.** Space Publications, *State of the Space Industry 1998,* 26, 32–34. **63.** Ibid., 58, 59; Peter B. de Selding, "Satellite Failures Put Big Squeeze on Underwriters," *Space News,* January 11, 1999, 1. **64.** Revenues for service providers climbed 23 percent in 1998 over the previous year to $26.2 billion. Satellite television led this surge with a rise of 30 percent to $17.6 billion. Anselmo, "Satellite TV Fuels Industry Growth," 32. **65.** *State of the Space Industry 1998,* 41–45; *State of the Space Industry 1999,* 37.

66. Peter B. de Selding, "Wary Investors Avoiding Satellite Deals," *Space News,* September 13, 1999, 1; Joseph A. Anselmo, "Iridium's Future Up in the Air," *Aviation Week & Space Technology,* August 23, 1999, 40, 41; Sam Silverstein and Warren Ferster, "ICO Goes Bankrupt," *Space News,* September 6, 1999, 1; Bruce A. Smith, "Is There a Market for Satellite Phones?" *Aviation Week & Space Technology,* May 8, 2000, 44, 45. **67.** Troy Thrash, "Little LEOs: Bigger Isn't Always Better," *Launchspace,* January/February 1999, 46, 47; Anthony L. Velocci Jr., "Orbcomm Nears Full Operational Status," *Aviation Week & Space Technology,* November 23, 1998, 46–48. **68.** Space Transportation Association, *State of the Space Industry 1998,* 50; William B. Scott, "Multimedia Satcom Competition Intensifies," *Aviation Week*

& Space Technology, April 13, 1998, 72–74; Mike Mills, "Orbit Wars," *Washington Post Magazine,* August 3, 1997, 10. **69.** In 1999, Congress began the work of eliminating cable's key advantage by allowing satellite television systems to carry local stations. Jack Egan, "For Satellite Television, the Limit Is the Sky," *U.S. New & World Report,* March 3, 1997, 54–56; *State of the Space Industry 1998,* 49; Joseph C. Anselmo, "House Bill Boosts Satellite TV, Eliminates Major Cable Advantage," *Aviation Week & Space Technology,* May 3, 1999, 51. **70.** Marco A. Caceres, *World Space Systems* (Fairfax, Va.: Teal Group, December 1998), 4.

71. John R. Copple to the Honorable Bill Frist, Senate Commerce Subcommittee on Science, Technology and Space, February 27, 1998. **72.** Firms such as SPIN-2 (a U.S.-Russian commercial partnership) do much of their business by e-mail and the rest by fax and phone. Imagery data is available within eight weeks of the SPIN-2 module landing, during which time the imagery is digitized. Pete Norris, SPIN-2, interview by author, Washington, D.C., June 25, 1998. **73.** Michael Mecham, "Commercial Imaging to Enter 1-Meter Era," *Aviation Week & Space Technology,* April 26, 1999, 84, 85; Joseph C. Anselmo, "OrbImage and Spot Link Up: Ikonos Readies for Launch," *Aviation Week & Space Technology,* September 20, 1999, 42. **74.** Warren Ferster, "U.S. Intelligence Slots $1 Billion for Civil Imagery," *Defense News,* April 12, 1999, 1. **75.** *State of the Space Industry 1998,* 54.

76. D-GPS uses receivers at known locations on the ground to calibrate the systems accuracy and then refine the degraded signal (from about one hundred meters). A GPS signal can be received by a fixed ground station with known geographic coordinates so that its inaccuracies are calculated. It is then retransmitted to a central station for correction, and that corrected information finally sent off to the user. Very high accuracies (usually within a meter) are possible by using the D-GPS augmentation method. Another method, called carrier tracking, though more difficult to accomplish, permits accuracies to within a few millimeters. Herring, "Global Positioning System," 49. See DoD New Release No. 139-98, "Additional Civil Coded Signals on Future Global Positioning System (GPS) Satellites," March 30, 1998; "New Study Charts GPS Growth Potential," *Military Space,* September 28, 1999, 2, 6. **77.** David F. Salisbury, "Tracking the Expanding Uses of GPS," *Aerospace America,* May 1997, 48–50; Peter B. de Selding, "France Begins to Equip 4,000 Buses with GPS Terminals," *Space News,* May 10, 1999, 28. **78.** National Defense Panel, *Transforming Defense: National Security in the 21st Century* (Washington, D.C.: National Defense Panel, December 1997), 38; Ben Iannotta, "Rockets Take Aim at Booming Market," *Aerospace America,* February 1998, 34–39. **79.** Frank Sietzen and Simon Mansfield, "Report Predicts Satellite Revenues of $171 Billion," Spacecast web site, http://www.spacer.com/spacecast/news/future-98i.html, May 26, 1998. **80.** Clayton Mowry, Satellite Industry Association, interview by author, Alexandria, Va., August 11, 1998.

81. William S. Cohen, *Annual Report to the President and the Congress* (Washington, D.C.: GPO, 1998), 67. **82.** The 1996 National Space Policy affirms this traditional understanding of space by declaring that key priorities for U.S. national security space activities are to improve "our ability to support military operations worldwide, monitor and respond to strategic military threats, and monitor arms control and non-proliferation agreements and activities." White House, *Fact Sheet: National Space Policy* (Washington, D.C.: National Science and Technology Council, September 19, 1996), 4. **83.** The Pentagon reportedly retains an interest in high-altitude reconnaissance balloons (stratospheric aerostats). Tethered balloons have been employed in drug monitoring operations. Deployed at sixty-five thousand feet, these reconnaissance balloons would produce imagery out to about 110 nautical miles. David A. Fulghum, "Balloons Studied for Intelligence Role," *Aviation Week & Space Technology,* November 27, 1995, 24. The U-2 photos proved beyond all doubt by 1961 that there was no "missile gap," a politically charged slogan used by presidential candidate John Kennedy in the 1960 race, which implied that his Republican opponent and the GOP had been soft on defense. See, for example, Thomas C. Reeves, *A Question of Character: A Life of John F. Kennedy* (New York: Free Press, 1991), 249, 365. **84.** Theresa Foley, "Corona Comes in from the Cold," *Air Force Magazine,* September 1995, 82–87; Dino A. Brugioni, "The Art and Science of Photoreconnaissance," *Scientific American,* March 1996, 78–85. **85.** "U.S. Department of State, *SALT II Agreement, Vienna, June 18, 1979* (Washington, D.C.: GPO, 1979), Selected Documents No. 12A, 53. See also Jeffrey T. Richelson, "High Flyin' Spies," *Bulletin of the Atomic Scientists,* September/October 1996, 48–54.

86. Friedman and Friedman, *Future of War,* 315; Burrows, *This New Ocean,* 530. **87.** Craig Covault, "Recon Satellites Lead Allied Intelligence Effort," *Aviation Week & Space Technology,* February 4, 1991, 25, 26. **88.** Craig Covault, "Secret Relay, Lacrosse NRO Spacecraft Revealed," *Aviation Week & Space Technology,* March 23, 1998, 26–28; Craig Covault, "Advanced KH-11 Broadens U.S. Recon Capability," *Aviation Week & Space Technology,* January 6, 1997, 24, 25; Philip J. Klass, "CIA Reveals Details of Early Spy Satellites," *Aviation Week & Space Technology,* June 12, 1995, 167–73. See also the website of the Federation of American Scientists, http://www.fas.org/spp/military/program/ imint/kh-12.htm. Curtis, *Space Satellite Handbook,* 107; Friedman and Friedman, *Future of War,* 318–21. **89.** Bernard Blake, ed., *Jane's Radar and Electronic Warfare Systems, 1995–96* (Coulsddon, Surrey: Jane's Information Group Limited, 1996), 674, 75; Phil Long, "Lost Satellite Believed to Be Super-Sensitive Vortex 2," *Miami Herald,* August 15, 1998, p. 1; Friedman and Friedman, *Future of War,* 321–24. See also http://www.fas.org/spp/military/ program/sigint/index.html. **90.** David A. Fulghum, "Growing Intelligence Operation Focuses on New Types of Signals," *Aviation Week & Space Technology,* August 2, 1999, 50–55; William B. Scott, "Composite Satellite to Detect Covert Nuclear Tests," *Aviation Week & Space Technology,* November 27, 1995, 48, 49.

91. Imagery satellites typically use data-relay satellites in geosynchronous orbit, such as the satellite data system or NASA's tracking and data relay satellites, to ensure that the information reaches the right ground stations in a timely and secure manner. **92.** Jack Kelley, "Bin Laden Was Stopped Seven Times," *USA Today,* February 24, 1999. **93.** Steven J. Bruger, "Not Ready for the First Space War: What about the Second?" *Naval War College Review,* Winter 1995, 77; Department of Defense, *Conduct of the Persian Gulf War* (Washington, D.C.: Department of Defense, April 1992), 169, 775, 777. **94.** "Radar Satellite Assesses Raids," *Aviation Week & Space Technology,* September 16, 1996, 26. **95.** David A. Fulghum, "Satellite Radars to Guide Missiles," *Aviation Week & Space Technology,* September 30, 1996, 33, 34.

96. Cohen, *Annual Report to the President and the Congress,* 1998, 67, 68; White House, *Fact Sheet: National Space Policy,* 4–6; Steven Lambakis, "Space Control in Desert Storm and Beyond," *Orbis,* Summer 1995, 417–33; Gil I. Klinger and Theodore R. Simpson, "Military Space Activities: The Next Decade," *Aerospace America,* January 1998, 45. **97.** Robert A. Ackerman, "Balkans Serve as Proving Ground for Operational Imagery Support," *Signal,* October 1999, 17; Walter Pincus, "Space Imagery Overhaul Aims at Better Data and Easier

Access," *Washington Post,* January 20, 1998, p. 7; "NIMA to Fill Imagery Gap," *Aviation Week & Space Technology,* December 11, 1995, 31. **98.** The U.S. military is counting on two or three big commercial players to succeed in this new communications market and are waiting to see what falls out in the competition that will take place over the next few years. Capt. Donald Slayton, Joint Chiefs of Staff, J-6 (Space Division), interview by author, the Pentagon, July 28, 1998. **99.** Deputy Under Secretary of Defense (Space), *Department of Defense Space Program: Executive Overview for FY 1999–2003* (Washington, D.C.: Department of Defense, February 1998), 17. DSCS users include the National Command Authorities, Diplomatic Telecommunications Service, White House Communications Agency, the commanders in chief of the armed services, and NATO authorities. **100.** Department of Defense, *Conduct of the Persian Gulf War,* K-25–K-26; Thomas S. Moorman Jr., "Space: A New Strategic Frontier," *Airpower Journal,* Spring 1992, 19.

101. Bruce A. Smith, "Milstar Balancing Cost, Mission Needs," *Aviation Week & Space Technology,* September 18, 1995, 50, 51; Mehuron, "Space Almanac," August 1998, 26; Cohen, *Annual Report to the President and Congress,* 1998, 70. **102.** Pat Cooper, "Military Expects System to Show the Big Picture," *Space News,* September 9–15, 1996, 10; Craig Covault, "'Info War' Advanced by Navy GBS Satcom," *Aviation Week & Space Technology,* March 23, 1998, 28, 29. **103.** Irving Lachow, "The GPS Dilemma," *International Security* 20, no. 1 (Summer 1995): 128, 29; Mark Hewish, "Multiple Uses of GPS," *International Defense Review,* Defense '95, 147; Herring, "Global Positioning System," 48. **104.** Rick Atkinson, *Crusade: The Untold Story of the Persian Gulf War* (Boston: Houghton Mifflin, 1993), 13–19; Lachow, "GPS Dilemma," 137. **105.** Department of Defense, *Conduct of the Persian Gulf War,* 569–70.

106. Statement by Gil Klinger, former acting deputy under secretary of defense for space, in Gigi Whitley, "Prospect of Commercial Satellite Navigation Service Stirs Debate," *Inside the Air Force,* February 27, 1998, 14. **107.** U.S. Space Command, *Long Range Plan: Implementing USSPACECOM Vision for 2020* (Colorado Springs: U.S. Space Command, 1998), 22, 23; Klinger and Simpson, "Military Space Activities," 45. The country also has several new commercial spaceports, some of which are collocated with existing government facilities in Virginia (Wallops Island), Florida (Cape Canaveral), California (Vandenberg Air Force Base), Alaska (Kodiak Island), and New Mexico (White Sands Missile Range). **108.** Tim Smart, "Lockheed, Boeing Get $3 Billion Contract," *Washington Post,* October 17, 1998, p. G1; Ionnatta, "Rockets Take Aim at Booming Market," 39; Joseph C. Anselmo, "Bigger Satellites Drive Plans for New Atlas, Athena Launchers," *Aviation Week & Space Technology,* April 13, 1998, 34, 55; Joseph C. Anselmo, "Interest Surging in Titan 4 Successors," *Aviation Week & Space Technology,* March 9, 1998, 24, 25. **109.** Gigi Whitley, "Budget Decision Restores Funds for Space Launch Improvements," *Inside the Air Force,* January 16, 1998, 11; "Weldon Calls for Redefinition on Cost of DoD Launch Sites," *Military Space,* November 10, 1997, 1. **110.** Cohen, *Annual Report to the President and the Congress,* 1998, 68.

Chapter 2

1. The inherent nature of the polity or state (as a living organism having purpose) and the eternal pursuit of power, prestige, and wealth have shaped for centuries the characteristics of peace and war. For an insightful and unparalleled treatment of this subject, see Raymond Aron, *Peace and War: A Theory of International Relations,* trans. Richard Howard and Annette Baker Fox (Garden City, N.Y.: Doubleday, 1966). **2.** National Defense Panel, *Transforming Defense,* 1. **3.** Zbigniew Brzezinski, "A Geostrategy for Eurasia," *Foreign Affairs* 76, no. 5 (September/October 1997): 51. **4.** Josef Joffe, "How America Does It," *Foreign Affairs* 76, no. 5 (September/October 1997): 13. **5.** See Aron, *Peace and War,* 47, 48. A noteworthy political philosopher and a brilliant theorist of international relations, Aron wrote that "all international politics involves a constant collision of wills, since it consists of relations among sovereign states which claim to rule themselves independently. So long as these units are not subject to external law or to an arbiter, they are, as such, rivals, for each is affected by the actions of the others and inevitably suspects their intentions."

6. Soft power garners international acceptance of norms and institutions that, according to Joseph Nye and William Owens, help shape the behavior of governments and rests on "the appeal of one's ideas or the ability to set the agenda in ways that shape the preferences of others." Joseph S. Nye Jr. and William A. Owens, "America's Information Edge," *Foreign Affairs* (March/April 1996): 20, 21; Joseph S. Nye Jr., *Bound to Lead: The Changing Nature of American Power* (New York: Basic Books, 1990). Today alliances are affected more by cultural and economic power. While the United States is the preeminent military power, it is also in a class of its own when it comes to wielding power through commerce, developing markets, spreading the liberal democratic ideas, offering cooperation. American English is the world's language, its universities are much sought after, and foreign peoples risk their lives to come to America. **7.** Joffe, "How America Does It," 16. According to former Speaker of the House of Representatives Newt Gingrich, there is no replacement for the United States: "If we do not lead the planet, there is no leader on the planet. We are in the classic sense a hegemon. We are the only military power that matters—if we decide to be decisive. No, that does not mean we can bully everybody. I said lead. I did not say dictate. I did not say dominate. I said lead." Excerpts from a speech delivered by Newt Gingrich, in "Viewpoint," *Aviation Week & Space Technology,* May 1, 1995, 86. **8.** Cohen, *Annual Report to the President and the Congress,* 1998, 1. **9.** Ibid., 1. **10.** National Defense Panel, *Transforming Defense,* 5–7.

11. Keith B. Payne, *Deterrence in the Second Nuclear Age* (Lexington: University Press of Kentucky, 1996). According to Payne, this age poses unique problems, not the least of which is deterring a number of nations that increasingly exhibit different types of behavior. **12.** Office of the Secretary of Defense, *Proliferation: Threat and Response* (Washington, D.C.: GPO, November 1997); Cohen, *Annual Report to the President and the Congress,* 1998, 2. **13.** National Defense Panel, *Transforming Defense,* 6. **14.** Jessica T. Mathews, "Power Shift," *Foreign Affairs,* January/February 1997, 50, 51. Some might even argue that the rise in the number of international arms-control agreements has diminished the likelihood of interstate conflict and increased the relative significance of these nontraditional threats. Although Mathews's arguments about the growing relevance of nonstate actors have merit, they ultimately rest upon a technological determinism to the exclusion of relevant political and strategic factors. **15.** See, for example, Robert O. Keohane and Joseph S. Nye Jr., "Power and Interdependence in the Information Age," *Foreign Affairs* 77, no. 5 (September/October 1998): 81–94. See also Carl H. Builder, *The Icarus Syndrome* (New Brunswick: Transaction, 1994), 241.

16. Paul Mann, "Info Technologies Transform National Security Doctrine," *Aviation Week & Space Technology,* November 23, 1998, 51, 52. **17.** Mathews, "Power Shift," 50–66. **18.** Peter F. Drucker argues that the nation-state has shown remarkable resilience in the face of the globalization of the world economy and the devolution of power to the individual. "So far," he writes, "there is no other institution capable of political integration and effective membership in the world's political community. In all probability, therefore, the nation-state will survive the globalization of the economy and the information revolution that accompanies it" (160). Peter F. Drucker, "The Global Economy and the Nation-State," *Foreign Affairs,* September/October 1997, 159–71. See also Keohane and Nye, "Power and Interdependence in the Information Age," 81–94. **19.** Builder, *Icarus Syndrome,* 241–46. After the emergence of satellites, solid-state electronics, and jet transport, according to Builder (244), "closed societies paid a growing price for their isolation; they denied the economic growth inherent in information technologies and their progeny—world commerce and community. The fate of the communist societies provides a striking example of the importance of discerning the dominant forces shaping the future." **20.** See Walter B. Wriston, "Bits, Bytes, and Diplomacy," *Foreign Affairs,* September/October 1997, 172–82. Wriston observed (172) that "instead of validating Orwell's vision of Big Brother watching the citizen, the third revolution enables the citizen to watch Big Brother. And so the virus of freedom, for which there is no antidote, is spread by electronic networks to the four corners of the earth."

21. National Defense Panel, *Transforming Defense,* 13. **22.** See, for example, Williamson Murray, "Thinking About Revolutions in Military Affairs," *Joint Forces Quarterly,* Summer 1997, 69–76. **23.** Alvin Toffler and Heidi Toffler, *War and Anti-War: Survival at the Dawn of the 21st Century* (Boston: Little, Brown, 1993), 48–63. For the next few decades, we are likely to see nations with varying levels of military capability and to be closely identified with agrarian, industrial, or information age warfare. According to the Tofflers (35–99), First Wave civilizations organize wealth creation and war making around their attachments to the land or territories. Second Wave civilizations emphasize the formation of systems for mass production, mass consumption, mass education, mass media and are characterized by the application of industrial techniques to production and mass-destruction techniques (to include nuclear arms) in warfare. Third Wave civilizations create and exploit "knowledge" (though perhaps a better word here is information) and have "brain-based" economies and war-fighting doctrines that emphasize speed, flexibility, smaller forces, information-based weapons, electronic infrastructures, de-massification, systems integration, and improvisation. **24.** See, for example, Ken Silverstein, "Buck Rogers Rides Again," *Nation,* October 25, 1999, 23. **25.** Lawrence Freedman, *The Revolution in Strategic Affairs,* (New York: Oxford University Press, 1998); Colin S. Gray, "RMAs and the Dimensions of Strategy," *Joint Forces Quarterly* 17 (Autumn/Winter 1997–98): 50–54; Phillip L. Ritcheson, "The Future of 'Military Affairs': Revolution or Evolution," *Strategic Review,* Spring 1996, 31–40; Paul Mann, "'Revisionists' Junk Defense Revolution," *Aviation Week & Space Technology,* April 27, 1998, 37, 38. See also Michael J. Vickers, "The Revolution in Military Affairs and Military Capabilities," in *War in the Information Age,* ed. Robert L. Pfaltzgraff Jr. and Richard H. Shultz Jr. (Washington, D.C.: Brassey's, 1997), 29–46.

26. Andrew F. Krepinevich and Steven M. Kosiak, "Smarter Bombs, Fewer Nukes," *Bulletin of the Atomic Scientists,* November/December 1998, 28, 29. See also Joseph S. Nye

Jr., "Nuclear Advice for South Asia," *Washington Post,* November 25, 1998, p. 21. **27.** Wriston, "Bits, Bytes, and Diplomacy," 177. **28.** Ambassador Ernest Preeg, cited in Douglas Johnston, ed., *Foreign Policy into the 21st Century: The U.S. Leadership Challenge* (Washington, D.C.: Center for Strategic and International Studies, 1996), 23. **29.** Johnston, ed., *Foreign Policy into the 21st Century,* 51. This notion that interdependence is behind a fundamental change in international power relationships is, in fact, an old one. Klaus Knorr wrote back in 1975 (207, 208) that "the exercise of international power and influence takes place in an increasingly interdependent world. Indeed, power and influence arise from and impinge on the structure of the interdependencies among actors. . . . Economic life has become interdependent and more or less integrated through the international flow of goods, services, capital, and technology, and through the medium of monetary relationships. This economic interaction affects not only national income and its distribution, the level of employment, occupational structure, and the rate of economic progress but it also indirectly impinges on class and influence structure, and hence on the political life of societies." Klaus Knorr, *The Power of Nations: The Political Economy of International Relations* (New York: Basic Books, 1975). In my eyes, it would be a mistake to conclude that the relevance of military power also has declined, as political-military events in the Middle East, Asia, and southeastern Europe have shown over the last thirty or more years. Growing interdependence, however, does make the calculations of power more challenging. To read more about the contending issues surrounding theories of economic integration see: James E. Dougherty and Robert L. Pfaltzgraff Jr., *Contending Theories of International Relations: A Comprehensive Survey,* 2d ed. (New York: Harper & Row, 1981), 417–67; Robert O. Keohane and Joseph S. Nye, *Power and Interdependence: World Politics in Transition* (Boston: Little, Brown, 1977); See also Drucker, "Global Economy and the Nation-State," 159–71. **30.** Stefan Barensky, "Europe's EELV is Here . . . and Its Whole Family Is Growing Up," *Launchspace,* January/February 1999, 35–38.

31. Antonio Rodota, interview by Norbert Lossau, in "Europe Could Be Number Two in Space Travel," *Die Welt* (Internet edition), 29 June 1998, in FBIS-WEU-98-181, July 1, 1998. **32.** Toffler and Toffler, *War and Anti-War,* 115. **33.** Speech by Robert D. Walpole, "North Korea's Taepo Dong Launch and Some Implications on the Ballistic Missile Threat to the United States," Center for Strategic and International Studies, Washington, D.C., December 8, 1998; David A. Fulghum, "North Korea Plans Booster Tests Soon," *Aviation Week & Space Technology,* November 30, 1998, 24, 25; "Pyongyang Prepared More Surprises to Shock the World," *South China Morning Post,* September 25, 1998; Bill Gertz, "N. Korean Missile Seen Posing Risk to U.S." *Washington Times,* September 16, 1998, p. 1. **34.** James R. Asker, "Growing Pains," *Aviation Week & Space Technology,* October 13, 1997, 17. **35.** Peter B. de Selding, "Newsmaker Forum: Bernard Molard," *Space News,* November 23–29, 1998, 22.

36. A. McLean and A. Swankie, "Helios 2—Myth or Reality?" *Space Policy* 14 (1998): 107–14. **37.** ITAR-TASS World Service, 1439 GMT, 19 November 1997, in FBIS-SOV-97-324, November 20, 1997. **38.** George Ojalehto and Henry Hertzfeld, "Growth, Cooperation Mark 1997 Space Activities," *Aerospace America,* July 1998, 4. **39.** Air Force Scientific Advisory Board, *New World Vistas: Air and Space Power for the 21st Century: Space Applications Volume* (Washington, D.C.: Air Force Scientific Advisory Board, 1996). **40.** John M. Logsdon, "Charting a Course for Cooperation in Space," *Issues in Science and Technology,* Fall 1993,

65–72; Kenneth Pedersen, "Thoughts on International Cooperation and Interests in the Post-Cold War World," *Space Policy,* August 1992, 205–20; George van Reeth and Kevin Madders, "Reflections on the Quest for International Cooperation," *Space Policy,* August 1992, 221–32.

41. See, for example, U.S. Crest, *Partners in Space: International Cooperation in Space: Strategies for the New Century* (Arlington, Va.: Center for Research and Education on Strategy and Technology, May 1993), 62–66. **42.** Cohen, *Annual Report to the President and the Congress,* 1998, 65; Jeff Erlich, "Israelis Will Get Missile Warnings from U.S.," *Defense News,* May 6–12, 1996, 24; Hamish McDonald, "U.S. Offers Insider Access to Spy Data," *Sydney Morning Herald,* August 1, 1998; Jeffrey M. Lenorovitz, "U.S., Russia to Share Missile Warning Data," *Aviation Week & Space Technology,* April 11, 1994, 24, 25. **43.** George I. Seffers, "Hamre Resists Proclamation of Year 2000 Conquest," *Defense News,* October 19–25, 1998, 22; M.J. Zuckerman, "U.S. Aims to Avert Y2K-Induced War," *USA Today,* November 13, 1998; David Hoffman, "Russia's Nuclear Force Sinks with the Ruble," *Washington Post,* September 18, 1998, p. 1; see also Anna Blundy, "Satellite 'Almost Started Nuclear Launch,'" *London Times,* September 23, 1998. **44.** "Share the High Ground?" *Aviation Week & Space Technology,* May 4, 1998, 21. **45.** Warren Ferster, "U.S., Europe to Merge Polar Satellite Efforts," *Space News,* November 23–29, 1998, 4.

46. See, for example, Stephen J. Hedges, "Double Edge of Exports," *Chicago Tribune,* July 20, 1998, p. 1, and Jeff Gerth and Eric Schmitt, "Chinese Said to Reap Gains in U.S. Export Policy Shift," *New York Times,* October 19, 1998, p. 1. **47.** Peter B. de Selding, "Chinese Satellite Technology Quest Shifts to Europe," *Space News,* November 23–29, 1998, 1. **48.** U.S. Congress, Office of Technology Assessment, *U.S.-Russian Cooperation in Space,* OTA-ISS-618 (Washington, D.C.: GPO, April 1995), 12, 19. Other areas of cooperation have included work on the *International Space Station,* U.S. astronaut presence on the Russian space station *Mir* (a record-setting ten U.S. and Russian space travelers gathered together on *Mir* in 1995), and cosmonaut riders on the U.S. space shuttle. See Kathy Sawyer, "'A Wonderful Dream Come True': Atlantis Docks Perfectly with Mir," *Washington Post,* June 30, 1995, p. A3. **49.** ESA Press Release, "ESA/Russia Cooperation: Another Step Forward," N42–97, Paris, 18 November 1997. **50.** Warren Ferster, "India, United States Near Data Agreement," *Space News,* April 1–7, 1996, 6.

51. Aleksandr Sharov, interview by Vladimir Ivanov, in "Russia Was, Is and Will Be a Great Space Power," *Rossiyaskaya Gazeta,* April 11, 1996, in FBIS-UST-96-023, June 25, 1996. **52.** The Pentagon threatened to sue WorldSpace in 1999 if it launched AmeriStar, mainly because it operates in the L-band frequency and is expected to disrupt military and civilian flight testing at eighty-six aircraft ranges using the same frequency. WorldSpace apparently has agreed to adjust the beams on its satellite to minimize potential interference. George I. Seffers, "Planned Satellite Launch Draws DoD Fire," *Defense News,* April 26, 1999, 4; "Military Protest 'Potential' Interference from Worldspace Satellites," *Military Space,* March 29, 1999, 4; "WorldSpace to Adjust Beams on AmeriStar," *Space News,* October 18, 1999, 2. **53.** NPO Yuzhnoye will provide the two-stage Zenit booster and RSC Energia will supply the upper stage and take responsibility for booster integration, launch operations, and range services. Kvaarner will operate the floating launch platform and command ship. William J. Broad, "Offering a Cheaper Ride to Orbit From the Middle of the Ocean," *New York Times,* June 16, 1998, p. C1; Bruce A. Smith, "Sea Launch Mission to Demonstrate

System," *Aviation Week & Space Technology,* March 22, 1999, 74–78. **54.** Craig Covault, "Russian Commercial Satcom Launched," *Aviation Week & Space Technology,* November 30, 1998, 28; see also "U.S. Space-Business Companies Widen Relations with Russian Firms," *Space Business News,* November 25, 1998, 3. **55.** Alasdair McLean, *Western European Military Space Policy* (Aldershot: Dartmouth, 1992), 107–8.

56. James R. Asker, "Hide 'N' Seek," *Aviation Week & Space Technology,* January 25, 1999, 29. **57.** Bill Gertz and Rowan Scarborough, "Inside the Ring: Nuclear Hide-And-Seek," *Washington Times,* July 9, 1999, p. 8. **58.** Peebles, *Corona Project,* 113–20, 140. The operational limitations placed on early reconnaissance satellites prevented them from playing any significant role in the detection of missiles on Cuba. In the early 1960s, the Corona satellites could not be used in time to be useful in fast breaking situations, nor did they have the maneuvering flexibility to enable precise flights over Cuba. **59.** Peebles, *Corona Project,* 174–79; Foley, "Corona Comes in From the Cold," 85. **60.** Brugioni, "Art and Science of Photoreconnaissance," 83.

61. Friedman and Friedman, *Future of War,* 314; Michael Dobbs, "Kissinger Offered China Satellite Data in 1973, Paper Shows," *Washington Post,* January 10, 1999, p. A2. **62.** See some early reports of the Serbs use of rubber tanks during Operation Allied Force to fool NATO targeters. "Serb CCD to Thwart Airpower," *London Times,* June 24, 1999. **63.** The START agreement seeks primarily to reduce ICBMs and associated equipment, heavy bombers equipped to deliver nuclear cruise missiles, and SLBMs and their associated systems. Recognizing the limitations of satellites, the negotiators of the treaty also provided for on-site inspections to assist in such activities as the verification of nuclear warhead levels. **64.** Warren Ferster, "Private Spacecraft Imagery Evolves as Treaty Tool," *Space News,* October 20–26, 1997, 3. **65.** Willliam J. Broad, "Spy Photos of Korea Missile Site Bring Dispute," *New York Times,* January 11, 2000, p. A8.

66. Richard Parker and Michael Zielenzige, "N. Korea Nuclear Facility Suspected," *Philadelphia Inquirer,* November 18, 1998, 1. **67.** Porcher l. Taylor III, "A Caveat on Spy Satellite Diplomacy," *Space News,* September 30–October 6, 1996, 13; see also Office of the Secretary of Defense, *Proliferation: Threat and Response* (Washington, D.C.: Office of the Secretary of Defense, April 1996), 24–28. During the late 1980s, Soviet intelligence satellites apparently unearthed in the late 1980s a Chinese program to develop biological weapons. William J. Broad, "Defector Tells of Soviet and Chinese Germ Weapons," *New York Times,* April 5, 1999, p. 1. **68.** Paul Bedard, "Spy vs. Spy," *U.S. News & World Report,* July 19, 1999; Walter Pincus and Vernon Loeb, "Spy Satellites," *Washington Post,* August 9, 1999, p. A13. **69.** Klass, "CIA Reveals Details of Early Spy Satellites," 173. **70.** Eric Schmitt, "Allies Check Satellite Pictures for Evidence of War Crimes," *New York Times,* May 19, 1999.

71. Richelson, "High Flyin' Spies," 54. **72.** *Executive Summary of the Report of the Commission to Assess the Ballistic Missile Threat to the United States,* July 15, 1997, 1, 4, 5, 22, 23. This Commission was led by former secretary of defense Donald Rumsfeld. **73.** According to a Russian source, "The troop subunits making secret preparations for the explosions at the Alpha nuclear test site in Pokharan (Rajasthan) in conjunction with scientists carefully synchronized their actions with the data provided by national space surveillance and Indian satellite flight monitoring systems. The 'schedule' for foreign reconnaissance satellites' appearances over Pokharan became known using space systems and it also proved possible to identify those areas under surveillance at particular times of the day." ITAR-

TASS report, "Satellites 'Missed' It," *Krasnaya Zvezda,* 19 May 1998, 3, in FBIS-TAC-98-139, 19 May 1998; see also Rahul Bedi, "India's Nuclear Test Preparations Avoid Detection," *Jane's Defence Weekly,* May 20, 1998; and William E. Burrows, "India Knew How to Hide from U.S. Eyes," *USA Today,* May 14, 1998. **74.** Cohen, *Annual Report to the President and the Congress,* 1998, 67. **75.** Klass, "CIA Reveals Details of Early Spy Satellites," 173; Nye and Owens, "America's Information Edge," 20.

76. Department of Defense, *Conduct of the Persian Gulf War,* 240; Phillip Finnegan, "Politics Hinders Joint Gulf Missile Defense," *Defense News,* March 22, 1999, 1. **77.** See, for example, Barbara Opall-Rome, "Israel Seeks Space Services From U.S.," *Space News,* January 24, 2000, 1. **78.** Friedman and Friedman, *Future of War,* 323. **79.** Pamela Hess, "DoD to Share Early Warning Data to Stem Y2K Failures in Russia, China," *Defense Information and Electronics Report,* June 5, 1998, 1. **80.** François Heisbourg, "Intelligence Failures Cast Doubt on 'Star Wars,'" *International Herald Tribune,* October 10–11, 1998.

81. Cited in Ivan A. Vlasic, "The Legal Aspects of Peaceful and Non-Peaceful Uses of Outer Space," in *Peaceful and Non-Peaceful Uses of Space: Problems of Definition for the Prevention of an Arms Race,* ed. Bhupendra Jasani (New York: Taylor & Francis, 1991), 38. **82.** State Department, *Summary of Foreign Policy Aspects of the U.S. Outer Space Program,* June 5, 1962, copy from the Lyndon Baines Johnson Library, 16; document declassified September 5, 1979. COPUOS was and remains an ineffective body; final agreement on space matters such as disarmament cannot be reached unless the major powers agree to it. **83.** McDougall, . . . *the Heavens and the Earth,* 179–94. In 1958, the opinion in the United States that space was to be used for nonmilitary purposes only reached its zenith—" 'space for peace' seemed like an unassailable proposition" (189). Since then, the idea that space must be demilitarized has remained favored only with the political Left in the United States and abroad. See also Vlasic, "Legal Aspects of Peaceful and Non-Peaceful Uses of Outer Space," 39, 40. **84.** Vlasic, "Legal Aspects of Peaceful and Non-Peaceful Uses of Outer Space," 37. **85.** Peter B. de Selding, "Developing Nations Push to Close Technology Gap," *Space News,* August 9, 1999, 6.

86. The activity of deep seabed mining is similar to the exploitation of space resources insofar as seabed exploration activities are not yet sufficiently economically viable to warrant commercial investment, despite development of prototype mining technologies. **87.** K.S. Jayaraman, "Newsmaker Forum: Udipi Ramachandra Rao," *Space News,* April 26, 1999, 30. Rao is the chairman of the UN Committee on Peaceful Uses of Outer Space. **88.** Beijing Xinhua Domestic Service, June 6, 1993, in FBIS-CH, June 15, 1993. **89.** See Oscar Schacter and Christopher C. Joyner, ed., *United Nations Legal Order* (Cambridge: Grotius Publications, 1995), 1:4. Of course, the authority of all international law is diminished if the law is not observed by, or has received negative votes from, affected state parties. **90.** Aron, *Peace and War,* 107.

91. Stephen Gorove, *Developments in Space Law* (Boston: Martinus Nijhoff, 1991), 256–57. **92.** Stephen Gorove observed that "in so far as space law is concerned, such a safety zone around an orbiting space object . . . could arguably be regarded as a violation of the ban on national appropriation in the Outer Space Treaty." Ibid., 282. **93.** Ibid., 266. **94.** See, for example, Colin S. Gray, "Space Is Not a Sanctuary," *Survival,* 25, no. 4 (July/August 1983): 196–206. For an opposing viewpoint, see Bruce M. DeBlois, "Space Sanctuary: A Viable National Strategy," *Airpower Journal* (Winter 1998): 41–57. **95.** Nathan C. Goldman,

American Space Law: International and Domestic (Ames: Iowa State University Press, 1988), 119–82; Gorove, *Developments in Space Law,* 3–15.

96. J.E.S. Fawcett, *Outer Space: New Challenges to Law and Policy* (Oxford: Clarendon Press, 1984), 54–62. **97.** Ibid., 51–54. **98.** Ibid., 61, 62. **99.** Gorove, *Developments in Space Law,* 155. **100.** Nandasiri Jasentuliyana, *Manual on Space Law* (Dobbs Ferry, N.Y.: Oceana Publications, 1979), 1:173–89.

101. Winston S. Churchill, "Will There Be War in Europe—And When?" in *Churchill and War,* ed. Michael Wolff, vol. 1 of *The Collected Essays of Sir Winston Churchill* (Bristol: Library of Imperial History, 1976), 436. See also Steven Lambakis, *Winston Churchill—Architect of Peace: A Study of Statesmanship and the Cold War* (Westport, Conn.: Greenwood Press, 1993), 7–12.

Chapter 3

1. See, for example, Russell Weigley, *The American Way of War: A History of United States Military Strategy and Policy* (Bloomington: Indiana University Press, 1973), in which one may read about how the United States' inclination to adopt strategies of annihilation that are particularly dependent on technology; also Carl H. Builder, *The Masks of War: American Military Styles in Strategy and Analysis* (Baltimore: Johns Hopkins University Press, 1989), who wrote that the air force, also the steward of America's space forces, "could be said to worship at the altar of technology. The airplane was the instrument that gave birth to independent air forces; the airplane has, from its inception, been an expression of the miracles of technology" (19). **2.** Isolationism (in its truest, most perfect form) characterizes the foreign policies of Albania during the cold war and North Korea since the 1950s. Japan and Sweden have practiced a form of cultural isolationism. Switzerland is known for its strategic isolationism. At no point in its history did the United States, the world's "melting pot," isolate itself from the affairs of the world, neither in the areas of culture or economics nor in the areas of foreign policy or strategic affairs. **3.** Felix Gilbert, *To the Farewell Address: Ideas of Early American Foreign Policy* (Princeton, N.J.: Princeton University Press, 1961); Paul Johnson, *A History of the American People* (New York: Harper Collins, 1997), 228–30. **4.** Allan R. Millett and Peter Maslowski, *For the Common Defense: A Military History of the United States of America* (New York: Free Press, 1994), 380–407. According to Paul Johnson, "The American system may have been near-pacifist in the interwar years, and isolationist in the 1930s, but it contrived to produce a generation of outstanding commanders . . . who not only organized victory in the field, in the air, and on the oceans, but also helped to establish American internationalism, and concern for the wellbeing of the entire globe." Johnson, *History of the American People,* 785, 86. **5.** This began in earnest during the Korean War, when President Truman committed the nation to funding military programs to develop capabilities to execute the grand containment strategy. The supplemental budget requests in 1951 served two purposes, according to Truman: "first, to meet the immediate situation in Korea, and, second, to provide for an early, but orderly, buildup of our military forces to a state of readiness designed to deter further acts of aggression." Cited in Millett and Maslowski, *For the Common Defense,* 513.

6. Cohen, *Annual Report to the President and the Congress,* 1998, 5. **7.** Edward S. Corwin, ed., *The Constitution of the United States of America: Analysis and Interpretation*

(Washington, D.C.: GPO, 1953), 284. **8.** William J. Clinton, *A National Security Strategy for a New Century,* Washington, D.C., White House, May 1997, 1. **9.** Accordingly, U.S. *written* policy is to "promote development of the full range of space-based capabilities in a manner that protects our vital security interests. We will deter threats to our interests in space and, if deterrence fails, defeat hostile efforts against U.S. access to and use of space. We will also maintain the ability to counter space systems and services that could be used for hostile purposes against our ground, air and naval forces, our command and control system, or other capabilities critical to our national security." White House, *A National Security Strategy for a New Century,* October 1998, 25, 26. **10.** William S. Cohen, *Annual Report to the President and the Congress* (Washington, D.C.: GPO, April 1997), 199. In March 1999, Deputy Defense Secretary John Hamre testified that the DoD considers space to be "an environment like the land, sea, and air within which military activities will be conducted to achieve U.S. national security objectives." One is left to wonder, however, how much space is "like" these other environments, especially when one considers the limits current national policy places on certain military uses of space for space-control and force-application purposes. See U.S. Senate, *Statement by Deputy Secretary of Defense John J. Hamre before the Senate Armed Services Committee, Strategic Forces Subcommittee,* copy, March 22, 1999, 9.

11. See, for example, the National Military Strategy, John M. Shalikashvili, *Shape, Respond, Prepare Now: A Military Strategy for a New Era* (http://www.dtic.mil/jcs/nms). Many of these challenges are also spelled out in National Defense Panel, *Transforming Defense,* 11–17. **12.** Ultimately, according to former Speaker of the House Newt Gingrich, "we intend to dominate the air, space and sea against anybody and have a mobilizable ground force capable of winning within a reasonable length of time." Speech by Newt Gingrich at the National Policy Conference of the Richard Nixon Center for Peace and Freedom in Washington, D.C., portions of which were reprinted in "Viewpoint," *Aviation Week & Space Technology,* May 1, 1995, 86. **13.** Cohen, *Annual Report to the President and the Congress,* 1998, 67. **14.** See, for example, David W. Ziegler, *Safe Heavens: Military Strategy and Space Sanctuary Thought* (Maxwell AFB, Ala.: Air University Press, June 1998). **15.** Maj. Gen. Robert Dickman, interview by author, Washington, D.C., July 27, 1998.

16. Ron Laurenzo, "Gansler: Forces Must Be There 'Within 24 Hours,'" *Defense Week,* September 7, 1999, 2. **17.** James D. Hessman and Gordon I. Peterson, "'Shape, Respond, and Prepare': Interview with Gen. Henry H. Shelton, Chairman of the Joint Chiefs of Staff," *Sea Power,* February 1999, 9. See also Bryant Jordan, "Overloaded," *Air Force Times,* August 30, 1999, 14. **18.** Cohen, *Annual Report to the President and the Congress,* 1998, 1–8. During the previous decade U.S. armed Services led international enforcements of economic sanctions, "no-fly" zones, and air campaigns to coerce and hold at bay an intransigent Saddam Hussein in Iraq and aggressive Serbs in the Balkans. Such small-scale operations have involved spasms of limited air and missile strikes to punish and coerce stubborn leaderships. On more than one occasion, the United States fired sea-launched and air-launched cruise missiles and let fly attack aircraft to destroy military targets in Iraq following Saddam Hussein's defiance of UN inspectors assigned to root out his weapons of mass destruction. Operation Allied Force (1999), aimed at confounding Serbian designs in Kosovo, was a more intensified air bombing campaign. **19.** According to one assessment, "Urban operations have historically required large numbers of troops while diluting technological advantages, making for extremely tough fighting. Urban structures and human densities vastly complicate targeting and maneuvering. Many of our current weapons are often ineffective in urban environments because of trajectory limitations, built-up areas, subterranean passages, and unobservable targets." National Defense Panel, *Transforming Defense,* 15. See also James R. Callard, "Viewpoint: Aerospace Power Essential in Urban Warfare," *Aviation Week & Space Technology,* September 6, 1999, 110: "Ground forces need the perspective of a city that is obtainable only from air and space if they are to operate there effectively." **20.** Satellite intelligence data, for example, and other forms of military cooperation are potential weapons in the war on drug trafficking in Colombia. Steven Lee Myers, "U.S. Pledges Military Cooperation to Colombia in Drug War," *New York Times,* December 1, 1998.

21. Office of the Secretary of Defense, *Proliferation: Threat and Response,* 1. **22.** National Defense Panel, *Transforming Defense,* 15, 16. **23.** Office of the Secretary of Defense, *Proliferation: Threat and Response,* 53. Presidential Decision Directive 60 is the U.S. government's current nuclear policy document. The top secret directive was signed by President Clinton in November 1997; its existence was reported December 7 of that year. Reportedly, it addresses the issue of nuclear retaliation against a chemical or biological attack. A *Defense News* story reports that "the PDD reflects the current reality, in which an attacker using weapons of mass destruction could face nuclear reprisal." According to U.S. government officials and analysts, "the new PDD leaves in place the policy of so-called negative security assurances. Countries that do not possess nuclear weapons, are not allied with a state that does, and are members in good standing of the Non-Proliferation Treaty are safe from U.S. nuclear reprisals under this policy." Jeff Erlich, "New U.S. Nuclear Policy Maintains Ambiguity," *Defense News,* Jan. 5–11, 1998, 4, 19. **24.** Many of the technologies and materials required to manufacture these weapons are readily available, multi-use, easy to use and learn about, and relatively inexpensive. More than twenty-five countries possess or are developing NBC weapons. Twenty-five countries have ballistic missiles, while others have programs to develop them. More than seventy-five countries possess cruise missiles. Briefing by Peter R. Lavoy, director for Counterproliferation Policy, Office of the Assistant Secretary of Defense for Strategy and Threat Reduction, "Countering the Proliferation and Use of WMD," September 13–16, 1998, given in Amman, Jordan. **25.** Office of the Secretary of Defense, *Proliferation: Threat and Response,* 58.

26. The director of the Central Intelligence Agency, George Tenet, testified in February 1999 that the Taepo Dong 1 ballistic missile had "demonstrated technology that with the resolution of some important technical issues would give North Korea the ability to deliver a very small payload to intercontinental ranges—including parts of the United States—although not very accurately." Cited in James Risen, "C.I.A. Sees a North Korean Missile Threat," *New York Times,* February 3, 1999, p. A6. Tenet stated in February 2000 congressional testimony, "We are all familiar with Russian and Chinese capabilities to strike at military and civilian targets throughout the United States. . . . Over the next 15 years, however, our cities will face ballistic missile threats from a wider variety of actors—North Korea, probably Iran, and possibly Iraq." George Tenet, *Hearing of the Senate Select Intelligence Committee,* February 2, 2000. **27.** Ballistic missile targets may be destroyed using hit-to-kill, or kinetic kill, vehicles, fragmentation warheads, nuclear warheads, or directed energy. **28.** Michael C. Sirak, "Army Plans Exercise with Commercial Pagers at Roving Sands Next Month," *Inside Missile Defense,* May 19, 1999, 1. **29.** Gen. Lester Lyles, director of the

Ballistic Missile Defense Organization, interview by author, Washington, D.C., August 1998. **30.** Defense Science Board, *Joint Operations Superiority in the 21st Century: Integrating Capabilities Underwriting Joint Vision 2010 and Beyond* (Washington, D.C.: Office of the Undersecretary of Defense for Acquisition and Technology, October 1998), 96–98. "Boost phase" is that portion of the flight of a ballistic missile or space vehicle during which the booster and sustainer engines operate. The "ascent phase" is that portion of the missile flight that follows boost phase and ends somewhere before the missile reaches apogee.

31. National Defense Panel, *Transforming Defense,* 13. See also Zalmay M. Khalilzad and John P. White, *The Changing Role of Information Warfare* (Santa Monica, Calif.: RAND, 1999); and John Donnelly and Vince Crawley, "Hamre to Hill: 'We're in a Cyberwar,'" *Defense Week,* March 1, 1999, 1. **32.** See, for example, Nye and Owens, "America's Information Edge," 20–36. **33.** See, for example, Ryan Henry and C. Edward Peartree, "Military Theory and Information Warfare," and Martin C. Libicki, "Halfway to the System of Systems," in *The Information Revolution and International Security,* ed. Ryan Henry and C. Edward Peartree (Washington, D.C.: Center for International and Strategic Studies, 1998), 105–27. **34.** GCCS, according to DoD, "provides nearly 700 locations with secret level functionality and increased capability. GCCS provides an enhanced common operational picture, force status, intelligence support, enemy order of battle, related facility information, and air tasking orders. . . . DoD will evolve toward more integrated and interoperable battle management systems through continued deployment of GCCS below the joint command level and into operational units." Cohen, *Annual Report to the President and the Congress,* 1998, 73. **35.** Arthur Money, assistant secretary of defense for Command, Control, Communications and Intelligence, speech before the Space Transportation Association, 15 October 1998, published in *Space Trans: The Newsletter of the Space Transportation Association* 8, no. 6 (November/December 1998): 1.

36. For a report on the performance of UAVs during Operation Allied Force, see Elizabeth Becker, "They're Unmanned, They Fly Low, and They Get the Picture," *New York Times,* June 3, 1999. According to a NATO officer, the allies received "fabulous imagery of Serb forces, refugee movements and battle-damage assessments." **37.** Cohen, *Annual Report to the President and the Congress,* 1998, 74. **38.** Ibid., 125; emphasis added. **39.** The national military strategy asserts that "a secure C4ISR (Command, Control, Communications, Computers, Intelligence, Surveillance and Reconnaissance) architecture must be designed and developed from the outset for rapid deployment and with joint and multinational interoperability in mind." **40.** Warren Ferster, interview by Keith Hall, *Space News,* January 11, 1999, 22.

41. Interoperability means people, procedures, and equipment operating together in parallel, effectively and efficiently enough under all the stressful conditions of the battlefield. Information sharing (such as early warning of missile launches), seamless interoperability with systems and weapons with coalition's partners, netted sensors and common doctrines and rules of engagement are all fundamental requirements, many of which involve the use of space. As a collector, disseminator, and a "networker" of common information, space is one of the essential ingredients of an agile world-class fighting force that must defend world interests jointly within its own organization and shoulder to shoulder with its coalition partners. Operation Allied Force caused some U.S. officials to voice their concern that the European allies risk becoming "junior partners" in future coalition military campaigns unless they make a concerted effort to acquire more advanced weapons. See William Drozdiak, "Allies Need Upgrade, General Says," *Washington Post,* June 20, 1999, p. 20. **42.** See, for example, excerpts from a speech delivered by Director of Central Intelligence George J. Tenet in "Viewpoint," *Aviation Week & Space Technology,* April 27, 1998, 74; Bradley Graham, "Authorities Struggle to Write the Rules of Cyberwar," *Washington Post,* July 8, 1998, p. 1; Winn Schwartau, *Information Warfare: Chaos on the Electronic Superhighway* (New York: Thunder's Mouth Press, 1994); Johan Benson, "Conversations with Jon Kyl," *Aerospace America,* October 1998, 12–14. **43.** "Banking and financial transactions are enormously dependent on global networks, and 95 percent of American wealth is digitally represented. Public switched networks (PSNs) that control telecommunications links on which these critical public and commercial infrastructures are themselves dependent have shown themselves vulnerable to digital attack and software manipulation. Further, a combination of cost concerns and the superiority of established commercial systems has created a situation in which 95 percent of all military communications travel over commercial systems." Henry and Peartree, "Military Theory and Information Warfare," 116. **44.** According to one prominent report on this subject, "Future opponents may attempt to bypass U.S. technological and material strength within a potential regional theater of operation and attack strategic vulnerabilities that may reside in the [U.S. homeland or in the homelands] of key U.S. allies." Roger C. Molander, Andrew S. Riddile, and Peter A. Wilson, *Strategic Information Warfare: A New Face of War* (Santa Monica, Calif.: RAND, 1996), 37; see also xiv-xvii. See interview with Jacques S. Gansler, undersecretary of defense for acquisition and technology, *The Year in Defense, 1998–1999,* 10. **45.** See Molander, Riddile, and Wilson, *Strategic Information Warfare,* 35, 36.

46. See, for example, Colin S. Gray, "The Influence of Space Power upon History," *Comparative Strategy* 15, no. 4 (1996): 293–308. **47.** White House, *Fact Sheet: National Space Policy,* 6. "These capabilities may also be enhanced by diplomatic, legal or military measures to preclude an adversary's hostile use of space systems and services." **48.** According to Keith Hall, director of the National Reconnaissance Office, "Up to now, space systems have not faced serious threats. As space capabilities grow and are major elements of national strength, adversaries will look for ways to diminish our advantage. Lessons from recent war games suggest *we must have policy formulated well in advance of hostilities.* If systems are attacked, what is our response?" Keith Hall, "Building Space Partnership," *Defense News,* September 29–October 5, 1997. **49.** White House, *Fact Sheet: National Space Policy,* 1, 2. **50.** "The control and utilization of space as a warfighting medium," accordingly, "will help to enable the United States to establish and sustain dominance over an area of military operations. Establishing such dominance will be a key to achieving success during a crisis or conflict." Cohen, *Annual Report to the President and the Congress,* 1997, 199. See also the *Quadrennial Defense Review,* which noted (in section 3) that "we must focus sufficient intelligence efforts on monitoring foreign use of space-based assets as well as develop the capabilities required to protect our systems and prevent hostile use of space by an adversary." National Defense Panel, *Transforming Defense,* noted the growing importance of space and stressed that the United States takes steps to fully exploit the opportunities of space and proactively address associated vulnerabilities (38). Cohen, *Annual Report to the President and the Congress,* 1997, 199, 200; see also Lambakis, "Space Control in Desert Storm and Beyond," 417–33.

51. Steven Lambakis, "Exploiting Space Control," *Armed Forces Journal International,* June 1997, 42–46. **52.** U.S. Space Command Director of Plans, *Long Range Plan,* viii. Estes's recommendation, however, did not have the support of Pentagon decision makers in 1997. "Space as Regional 'AOR' Seen Unlikely to Appear in Command Plan Revision," *Inside the Pentagon,* September 4, 1997, 1. See also Douglas Berenson, "U.S. Space Command Wants Space Declared Its 'Regional" Area of Ops," *Inside the Pentagon,* February 20, 1997, 1. **53.** See, for example, a briefing presented by Christopher D. Lay, "Can We Control Space?" Electronic Industries Association Fall Conference, October 9, 1997. **54.** Joint Chiefs of Staff, *Concept for Future Joint Operations: Expanding Joint Vision 2010* (Washington, D.C.: Joint Chiefs of Staff, May 1997), 60. See also Defense Science Board, *Joint Operations Superiority in the 21st Century.* **55.** "Joint Vision 2010 achieves full-spectrum dominance through the five tenets of dominant maneuver, precision engagement, full-dimensional protection, focuses logistics, and information superiority. These concepts are well rooted in proven principles beginning with the classical principles of war: offensive mass, economy of force, movement, unity of command, security, surprise, and simplicity. It also builds on a series of enduring operational themes: increased precision, lethality, protection, command and control, and awareness; and decreased time, casualties, and collateral damage." Defense Science Board, *Joint Operations Superiority in the 21st Century,* 12, 17, 19.

56. Gil Siegert, defense space consultant, interview by author, Washington, D.C., July 1998. **57.** Henry H. Shelton, "Operationalizing Joint Vision 2010," *Airpower Journal,* Fall 1998, 102–7. See also Elaine M. Grossman, "Information Operations to Play Key Role in New 'Joint Vision 2020,'" *Inside the Pentagon,* March 23, 2000, 1. **58.** High-level Pentagon officials believe that in future conflicts we will see "100 percent utilization" of highly accurate munitions. Vice chief of the U.S. Air Force, Gen. Lester Lyles, cited the "mandate to minimize collateral damage" as the reason for this dependence on precision weapons. Timothy Hoffman, "Vice Chief Cites Importance of Space," *U.S. Air Force ONLINE News,* September 1, 1999 (http://www.af.mil/newspaper/v1_n25_s5.htm). See also Bryan Bender, "USAF Looks to Phase Out Laser-Guided Bombs," *Jane's Defence Weekly,* November 17, 1999. **59.** See, for example, Department of the Air Force, *Global Engagement: A Vision for the 21st Century Air Force,* (Washington, D.C.: Department of the Air Force, 1996); Department of the Air Force, *Organization and Employment of Aerospace Power: Air Force Doctrine Document 2* (Washington, D.C.: Department of the Air Force, September 1998). **60.** Edward G. Anderson III, "The New High Ground," *Armed Forces Journal International,* October 1997, 68; see also William Gregory, "Down to Earth: U.S. Army Increases Attention to Space Issues and Technologies," *Armed Forces Journal International,* December 1997, 12. See also Hunter Keeter, "Costello: DoD to Lose Majority in Space Operations," *Defense Daily,* March 3, 1999, 2.

61. See, U.S. Army Space and Missile Defense Command, Vision 2010 (Huntsville, Ala.: USASMDC, October 1997), 1–4. The U.S. Army prides itself on being the first to launch a U.S. satellite into orbit in 1958. **62.** Department of the Navy, *Forward . . . From the Sea* (Washington, D.C.: Department of the Navy, 1994), 8. **63.** See Colin S. Gray, "Vision for Naval Space Strategy," U.S. Naval Institute *Proceedings,* January 1994, 63–68; Randall G. Bowdish and Bruce Woodyard, "A Naval Concepts-Based Vision for Space," U.S. Naval Institute *Proceedings,* January 1999, 50–53. It is worth noting that when on August 18, 1998, at the annual Naval War College Convocation ceremony, Vice Adm. Arthur K.

Cebrowski (USN), newly installed president of the college, delivered a speech wherein he reviewed the impact of the information age on warfare without once mentioning the role of space. This is astounding, especially in light of the fact that Admiral Cebrowski was former director, Navy Space, Information Warfare, and Command and Control. See his speech in "President's Notes," *Naval War College Review,* Winter 1999, 4–13. **64.** Gray, "Vision for Naval Space Strategy," 63. See also Harold Kennedy, "Navy Agency Wants to Boost Awareness of Space Programs," *National Defense,* June 2000, 20. **65.** The basis for much of the information that follows was drawn from an excellent document produced by the Under Secretary of Defense for Acquisition and Technology, *Department of Defense Space Program: Executive Overview for FY1999–2003* (Washington, D.C.: Department of Defense, February 1998), 12–28.

66. Fulghum, "Growing Intelligence Operation," 50–55; Bryan Bender, "USA to Limit Sale of Satellite Imagery on Security Grounds," *Jane's Defence Weekly,* March 8, 2000. One of the biggest problem facing analysts is, and will be information overload. New technologies and methods are required for sorting through the vast amount of data to identify what is important. **67.** I would like to thank Daniel Goure of the Center for Strategic and International Studies for his ideas in this area. **68.** See then–Lt. Gen. Thomas S. Moorman Jr., USAF, who, as chairman of the DoD Space Launch Modernization Study, published *Space Launch Modernization Plan,* Executive Summary (Washington, D.C.: GPO, April 1994). **69.** Brian Berger, "Reusable Launch Vehicles a Decade Away, NASA Says," *Space News,* June 21, 1999, 3; Craig Covault, "New Shuttle Data Center Speeds Orbiter Processing," *Aviation Week & Space Technology,* August 30, 1999, 50. **70.** A string of expendable launch vehicle failures in 1998 and 1999 has raised the Pentagon's estimation of the value of NASA's reliable space shuttle for launching military payloads. Sig Christenson, "Future Military Missions May Utilize Space Shuttle," *San Antonio Express News,* May 21, 1999.

71. Ferster, interview by Hall, 22; and Johan Benson, "Conversations with Keith Hall," *Aerospace America,* January 1999, 16–18. **72.** Craig Covault, "Space-Based Radars Drive Advanced Sensor Technologies," *Aviation Week & Space Technology,* April 5, 1999, 49, 50; Jeremy Singer, "Fiber Optics Boom Complicates NSA Satellite Strategy," *Space News,* June 12, 2000, 1. **73.** Thomas S. Moorman Jr., "The Explosion of Commercial Space and the Implications for National Security," *Airpower Journal,* Spring 1999, 6–20. See also Robert Wall, "USAF Readies Secure Satcom Designs," *Aviation Week & Space Technology,* August 30, 1999, 60. **74.** Office of the Under Secretary of Defense for Acquisition and Technology, *Department of Defense Space Program,* 16–21. See also the website of the Federation of American Scientists at www.fas.org/spp/military/program. **75.** For a prescient discussion of many of these issues, see Commission on Integrated Long-Term Strategy, *The Future Security Environment,* October 1988, 70, 71. The issue of access to bases and the ability of U.S. forces to perform fundamental missions overseas has significant relevance and implications for space-based force application weapons. According to this report (59), loss of overseas bases will mean "more money for systems that can perform those missions without overseas bases may be required. Our forces' flexibility and mobility would become even more important, and technological innovation might create or enhance systems with those attributes. More communications relay satellites and more systems with ultra-long ranges may be necessary, which may force a restructuring of our Navy and Air Force."

76. Ivan Bekey, "Force Projection from Space," in Air Force Science Advisory Board,

New World Vistas, 83; see also a perspective on the potential threat to the United States from space combat platforms in Kenneth Roy, "Ship Killers: From Low Earth Orbit," U.S. Naval Institute *Proceedings,* October 1997, 40–43. **77.** See, for example, Robert Walls, "USAF Weighs Multi-Role ICBM," *Aviation Week & Space Technology,* October 18, 1999, 34. **78.** David A. Fulghum and Robert Wall, "Weapons, Intelligence Targeted in Probe," *Aviation Week & Space Technology,* July 29, 1999, 69; Friedman and Friedman, *Future of War,* 341. Some of the other programs undertaken by the U.S. services and NASA to demonstrate hypersonic technologies include the U.S. Air Force's Hypersonic Technology Program, NASA's Hyper-X research vehicle and its Future-X focus on a space plane demonstrator, and the U.S. Navy's Fasthawk program to build the next-generation Tomahawk land-attack cruise missile and the High Speed Strike System, a hypersonic missile. See David G. Wiencek, *The Fast Things in the Sky: Hypersonic Missiles and Aerial Vehicles; A Technology for the 21st Century* (Lancaster, UK: Centre for Defence and International Security Studies, 1998). See also Craig Covault, "'Global Presence' Objective Drives Hypersonic Research," *Aviation Week & Space Technology,* April 5, 1999, 54–56, and John A. Tirpak, "Mission to Mach 5," *Air Force Magazine,* January 1999, 28–33. **79.** This case was cited by James Oberg in Bryant Jordan, "Will the Air Force Lose Its Space Program," *Air Force Times,* February 8, 1999, 26. **80.** See, for example, Krepinevich and Kosiak, "Smarter Bombs, Fewer Nukes," 26–32. The Defense Science Board cited in 1998 operational advantages provided by long-range aviation. "The ability to strike from global ranges reduces some of the constraints associated with potential theater basing restrictions and reduces the vulnerability of the force to attack." Global space strike capabilities, given adequate weapons and reconnaissance constellations, could improve on some these operational advantages, especially when one considers that what is sought is rapid initial response (satellites on-orbit are on-call for missions), reduced numbers of forward-based forces, an ability to operate from bases beyond range of enemy strikes, an ability to hit mobile and fixed targets, and an ability to reinforce forward-presence enabling forces. More of a challenge to future defense planners may be trying to develop a space platform having a large "payload capacity" and capable of being replenished (with energy) or restocked with projectiles. Defense Science Board, *Joint Operations Superiority in the 21st Century,* 23.

81. Although currently there is no funding to develop the SOV and SMV concepts, some congressmen are on record supporting the eventual development and acquisition of a military space plane. See Defense Science Board, *Joint Operations Superiority in the 21st Century,* 30–32. See also Marilyn Haddrill, "Air Force Eyes Further SMV Development," *Space News,* August 17–23, 1998, 18. **82.** Gen. Bernard Schriever (ret.), USAF, believes that deploying space weapons on demand would contribute to deterrence and military effectiveness. Interview by author, Washington, D.C., August 28, 1998. **83.** Richard Lardner, "Pentagon Eyes Info Operations in Bid to Defeat Deeply Buried Targets," *Inside the Air Force,* September 3, 1999, 1. **84.** Defense Science Board, *Joint Operations Superiority in the 21st Century; Integrating Capabilities Underwriting Joint Vision 2010 and Beyond* (Washington, D.C.: Office of the Under Secretary of Defense for Acquisition and Technology, October 1998), 2:32–34. According to the report, "The successful development of this system [an orbiting vehicle carrying long thin rods in highly elliptical orbits] would provide the United States with a capability to strike targets anywhere on the globe within six to nine hours depending upon constellation size and orbital parameters. In lieu of forward basing, ballistic delivery of precision weapons from space is the only feasible way to assure prompt attack of targets anywhere on the globe within the opening hours of the war and without using systems currently countable under the START Treaty." **85.** Although not a very significant point, there is some question as to whether a ballistic missile ought to be considered a space weapon. A ground-based ballistic missile (as opposed to a fractional orbital bombardment system) does not exploit orbit in any significant way to reach its target. If it is not based in space, it seems only logical that a weapon system would have to rely for its performance on the physics of the space environment, or be designed to achieve the interception (e.g., for reconnaissance or destruction) of a space object, to qualify as a true *space* weapon. We do not consider ground-launched arrows or artillery air weapons, even though they travel through the air medium to reach their targets. Why must we consider a ballistic missile a space weapon?

86. Long-wave lasers use a chemical reaction to generate energy that may be directed and focused against a target. Shorter wavelength lasers would be more powerful (and more likely to penetrate the atmosphere), and generate a coherent electrical energy from a nuclear explosion. X-ray lasers, although never shown to be technically feasible, would self-destruct after generating several small explosions and releasing concentrated energy against a target. Particle beam weapons would use chemical or nuclear reactions to generate a beam that would destroy the internal workings of the target. And finally, chemically or nuclear-generated high power microwave (HPM) weapons would focus radiation to disrupt electronic circuitry. Friedman and Friedman, *Future of War,* 351–55; BGL, "Directed Energy Weapons—Where Are They Headed?" *Physics Today,* August 1983, 17–20; Clarence A. Robinson Jr., "Beam Weapons Technology Expanding," *Aviation Week & Space Technology,* May 25, 1981, 40–47; "Pentagon Studying Laser Battle Stations in Space," *Aviation Week & Space Technology,* July 28, 1980, 57–62. **87.** See, for example, Bekey, "Force Projection from Space," 83–87. **88.** See, for example, William B. Scott, "Advances Enable Space Vehicles, Lasers," *Aviation Week & Space Technology,* April 5, 1999, 50. **89.** Suzann Chapman, "Space Junk," *Air Force Magazine,* November 1996, 39; see also Judy Donnelly and Sydelle Kramer, *Space Junk: Pollution Beyond the Earth* (New York: Morrow Junior Books, 1990). AlliedSignal Technical Services Corporation recently completed a three-year conversion of Space Command's central data library of satellites and orbital debris putting it into an electronic format. The conversion involved fifteen million pages of documentation, 1,800 technical manuals, 600,000 drawings and 300 specialized documents. See "Space Control Shuns Paper," *Aviation Week & Space Technology,* August 30, 1999, 53. **90.** In a 1985 orbital test, the F-15 homing kill vehicle destroyed an old satellite, casting about 285 trackable pieces of space junk.

91. Air Force Space Command, "Fact Sheet: Ground-Based Electro-Optical Deep Space Surveillance" (Peterson Air Force Base: Directorate of Public Affairs, May 1997). **92.** U.S. Space Command, *Space Surveillance,* http://www.spacecom.af.mil/usspace/space.htm. **93.** The commander in chief of U.S. Space Command, Gen. Richard Myers, sees surveillance of space as the "No. 1" space-control priority. Warren Ferster, "Profile: Gen. Richard Myers," *Space News,* May 17, 1999, 22. See also Howell M. Estes III, "Protecting U.S. Assets in Space," *International Space Industry Report,* June 8, 1998, 30. William B. Scott, "New Satellite Sensors Will Detect RF, Laser Attacks," *Aviation Week & Space Technology,* August 2, 1999, 57, 58. **94.** William B. Scott, "Cincspace Wants Attack Detectors on Satellites," *Aviation Week &*

Space Technology, August 10, 1998, 22–24. **95.** John J. Hamre, Deputy Secretary of Defense, *Statement before the Senate Armed Services Committee, Strategic Forces Subcommittee,* March 22, 1999, 32. Strictly speaking, all trade (trade over land, on the seas, in the air) requires global cooperation, but commercial interests have not stopped attack operations in these other environments. Also, the emphasis on tactical, or temporary, denial of space to an enemy is a suspiciously politically correct approach to a rather serious, even life-threatening problem. Non-lethal, non-destructive means may work under certain limited circumstances. But the fact remains, denial in a crisis or war may require destruction.

96. Stares, *Militarization of Space,* 81, 96, 107–9, 127. **97.** Warren Ferster, "U.S. Army to Study Making Satellite Killer Less Lethal," *Defense News,* October 26–November 1, 1998, 38. **98.** Bryan Bender, "Russia Concerned About U.S. Space Laser Test," *Defense Daily,* October 17, 1997, 3. More than likely, the numbers quoted here refer to the spread of laser technologies capable of temporarily blinding satellites. The number of countries capable of destroying satellites is much lower. **99.** Michael A. Dornheim, "Laser Engages Satellite, with Questionable Results," *Aviation Week & Space Technology,* October 27, 1997, 27; Daniel G. DuPont, "Army Continuing to Refine Laser's Ability to Track, Defeat Satellites," *Inside the Army,* November 30, 1998, 1, 8–11. **100.** David A. Fulghum, "USAF Aims Laser at Antimissile Role," *Aviation Week & Space Technology,* August 14, 1995, 24.

101. John A. Tirpak, "Military Lasers High and Low," *Air Force Magazine,* September 1999. **102.** The fact that the U.S. Space Command's *Long Range Plan* repeatedly makes reference to "the notion of weapons in space is not consistent with national policy" is a clue that not much is being done in the space-control area, even in the classified world. **103.** Ferster, "Profile: Gen. Richard Myers," 22. The general indicated that improvements still are needed, including development of a capability to derive accurate coordinates for precision guided munitions from national and in-theater imagery without having to turn first to a central site in the United States. See also Bryan Bender, "Allies Still Lack Real-Time Targeting," *Jane's Defence Weekly,* April 7, 1999; Craig Covault, "Military Space Dominates Air Strikes," *Aviation Week & Space Technology,* March 29, 1999, 31–33; Jan Wesner Childs, "'Space Age' Hits Albania," *European Stars and Stripes,* June 2, 1999, 5. **104.** Bob Smith, "The Challenge of Spacepower," Speech before the Fletcher School/Institute for Foreign Policy Analysis, Cambridge, Mass., November 18, 1998, unpublished copy, 3. **105.** Greg Caires, "Estes: AF Must Integrate Air and Space or Risk Splitting Service, *Defense Daily,* July 30, 1998, 3. Estes is also on record stating that, should the air force not undertake a major restructuring of its budget, the service can "kiss the space mission goodbye." Cited in David Atkinson, "Estes Says Air Force Budget Must Be Realigned for Space Ops," *Defense Daily,* January 13, 1999, 1. See also Office of Assistant Secretary of Defense for Public Affairs, "Air Force Announces Force Structure Changes," News Release No. 088-99, March 5, 1999.

106. Schriever interview. General Schriever, widely considered to be the father of the United States' space and missile force, made the following observation back in 1957: "In the long haul our safety as a nation may depend upon our ability to achieve 'space superiority.' Several decades from now the important battles may not be sea or air battles, but space battles, and we should be spending a certain fraction of our national resources to ensure that we do not lag behind in obtaining space supremacy." There are, of course, many who would argue otherwise, especially those who still wear the air force uniform. Lester Lyles, former director of the Ballistic Missile Defense Organization and then lieutenant general, said in 1998, "I don't envision, nor do I desire that we get to a separate space force. We need to work as a team. Think of space as an extension of the air. People need to realize its benefit, that they utilize it and take advantage of it, work it into warplanning and other activities. We need to see space become a seamless, invisible part of the way we operate today (as opposed to seeing it become completely different and separate). The Air Force is trying to see that every one gets air and space education." **107.** Col. Simon "Pete" Worden, interview by author, Washington, D.C., August 4, 1998. Worden argued that JV2010 basically ignored space, even as it talked about the Revolution in Military Affairs. See also Grossman, "Information Operations to Play Key Role," 1. **108.** I wish to thank Christopher Lay of SAIC and Daniel Goure of the Center for Strategic and International Studies for sharing their thoughts on this subject. **109.** The Defense Science Board has concluded that the SBL may be able to target low-flying aircraft and ground-based missile launchers, in addition to doing laser designation for precision-guided bombs and attacking satellites in LEO. "Zapper," *Aviation Week & Space Technology,* May 24, 1999, 27. **110.** Johan Benson, interview by Daniel Hastings, *Aerospace America,* July 1998, 14–16. See also Moorman, "Explosion of Commercial Space," 18. Moorman writes that the Pentagon and the air force will have great difficulty in funding new space requirements. "The problem," he notes, "lies in affording new initiatives while maintaining basic space services in the face of a flat or declining DOD budget." New initiatives, according to Moorman, might entail space asset protection programs, development of new technologies to contribute to the development of a space-based radar to supplement or replace the AWACS and JSTARS, and programs to develop satellite negation systems to protect U.S. forces.

111. For further expansion of these ideas, see Gray, "Influence of Space Power upon History," 293–308.

Part 2

1. Joint Chiefs of Staff, *Concept for Future Joint Operation,* 15.

Chapter 4

1. Figures are in fiscal year 1998 constant dollars. Tamar A. Mehuron, ed., "Space Almanac," *Air Force Magazine,* August 1999, 31. **2.** U.S. Space Command, "Satellite Box Score," February 17, 1999, http://www.spacecom.af.mil/usspace/boxscore.htm; Estes, "Protecting U.S. Assets in Space," 30. **3.** Paul Mann, "NASA, Air Force Seek Funds for New Orbiter, Expendables," *Aviation Week & Space Technology,* March 3, 1986, 15. **4.** "USAF May Increase Titan 34D-7 Production," *Aviation Week & Space Technology,* February 17, 1986, 26. See also "Delta Launchers," *Aviation Week & Space Technology,* February 17, 1986, 22. **5.** According to one European space official at the time, "With the Delta launch failure that followed the Titan 34D explosion and the space shuttle tragedy, Ariane remains as the only Western launcher currently on an operational status." The official overlooked the availability of the U.S. Atlas booster, although the point made is well taken. "Europeans Plan Eight Missions/Year for Ariane," *Aviation Week & Space Technology,* May 12, 1986, 22.

6. "Titan Solid Booster Failure Caused Vandenberg Accident," *Aviation Week & Space Technology,* May 5, 1986, 24, 25. At the time, then Maj. Gen. Kutyna was USAF director of

Space Systems, Command, Control and Communications. **7.** "Several U.S. Military Spacecraft Operating on Final Backup Systems," *Aviation Week & Space Technology,* March 30, 1987, 22, 23. For a review of technical lessons from past launch failures, see Charles Gunn, "Failures Are Not an Option," *Launchspace,* January/February 1999, 24–29. **8.** Warren Ferster, "Failures Trigger Industry Scrutiny," and "Failures at a Glance," *Space News,* May 17, 1999, 3; Craig Covault, "Reviews Advance as New Satellite Fails," *Aviation Week & Space Technology,* May 24, 1999, 61, 62; David Atkinson, "White House to Take on Space Review Role," Defense Daily, May 18, 1999, 4; Kathy Sawyer, "Rocket Failures Shake Faith in Space Industry," *Washington Post,* May 11, 1999, p. 1; Craig Covault, "Titan, Delta Failures Force Sweeping Reviews," *Aviation Week & Space Technology,* May 10, 1999, 28–30; William B. Scott, "Launch Failures Cripple U.S. Space Prowess," *Aviation Week & Space Technology,* May 3, 1999, 31, 32; Tom Raum, "Lawmakers Urge U.S. to Upgrade Satellite Launch Capabilities," *Washington Times,* July 10, 1999, p. C10; Gigi Whitley, "NRO Chief Details Organization, Membership of Space Launch Panel," *Inside the Air Force,* July 9, 1999, 5. **9.** Theresa M. Foley, "Loss of Satellite Hinders Weather Forecasting System," *Aviation Week & Space Technology,* May 12, 1986, 23, 24; James R. Asker, "First Goes-Next Set for Launch," *Aviation Week & Space Technology,* April 4, 1994, 73; and James R. Asker, "Satellites Spawn New Weather Data," *Aviation Week & Space Technology,* May 1, 1995, 27. **10.** According to William Perry, prior to his becoming secretary of defense, "Because C3I was so significant [in the war], it will be carefully studied by other nations, not only to learn how to emulate it, but also how to counter it. Many of the C3I systems used in Desert Storm could be degraded by foreseeable countermeasures. . . . C3I got a "free-ride" to some extent, because inexplicably they were not subjected to countermeasures. But this is unlikely to happen again." William J. Perry, "Desert Storm and Deterrence," *Foreign Affairs* 70, no. 4 (Fall 1991): 78, 79; "NRO Worries that Techniques in Yugoslavia May Have Been Exposed," *Aerospace Daily,* July 9, 1999.

11. Colin S. Gray, "Space Power Survivability," *Airpower Journal* 7, no. 4 (Winter 1993): 27–42. **12.** Dietrich Rex, "Will Space Run Out of Space? The Orbital Debris Problem and Its Mitigation," *Space Policy* 14 (1998): 95–105; Leonard David, "Low-Power Lasers Might Counteract Orbital Debris," *Space News,* October 21–27, 1996, 6. **13.** Air Force Scientific Advisory Board, *New World Vistas,* 81. **14.** Alan Ladwig, senior adviser to the NASA administrator, interview by Richard W. Scott Jr. **15.** Catherine MacRae, "4.7 Billion Space Weather Architecture Transition Plan Completed," *Inside the Pentagon,* November 18, 1999, 23–25; William B. Scott, "Rad-Hard Chip Industry Recovering," *Aviation Week & Space Technology,* November 9, 1998, 90–93; "High-Energy Threat in Space Detailed," *Aviation Week & Space Technology,* August 22, 1994, 24; Bruce D. Nordwall, "Natural Hazards Threaten Milsats," *Aviation Week & Space Technology,* April 5, 1999, 59, 60.

16. "Space Debris, Watch Out," *Military Space,* October 25, 1999, 1; Brian Berger, "Earth Heads for Comet Tempel-Tuttle's Wake," *Space News,* November 2–8, 1998, 8; William B. Scott, "Leonids Shower Triggers New Look at Space Debris," *Aviation Week & Space Technology,* January 4, 1999, 51, 52, 55; Paula Shaki, "Mild Leonid Meteor Activity Catches Few by Surprise," *Space News,* November 23–29, 1998, 6; Ben Ianotta, "U.S. Officials Warn Satellite Operators of Leonid Storm," *Space News,* November 15, 1999, 6. **17.** The www.heavens-above.com Internet web site advertises a capability to locate any satellite at any given moment. **18.** Cited in Leonard David, "New Software Enables Amateurs to Track Satellites," *Space News,* August 12–18, 1996, 8. See also Alan Thomson, "Satellite Vulnerability:

A Post-Cold War Issue?" *Space Policy,* February 1995. **19.** Vernon Loeb, "Hobbyists Track Down Spies in Sky," *Washington Post,* February 20, 1999, p. 1. **20.** Thomson, "Satellite Vulnerability," 19–26.

21. Joseph C. Anselmo, "Satellite Builders Fear Export Nightmare," *Aviation Week & Space Technology,* February 22, 1999, 24; see also Michael Hirsh, "The Great Technology Giveaway? Trading with Potential Foes," *Foreign Affairs,* September/October 1998, 2–9. **22.** Department of Defense, *Military Forces in Transition* (Washington, D.C.: Office of the Secretary of Defense, 1991), 44; U.S. Congress, Office of Technology Assessment, *Anti-Satellite Weapons, Countermeasures, and Arms Control,* OTA-ISC-281 (Washington, D.C.: GPO, September 1985), 52. The Soviet ASAT was a *co-planar system.* Launches had to be timed to coincide with the passage of the target's ground track [orbital plane] over the launch site at Tyuratam Cosmodrome. See also William J. Durch, *Anti-Satellite Weapons, Arms Control Options, and the Military Use of Space,* Center for Science and International Affairs, John F. Kennedy School of Government, Harvard University, 1984, 11–13. **23.** Aaron Karp, *Ballistic Missile Proliferation the Politics and Technics,* SIPRI (Oxford: Oxford University Press, 1996), 17; see also W. Seth Carus, *Ballistic Missiles in Modern Conflict,* Center for Strategic and International Studies (Westport, Conn.: Praeger, 1991), 20–21, 26; Maj. Thomas A. Torgerson, USAF, *Global Power Through Tactical Flexibility Rapidly Deployable Space Units,* (Maxwell AFB, Ala.: Air University Press, June 1994), 11–15. **24.** See National Intelligence Council, *Foreign Missile Developments and the Ballistic Missile Threat to the United States Through 2015,* mimeo, September 1999; Paul Beaver, "China Prepares to Field New Missile," *Jane's Defence Weekly,* February 24, 1999; Dov S. Zakheim, "Old Rivalries, New Arsenals: Should the United States Worry? *IEII Spectrum,* March 1999, 29–31. **25.** A rogue state might be able to acquire a long-range ballistic missile to attack the United States, but to do so without being detected is unlikely according to U.S. Army general Henry Shelton, chairman of the Joint Chiefs of Staff. "Rogue Country Could Secretly Buy Long-Range Missile, Shelton Says," *Aerospace Daily,* June 8, 1998, 379.

26. John Pike and Eric Stambler, "Anti-Satellite Weapons and Arms Control," in *Encyclopedia of Arms Control and Disarmament,* ed. Richard Dean Burns (New York: Charles Scribner's Sons, 1993), 2:998. **27.** Thomson, "Satellite Vulnerability," 25. **28.** G. Harry Stine, *Confrontation in Space,* (Englewood Cliffs, N.J.: Prentice-Hall, 1981), 82, 83. **29.** Thomson, "Satellite Vulnerability," 25, 26. **30.** The Aspen Strategy Group, *Anti-Satellite Weapons and U.S. Military Space Policy* (Washington, D.C.: University Press of America, 1986), 28. See also Yevgeni Velikhov, Roald Sagdeev, and Andrei Kokoshin, *Weaponry in Space* (Moscow: Mir Publishers, 1986), 99.

31. Lambakis, "Space Control in Desert Storm and Beyond," 429, 430; Steven Lambakis, "The United States in Lilliput: The Tragedy of Fleeting Space Power," *Strategic Review* (Winter 1996): 35–36; Maj. David W. Ziegler, USAF, *Safe Heavens—Military Strategy and Space Sanctuary Thought* (Maxwell AFB, Ala.: Air University Press, 1998), 24. **32.** Durch, *Anti-Satellite Weapons,* 14; Stares, *Militarization of Space,* 108. **33.** JAYCOR, *Analysis of the Feasibility of Degrading U.S. National Technical Means and Affecting Associated Communications by the Effects of a High Altitude Nuclear Detonation,* report prepared for the Defense Nuclear Agency, DNA001–91–C-0030 (Vienna, Va.: JAYCOR, 23 December 1994), 2–4, 9. **34.** Bill Gertz and Rowan Scarborough, "Russian ASAT," *Washington Times,* June 18, 1999, p. 9; and John Donnelly, "New Chips to Make Satellites Nuke-Proof," *Defense Week,* December 7, 1998, 1; Roger Fontaine, "EMPs No Longer Science Fiction," *Washington Times,* July 14, 1997, p. 10. **35.** Robert Wall, "Directed Energy Arms Sought by China," *Aviation Week & Space Technology,* November 9, 1998, 35; Bill Gertz, "Chinese Army Is Building Lasers," *Washington Times,* November 3, 1998, pp. A1, A9; Select Committee on U.S. National Security and Military/Commercial Concerns with the People's Republic of China, *U.S. National Security and Military/Commercial Concerns with the People's Republic of China* (Washington, D.C.: GPO, 1999), xiii, 11, 12 (hereafter cited as the Cox Committee Report); Robert Wall, "Panel Says Spying Aided Conventional Weaponry," *Aviation Week & Space Technology,* May 31, 1999, 31, 32.

36. Paul B. Stares, *Space and National Security* (Washington, D.C.: Brookings Institute, 1987), 75–77; Durch, *Anti-Satellite Weapons,* 10. **37.** Steven R. Peterson, *Space Control and the Role of Antisatellite Weapons* (Maxwell AFB, Ala.: Air University Press, May 1991), 38; Department of Defense, *Military Forces in Transition,* 44. **38.** Lyles interview; cited in Robert Wall, "Intelligence Lacking on Satellite Threats," *Aviation Week & Space Technology,* March 1, 1999, 54. **39.** Arthur F. Manfredi, interview by Richard W. Scott Jr.; John Sheldon, James Kiras, and Colin S. Gray, "Any Time, Any Place—Even Space? The Use of Special Operations Forces for Space Control and Space Denial" (unpublished research project proposal, University of Hull, England, June 1998), 2. **40.** Air Force Scientific Advisory Board, *New World Vistas,* 80.

41. Durch, *Anti-Satellite Weapons,* 13. **42.** Gerald R. Hust, *Taking Down Telecommunications* (Maxwell AFB, Ala.: Air University Press, 1994), 10. **43.** See Richard J. Newman, "The New Space Race," *U.S. News and World Report,* November 8, 1999, 30: Newman writes that the Pentagon believes the Chinese may have a GPS interdiction capability by 2010, and that Russian businesses are advertising the sale of hand-held GPS jammers having a range of 150 miles. See also Thomas G. Seigel, "The Application of Space to Military and Naval Operations," *Naval War College Review,* Winter 1994, 119; Bryan Bender, "GPS Rethink Likely After Operation 'Allied Force,'" *Jane's Defence Weekly,* April 28, 1999. **44.** Wall, "Intelligence Lacking on Satellite Threats," 54. **45.** Victor Villhard, White House Office of Science and Technology, interview by Richard W. Scott Jr.

46. Col. Marc C. Johansen (ret.), USAF, interview by Richard W. Scott Jr. **47.** Damon R. Wells, State Department telecommunications expert, interview by Richard W. Scott Jr. See also Hust, *Taking Down Telecommunications,* 19. **48.** Kamran Khan and Pamela Constable, "Bomb Suspect Details Anti-U.S. Terror Force," *Washington Post,* August 19, 1998, pp. A1, A25. **49.** Manfredi, interview by Scott. **50.** James Hackett, "Clinton's Unilateral Disarmament Move," *Washington Times,* October 29, 1997, p. 19.

51. Lyles interview. **52.** Christopher Lay, interview by author, Arlington, Va., June 3, 1998. **53.** See John M. Collins, *Military Space Forces: The Next Fifty Years* (Washington, D.C.: Congressional Research Service, October 12, 1989), 58. This discussion of countermeasures used information from the following sources: Congressional Research Service, *U.S. Military Satellites and Survivability,* Library of Congress, February 26, 1986, CRS-1–CRS-10; Colin S. Gray, *American Military Space Policy Information Systems, Weapon Systems and Arms Control* (Cambridge, Mass.: Abt Books, 1982), 70–71; Stares, *Space and National Security,* 84–85; Thomson, "Satellite Vulnerability," 27–29. Petersen, *Space Control and the Role of Antisatellite Weapons,* 72. Advisory Group for Aerospace Research and Development, *Tactical Applications of Space Systems* (Neuilly sur Seine, France: U.S. Department of Commerce, National Technical Information Service, October 1989), 25–30; JAYCOR, *Analysis of the Feasibility;* Pierre Lellouche, *Satellite Warfare: A Challenge for the International Community* (New York: United Nations Institute for Disarmament Research, 1987), 13–14; Arthur F. Manfredi, "U.S. Space Systems Survivability," *National Defense,* September 1988, 45–47; Durch, *Anti-Satellite Weapons,* 14, 15; Friedman and Friedman, *Future of War,* 365–66. **54.** JAYCOR, *Analysis of the Feasibility,* 35. **55.** Mowry interview.

56. Cited in Atkinson, "Estes Says Air Force Budget Must Be Realigned," 1. SPIN-2 executives, on the other hand, are not concerned about the potential vulnerability of their systems or survivability in a hostile environment. They do expect the United States to step in for protection. They would accept losses as a business risk. Moreover, their operations are insured. Norris interview. **57.** The launch sites are also at the mercy of the weather, as Hurricane Floyd reminded NASA's managers in 1999. See Craig Covault, "Floyd Disrupts Space, Aviation Operations," *Aviation Week & Space Technology,* September 20, 1999, 38, 39; and "Minimize Threat to Shuttles," *Space News,* October 4, 1999, 12. **58.** Lt. Gen. Lance Lord, USAF, "Range Modernization: The Rest of the Story," *Space News,* March 22, 1999, 13. **59.** Collins, *Military Space Forces,* 60. **60.** Robert Wall, "Darpa to Demonstrate new Satellite Concept," *Aviation Week & Space Technology,* December 6, 1999, 30, 31.

61. John H. Campbell, associate director of flight projects for Hubble Space Telescope, Goddard Space Flight Center, interview by Richard W. Scott Jr., Washington, D.C., April 1999. **62.** Scott, "Cincspace Wants Attack Detectors," 22–24; Estes, "Protecting U.S. Assets in Space," 30. **63.** Albert Wohlstetter and Brian G. Chow, *Self-Defense Zones in Space,* Report prepared by Pan Heuristics, R&D Associates, contract No. MDA903–84–C-0325, July 15, 1986; Aspen Strategy Group, *Anti-Satellite Weapons,* 22, 26. Col. Thomas Oldenberg (ret.), USAF, interview by Richard W. Scott Jr, Rosslyn, Va., July 29, 1998. **64.** For a fuller discussion of classic problems associated with ASAT arms control, see Ronald Reagan, *Report to Congress on U.S. Policy on ASAT Arms Control,* H. Doc. 98-197, 98th Cong., 2d sess. (Washington, D.C.: GPO, 1984), 11–15; see also Patrick J. Garrity, Raymond A. Gore, Robert E. Pendley, and Joseph F. Pilat, *Monitoring Space Weaponry: Detection and Verification Issues,* LA-11278 (Los Alamos, N.M.: Center for National Security Studies, Los Alamos National Laboratory, April 1988); Paul B. Stares, "The Problem of Non-Dedicated Space Weapon Systems," in Jasani, *Peaceful and Non-Peaceful Uses of Space,* 147–55. **65.** See, for example, Aspen Study Group, *Anti-Satellite Weapons and U.S. Military Space Policy* (Washington, D.C.: University Press of America, 1986); Paul B. Stares, "Arms Control in Space," in Stares, *Space and National Security,* 142–86; Donald L. Hafner, "Averting a Brobdingnagian Skeet Shoot: Arms Control Measures for Anti-Satellite Weapons," *International Security* 3, no. 5 (Winter 1980/81): 41–60; Richard L. Garwin, Kurt Gottfried, and Donald L. Hafner, "Antisatellite Weapons," *Scientific American* 250, no. 6 (July 1984): 45–55; Union of Concerned Scientists, *Anti-Satellite Weapons: Arms Control or Arms Race?* (Cambridge, Mass.: Union of Concerned Scientists, 1984).

66. Cited in Grossman, "Information Operations to Play Key Role," 1. **67.** Leading defense officials are concerned that the United States does not have the intelligence, surveillance, and reconnaissance (ISR) assets to fight two major wars simultaneously. See Frank Wolfe, "Hamre: Pentagon Needs to Examine Use, Force of ISR Assets," *Defense Daily,* September 10, 1999, 3. **68.** *Contrails: The Air Force Cadet Handbook* (Colorado Springs: U.S.

Air Force Academy, 1974), 20:17, cited in Thomas D. Bell, *Weaponization of Space: Understanding Strategic and Technological Inevitabilities* (Maxwell AFB, Ala.: Air University Press, 1999), 6. **69.** Cited in Bill Gertz and Rowan Scarborough, "Future War," *Washington Times,* October 22, 1999, p. 9. **70.** Future reconnaissance satellites may be built with serviceability features, to include fuel ports. Concepts involving autonomous maintenance and resupply satellites are under exploration by the Defense Advanced Research Projects Agency. See Paula Shaki Trimble, "Robotic Spacecraft May Service Future U.S. Spy Satellites," *Space News,* October 25, 1999, 4.

71. The United States' former Space Command chief noted, "Any threat to our use of space is a threat to our nation's security. Here's where the United States military must play an important role. As we have protected national and economic security on land, sea and air for more than 200 years, we must be prepared to defend our interests in space tomorrow." Estes, "Protecting U.S. Assets in Space," 30. **72.** Cited in William Scott, "Space Ops Threatened by Launch Failures," *Aviation Week & Space Technology,* May 17, 1999, 26. Russia's sales of GPS-jamming equipment and the disputes of geosynchronous orbital slots confirm that space is a place where confrontation will take place. William B. Scott, "Space Chief Warns of Threats to U.S. Commercial Satellites," *Aviation Week & Space Technology,* April 12, 1999, 51.

Chapter 5

1. Cited in Burrows, *This New Ocean,* 332. **2.** See, for example, the testimony of Andrew F. Krepinevich Jr., *Hearing of the Emerging Threats and Capabilities Subcommittee of the Senate Armed Services Committee,* March 5, 1999 (Washington, D.C.: Federal News Service, 1999). **3.** The United States Commission on National Security/21st Century, *New World Coming: American Security in the 21st Century, Supporting Research and Analysis,* (Arlington, Va.: The Commission, 1999). **4.** Gregor Ferguson, "Australia Seeks to Build Satcom," *Defense News,* January 19–25, 1998, 10; Gregor Ferguson, "Australian Defence Force Enters Satellite Partnership," *Defense News,* November 15, 1999, 60. **5.** Gregor Ferguson, "Australians Eye Surveillance Satellites," *Space News,* April 3–9, 1995, 3.

6. The Nurrangar facility, which used to perform the early warning mission using Defense Support Program satellites, was shut down on October 12, 1999, because its technology had become obsolete. Gregor Ferguson, "Australia, U.S. Discuss SBIRS Collaboration Effort," *Space News,* November 1, 1999, 16. Upgraded systems at Pine Gap will be able to detect launches out of North Korea. See "Satellites Put U.S. Base in the Outback Out of Business," *San Diego Union Tribune,* December 17, 1999. **7.** Marco Caceres, "CBERS (Ziyuan)," *World Space Systems Briefing* (Fairfax, Va.: Teal Group, January 1999). **8.** Frank Dirceu Braun, "Alcântara: Ready and Waiting," *Launchspace,* May/June 1999, 35, 36. **9.** David Pugliese, "Canada Pursues First Dedicated Military Satcom System," *Defense News,* May 18–24, 1998, 18. **10.** Marco Caceres, "Radarsat," *World Space Systems Briefing* (Fairfax, Va.: Teal Group, January 1999).

11. Current U.S. policy is that U.S. companies may not sell radar satellite imagery with a resolution better than five meters to customers other than the United States. Warren Ferster, "NASA Questions Wisdom of Launching Radarsat-2," *Space News,* February 22, 1999, 1; Brian Berger and David Pugliese, "Canada Plans to Control Radarsat-2 Data Access," *Space News,* June 21, 1999, 1; and Warren Ferster, "U.S., Canada Reach Radarsat 2 Accord," *Space News,* October 18, 1999, 1. According to Foreign Minister Lloyd Axworthy, "We need to take steps to ensure that photographs taken by these satellites are not used against Canada and its allies." Defense Minister Arthur Eggleton observed that "as modern remote-sensing satellites can produce imagery whose quality approaches that obtained from specialized intelligence satellites, we must ensure that the data produced by Canadian satellites cannot be used to the detriment of our national security and that of our allies." **12.** David Pugliese, "Canadian Space Assets Key to NORAD's Future," *Space News,* February 15, 1999, 6; David Pugliese, "Canada Mulls Backing Missile Shield," *Ottawa Citizen,* May 16, 1999, 1. **13.** Cited by Peter B. de Selding, "China Makes Space Top Priority," *Space News,* October 14–20, 1996, 6. **14.** Zhu Yilin, "Fast-track development of space technology in China," *Space Policy,* May 1996, 139–42. For an excellent overview of the Chinese space program, see Joan Johnson-Freese, *The Chinese Space Program: A Mystery Within a Maze* (Malabar, FL: Krieger, 1998). **15.** Zhu Yilin, "Fast-track development of space technology in China," 141; see also Cox Committee Report, 10–20; Peter B. de Selding, "China Courting Deals with Non-U.S. Firms," *Space News,* November 15, 1999, 1; Bill Gertz, "China Plots Winning Role in Cyberspace," *Washington Times,* November 17, 1999, p. 1; Bill Gertz, "China Recruits Spies for Science," *Washington Times,* October 11, 1999, p. A1.

16. Cox Committee Report, 199–204. **17.** "Long March 4B Lays Missile Groundwork," *Space News,* May 24, 1999, 2; Bill Gertz, "China Prepared to Test ICBM with Enough Range to Hit U.S." *Washington Times,* November 12, 1998, p. 1. **18.** Craig Covault, "Manned Program Advances Chinese Space Technology," *Aviation Week & Space Technology,* November 29, 1999, 28–30; Phillip Clark, "Chinese designs on the race for space," *Jane's Intelligence Review,* April 1997, 178–82; Johnson-Freese, *Chinese Space Program,* 98–100. **19.** Johnson-Freese, *Chinese Space Program,* 93. **20.** Ibid., 95; Marco Caceres, "Sinosat," *World Space Systems Briefing,* February 1999; Cox Committee Report, 219, 220.

21. Fan Qing and Liu Jun, Beijing Xinhua Domestic Service in Chinese 0825 GMT, 18 July 1998, in FBIS-CHI-98-202, July 21, 1998. Accord to Fan and Liu, "Northwest China occupies a third of China's total territory. Because the troops' defense line is long and they are relatively far apart from one another, it was very difficult to ensure telecommunications. Taking into account technological progress in global telecommunications and to rectify backwardness in China's telecommunications, in recent years, the broad masses of officers, men engineers, and technicians of the Lanzhou Military Region Telecommunications Department have accelerated in modernizing military telecommunications technology and considerably increased the overall comprehensive capability to guarantee telecommunications." **22.** Bill Gertz, "China's Military Links Forces to Boost Power," *Washington Times,* March 16, 2000, p. 1. **23.** Johnson-Freese, *The Chinese Space Program,* 97–98. **24.** Chou Kuan-wu, "China's Reconnaissance Satellites," *Kuang Chiao Ching* (Hong Kong) 306 (March 16, 1998): 36–40, in FBIS-CHI-98-098, Article ID: drchi04081998000230, April 8, 1998. **25.** Ibid. The author writes that "politically, reconnaissance satellites help establish China's position as a major power." Concerning the capability of China's reconnaissance satellites, he writes that "the outside world knows precious little about the resolution power of China's reconnaissance satellites. The stellar cameras on China's reconnaissance satellites are used for the precise measurement of the position of satellites and ground targets." See also Joseph C. Anselmo, "China's Military Seeks Great Leap Forward," *Aviation Week &*

Space Technology, May 12, 1997, 68; Phillip Clark, "Global activity grows in the hands of a few," *Jane's Defence Weekly,* July 30, 1997.

26. Curtis, *Space Satellite Handbook,* 92. **27.** Richard D. Fisher Jr., "China Rockets Into Military Space," *Asian Wall Street Journal,* December 29, 1998; Marco Caceres, "Twin-Star," *World Space Systems Briefing,* December 1998; Wang Chiung-hua, "The Military Importance of Communist China Plans to Use Global Positioning System to Improve Precision of Missiles," *Chung-Yang Jih-Pao,* 24 December 1997, 10, in FBIS-CHI-97-361, December 27, 1997. **28.** Cox Committee Report, 209, 210; see also "Space Technology Could Beat US Defences, Scientist Says," *South China Morning Post,* November 22, 1999, 1. **29.** Rodota, interview by Lossau. **30.** Peter B. de Selding, "Europe Looks for Alternatives to Ariane," *Space News,* May 10, 1999, 8; Michael A. Taverna, "Europe Begins Plotting Strategy for RLVs," *Aviation Week & Space Technology,* April 12, 1999, 73–75; Michael A. Taverna, "Upgrades Aim to Keep Ariane 5 Competitive," *Aviation Week & Space Technology,* December 13, 1999, 61.

31. Peter B. de Selding, "European Space Agency Simplifies Imagery Sales Policy," *Defense News,* April 12, 1999, 32. **32.** Peter B. de Selding, "European Group Proposes NIMA-Like Agency," *Space News,* May 17, 1999, 3. **33.** Curtis, *Space Satellite Handbook,* 76. **34.** Craig Covault, "European Hot Bird Satcom Advances On-Board Switching," *Aviation Week & Space Technology,* October 19, 1998, 44. Marco Caceres, "Eutelsat," *World Space Systems Briefing,* April 1997. **35.** The eighty-satellite Skybridge LEO constellation is scheduled to begin in 2002 and will provide more than two hundred gigibits-per-second of overall system capability. See Peter B. de Selding, "Investments Boost Broadband Systems," *Space News,* October 11, 1999, 10; see also Michael A. Taverna, "European Mergers Boost Common Broadband Satellite System Plans," *Aviation Week & Space Technology,* November 8, 1999, 94, 95.

36. Michael A. Taverna, "Europe Launches Satnav Project," *Aviation Week & Space Technology,* July 5, 1999, 25; Rodota, interview by Lossau, in FBIS-WEU-98-181, June 30, 1999; Peter B. de Selding, "ESA Gets Increase for Satellite Navigation; Other Programs Cut," *Space News,* May 24, 1999, 1; Michael A. Taverna, "Germany to Play Key Role in GNSS-2," *Aviation Week & Space Technology,* February 22, 1999, 31. **37.** The ten members of the WEU are the United Kingdom, Belgium, France, Italy, Spain, Portugal, Luxembourg, the Netherlands, Germany, and Greece. **38.** de Selding, "Newsmaker Forum: Bernard Molard," 22. **39.** Peter B. de Selding, "France Optimistic on European Military Space Effort," *Space News,* December 20, 1999, 6; see also Michael A. Taverna, "Helios Success Fails to Quell Recon Discord," *Aviation Week & Space Technology,* December 13, 1999, 41, 42. **40.** Cited in Peter B. de Selding, "Europe Decries Reliance on U.S. Satellites," *Defense News,* May 17, 1999, 38; Jacques Baumel, "In Favor of a True European Defense System—Europe's Defense System is Lagging 10 Years Behind and There Is an Urgent Need to Make Up for Lost Time," *Le Figaro,* April 22, 1999, 2, in FBIS, Document ID: FTS 19990422000622, April 22, 1999.

41. Peter B. de Selding, "Expected Military Demand Drives Spot 5 Sensor Deal," *Space News,* February 1, 1999, 4; Peter B. de Selding, "Alcatel, Matra Vie for Spot Follow-on Contract," *Space News,* October 19–25, 1998, 6; Ben Iannotta, "Spot Image Plans Strategy of Diversity, Efficiency," *Space News,* October 11, 1999, 16. **42.** Pierre Sparaco, "French Satellite Details Air Strike Damages," *Aviation Week & Space Technology,* April 12, 1999, 26, 27;

Peter B. de Selding, "French Seek to Speed Helios Data Transfer," *Defense News,* June 21, 1999, 3. **43.** France would sell satellites to friendly countries and not for commercial use. Peter B. de Selding, "France Offers to Sell Spy Satellites," *Defense News,* March 13–19, 1995, 4; McLean and Swankie, "Helios 2—Myth or Reality?" 107–14; Marco Caceres, "Helios," *World Space Systems Briefing,* January 1999; "First Helios 2 Satellite to Launch in 2003," *Defense News,* October 11, 1999, 2; Michael A. Taverna, "Helios 2 Award Goes to MMS," *Aviation Week & Space Technology,* November 15, 1999, 32. **44.** Christian Lardier, "Review of French Military Space Program," *Cosmos/Aviation International,* March 19, 1999, 48, in FBIS, Document ID: FTS19990417000521, April 17, 1999. **45.** Michael A. Taverna, "Euro Milsatcom Accord Inked," *Aviation Week & Space Technology,* December 13, 1999, 41.

46. Anatol Johansen, "New Radar Satellites to Ease European Reconnaissance Woes: European Weak in Space Reconnaissance: Americans Less than Helpful," *Duesseldorf Handelsblatt,* May 5, 1999, 60, in FBIS, Document ID: FTS 19990521000882, May 21, 1999; Peter B. de Selding, "Europeans Take Fresh Look at Spy Satellites: Germany Pens Contracts for Defense Radar Work," *Space News,* November 16–22, 1998, 3. **47.** Dirk Weber, "Armed Forces Satellite Communications in Action," *Wehrtechnik,* January 1997, 51, 52, in FBIS-EST-97-005, January 1, 1997. Organizational charters specify that German armed forces may use Eutelsat, Inmarsat, Intelsat, the Russian-led Intersputnik, and the Kopernikus system only for humanitarian operations. **48.** Richard Barnard, "One on One: A.P.J. Abdul Kalam," *Defense News,* April 10–16, 1995, 54. Parroting the vision of the U.S. Air Force, India's Air Force chief is on record as stating that there is a growing necessity to "progress from an air force to an air and space force by deploying space-based systems." "Indian Air Chief: War Success to Depend on Satellites," *Asian Age,* 8 October 1998, 3, in FBIS, Document ID: FTS19981008000873, October 8, 1998; K.S. Jayaraman, "Indian Scientists Attack Malaria from Space," *Space News,* November 15, 1999, 24. **49.** Vivek Raghuvanshi, "India Exerts Military Control of Satellite," *Defense News,* June 17–23, 1996, 10. **50.** "Indian Agency Eyes Expanded Imagery, Launcher Markets," *Space News,* August 2, 1999, 10; Caceres, "IRS," *World Space Systems Briefing,* February 1999; Michael Mecham, "Cost-Conscious Indians Find Profits in Imaging Satellites," *Aviation Week & Space Technology,* August 12, 1996, 59, 60; Debra Polsky Werner, "Eosat Counts on Indian Imagery," *Space News,* February 6–12, 1995, 10; "India to Launch 'Indigenous' Ocean Scanning Satellite," *Hindustan Times,* 19 November 1998, 22, in FBIS, Document ID: FTS19981119000703, November 19, 1998.

51. Warren Ferster and Peter B. de Selding, "Indians Dive Into Mobile Market," *Space News,* January 23–29, 1995, 1; Michael Mecham, "India Builds a 'Crown Jewel,'" *Aviation Week & Space Technology,* August 12, 1996, 56, 57; Caceres, "Insat," *World Space Systems Briefing,* March 1999; Caceres, "Gramsat," *World Space Systems Briefing,* July 1998. **52.** K.S. Jayaraman, "Indian Firm Prepares for Regional GPS Boom," *Space News,* April 27–May 3, 1998, 10. **53.** K.S. Jayarman, "India to Shift Launches, Construction to Industry," *Space News,* February 7, 2000, 1. **54.** Vivek Raghuvanshi, "India to Develop Extensive Nuclear Missile Arsenal," *Defense News,* May 24, 1999, 14; Barry Bearak, "India Tests Missile Able to Hit Deep Into Neighbor Lands," *New York Times,* April 12, 1999, p. A3; and Anon, "Indian Scientists Said Working on Space Plane 'Avatar,'" *Deccan Herald,* 8 June 1998, in FBIS-NES-98-159, June 8, 1998; K.S. Jayaraman, "PSLV to Serve Global Market by 2000, Officials Say," *Space News,* August 9, 1999, 12. **55.** "Indonesia Not to Develop Military

Products," *Medan Waspada* (Internet edition), 11 December 1998, in FBIS, Document ID: FTS19981210001702, December 10, 1998.

56. Amnon Barzilay, "IDF to Expand Use of Satellites," *Ha'aretz,* 13 July 1998, A4, in FBIS-NES-98-194, July 13, 1998. Regarding the threat to Israel's existence posed by Iraq's nuclear weapons program, one former senior intelligence officer said that "the main difference between the situation in the Gulf war and now is that Israel now has an Ofeq satellite capable of taking high definition pictures of two-meter objects in Iraq." Yosi Melman, "The Iraqi Section Has Grown," *Ha'aretz,* 2 February 1998, B3, in FBIS-TAC-98-033, February 2, 1998. **57.** Barbara Opall-Rome, "Dual-Use Eros Expected to Boost Commercial Data Quality," *Space News,* September 20, 1999, 8; Gerald M. Steinberg, "Middle East space race gathers pace," *International Defense Review,* October 1995, 20–23; Caceres, "Eros," *World Space Systems Briefing,* January 1998; A. Peer, "'Revolutionary' Satellites Ready for Launch in Few Months," *Hamodi'a,* 30 April 1999, 18, in FBIS, Document ID: FTS19990430001184, April 30, 1999. **58.** James Bruce, "Israel's Amos-1 launches ABM early warning bid," *Jane's Defence Weekly,* March 6, 1996, 23. **59.** Stefan Barensky, "Small Launch Services: Little Trees in a Big Forest," *Launchspace,* January/February 1999, 30, 31. **60.** Japan's general space budget rose in 1999, boosted by funds allotted for the planned reconnaissance satellite constellation. "Outline of FY97 Defense White Paper Noted," *Yomiuri Shimbun,* 29 May 1997 (morning edition), 3, in FBIS-EAS-97-150, May 30, 1997; Paul Kallender, "Stimulus Package Bolsters Japan's Space Programs," *Space News,* January 25, 1999, 6.

61. "Japan Will Buy U.S. Parts for Spy Satellites," *Military Space,* October 12, 1999, 1; Naokio Usui and Warren Ferster, "Agreement Opens Japanese Program to U.S. Firms," *Space News,* October 11, 1999, 1; "U.S. Military Will Cooperate with Japan Satellite Program," *Military Space,* August 2, 1999, 2; "First Spy Satellites to Be Launched in 2002," *South China Morning Post* (Internet edition), November 7, 1998. Officials in the prime minister's cabinet do not believe that the satellites would violate a 1969 parliamentary resolution barring Japan from activities in space that were not for peaceful purposes. See also Y. Noki, "Summing up the Pluses and Minuses of Japan's Reconnaissance Satellite Development," *Gunji Kenkyu,* December 1998, 60–74, in FBIS, Document ID: FTS19981212000029, December 12, 1998; "Spy Satellite Development Going Domestic," *Nikkei Sangyo Shimbun* (Nikkei Telecom Database version), 19 May 1999, 28, in FBIS, Document ID: FTS19990523000156, May 23, 1999; James Morrison, "Sky Spy for Japan," *Washington Times,* September 30, 1999; Calvin Sims, "U.S. and Japan Agree to Joint Research on Missile Defense," *New York Times,* August 17, 1999. **62.** "Outline of Space Development Policy," *Uchu Kaihatsu Iinkai Geppo,* March 1996, 53–75, in FBIS-JST-97-002, March 1, 1996; Paul Kallendar, "Japan Takes Steps to Manufacture Satellites," *Space News,* April 7–13, 1997,.6; Paul Kallendar, "Japan Advances Agenda for High-Speed Data Satellite," *Space News,* October 7–13, 1996, 8. **63.** Michael Mecham and Eiichiro Sekigawa, "Japanese Reschedule More H-2A Launches," *Aviation Week & Space Technology,* July 19, 1999, 56, 57; Eiichiro Sekigawa, "Space Funding Supports H-2, International Station Work," *Aviation Week & Space Technology,* February 23, 1998, 81; Eiichiro Sekigawa, "Japan's H-2A Launcher Prompts Manifest Changes," *Aviation Week & Space Technology,* June 14, 1999, 212; Terrey H. Quindlen, "Japan's J1 Replacement to Use Russian, U.S. Technology," *Space News,* September 13, 1999, 8. **64.** Simon Saradzhyan, "Russia Earns Cash Training Amateurs at Star

City," *Space News,* July 12, 1999, 6. **65.** Simon Saradzhyan, "Russian Deputy Premier: Abandon Mir for Sake of Military Spacecraft," *Defense News,* June 28, 1999, 10; "Novosti" newscast presented by Aleksandra Buratayeva, Russian Public Television First Channel Network, 1500 GMT, 20 January 1998, in FBIS-SOV-98-020, January 20, 1998; Simon Saradzhyan, "Failure to Boost '99 Space Spending Clouds RKA's Future," *Space News,* February 1, 1999, 10; Andrey Yarushin, "Service Life of Most Russian Spacecraft 'Already Ended,'" ITAR-TASS, 1121 GMT, 25 November 1998, in FBIS, Document ID: FTS19981125000538, November 25, 1998; Simon Saradzhyan, "Russian Chiefs to Shrink Space Control Network," *Defense News,* June 14, 1999, 28; Joseph C. Anselmo, "Discovery Visits Space Station; Russia to Pull Crews Off Mir," *Aviation Week & Space Technology,* June 7, 1999, 27.

66. Saradzhyan, "Russian Deputy Premier: Abandon Mir for Sake of Military Spacecraft," 10. **67.** Simon Saradzhyan, "Russia to Make Spy Satellites a Top Budget Priority in 2000," *Space News,* October 25, 1999, 3; "Russia Loses Space Reconnaissance Capability," *Military Space,* July 19, 1999, 1. Mikhail Osokin, "Segodnya," *NTV,* 1500 GMT, 22 May 1999, in FBIS, Document ID: FTS19990522000740, May 22, 1999; Michael R. Gordon, "Russian Military Loses Satellites," *New York Times,* November 22, 1996, p. A1; Craig Covault, "New Russian Recon Keyed to Area Surveillance," *Aviation Week & Space Technology,* February 23, 1998, 37; Phillip S. Clark, "Russia's latest spy satellite," *Jane's Intelligence Review,* February 1996, 71–74; Covault, "Military Space Dominates Air Strikes," 33. **68.** Phillip Clark, "Russian and Ukrainian satellites to observe the oceans," *Space Policy,* 15 (1999), 13–17. **69.** Michael A. Taverna, "Proton Failure Upsets Russian Satcom Program," *Aviation Week & Space Technology,* November 22, 1999, 58, 59; Peter B. de Selding, "Russia Eyes West's Technology," *Space News,* September 27, 1999, 1; "Financing Plan Set for Russian Telecoms," *Aviation Week & Space Technology,* September 13, 1999, 29; Natalya Efrussi, *TV-6,* 0700 GMT, 23 April 1999, in FBIS, Document ID: FTS19990423000957, April 23, 1999; Caceres, "Gonets," and "Yamal," *World Space Systems Briefing,* March 1999 and October 1998 respectively. **70.** Geoffrey E. Perry, "Operational trends in Russian navigation satellite systems," *Space Policy,* February 1997, 5–7; "Moscow Approves Civilian Use of Global Satellite Network," *Interfax,* 0953 GMT, 2 October 1997, in FBIS-SOV-97275, October 2, 1997; Simon Saradzhyan, "Russia Plans to Refurbish Glonass Satellite Fleet," *Space News,* January 31, 2000, 6.

71. Craig Covault, "Russians Advance with Angara, Proton and Soyuz," *Aviation Week & Space Technology,* December 13, 1999, 70–73. **72.** Craig Covault, "Russian Space Program Decline Accelerating," *Aviation Week & Space Technology,* January 15, 1996, 26; Craig Covault, "Promise and Peril Mark Russian Launch Surge," *Aviation Week & Space Technology,* July 13, 1998, 78, 79; "Russian Submarine Launches Satellite," *London Times,* July 8, 1998; Phillip Clark, "Russian proposals for launching satellites from the oceans," *Space Policy,* 15 (1999), 9–12. **73.** Simon Saradzhyan, "Russians Strive to Bring Launches Back to Home Soil," *Defense News,* May 17, 1999, 34. **74.** Steven J. Zaloga, "Red Star Wars," *Jane's Intelligence Review,* May 1997, 205–8; James Oberg, "Secret Soviet Spacecraft," *Popular Science,* October 1998, 82–86. See also Mikhail Rebrov, "Russian Weapons: 21st Century Designs: Plasma Weapons: Fantasy or Reality?" *Krasnaya Zvezda,* 18 May 1996, 6, in FBIS-UMA-96-123-S, June 25, 1996, 21–24; William B. Scott, "Russian Politics May Stymie Laser/Optics Collaboration," *Aviation Week & Space Technology,* March 21, 1994, 49,

50; Gertz and Scarborough, "Russian ASAT," 9. **75.** Lord Reay, "UK space policy," *Space Policy,* November 1991, 307–15.

76. Harry R. Davis and Robert C. Good, *Reinhold Niebuhr on Politics* (New York: Charles Scribner's Sons, 1960), 265. **77.** William Mitchell, *Winged Defense: The Development and Possibilities of Modern Air Power—Economic and Military,* (Port Washington, N.Y.: Kennikat Press, 1925), 3, 4. **78.** Paula Shaki, "Firm Poised to Meet Demand for Easy Access to Imagery," *Space News,* April 5, 1999, 7; Vernon Loeb, "Sharp Eye in the Sky Lets Nations Spy—For a Price," *Washington Post,* May 10, 2000, p. 3; Catherine MacRae, "Space Architect Reviewing 'Stakeholder' Proposals for Future Security," *Inside the Pentagon,* May 18, 2000, 21, 22. **79.** Vernon Loeb and David Vise, "Satellite Snaps," *Washington Post,* November 5, 1999, p. A31. **80.** According to CIA Director George Tenet, the growing availability of commercial high-resolution satellite imagery "will have major implications for denial and deception and surprise—both on our part and on the part of the enemy." See Robert Holzer, "Commercial Imagery Could Aid Rogue Nations," *Space News,* April 3, 2000, 6; and Jason Bates, "Policy-Makers Not Ready for New Imagery," *Space News,* April 3, 2000, 1.

81. Joseph C. Anselmo, "Commercial Satellites Zoom in on Military Imagery Monopoly," *Aviation Week & Space Technology,* September 22, 1997, 75. **82.** SPOT Image Advertisement, *Defense News,* June 16–22, 1997, 63. **83.** Pedro Rustan, "Commercial Imagery: Nice But No Substitute for Recon," *Aviation Week & Space Technology,* August 30, 1999, 66. For a good discussion of these issues, see Vipin Gupta, "New Satellite Images for Sale," *International Security* 20, no. 1 (Summer 1995): 94–125. See also Bob Preston, *Plowshares and Power: The Military Use of Civil Space* (Washington, D.C.: National Defense University Press, 1994), 25–54. **84.** Michael Mecham, "Imagery a Vital Planning Tool," *Aviation Week & Space Technology,* December 7, 1998, 64–69. **85.** Jefferson Morris, "Space Imaging," *Launchspace,* October/November 1999, 40, 41; and "Remote Sensing," *Aviation Week & Space Technology,* October 11, 1999, 21.

86. Frank G. Klotz, *Space, Commerce, and National Security* (New York: Council on Foreign Relations, 1998), 14. See also Preston, *Plowshares and Power,* 35–42. Imagery of General Schwartzkopf's "Hail Mary" could have been seen in enough time for Iraqi forces to attack the redeployed U.S. forces using chemical weapons or to exploit the departures of the VII and XVIII Corps to break through thinned out lines to attack coalition airfields and supply lines in Saudi Arabia. According to Preston, even satellites with a poor revisit rate (that is, given their orbital parameters and imaging capabilities, they must wait a couple of weeks to view the same objects a second time) would have provided an opportunity to discover the movement. **87.** Cited in McLean, *Western European Military Space,* 101. **88.** See, for example, Berger and Pugliese, "Canada Plans to Control Radarsat-2 Data Access," 1. **89.** "Analysis 3—Commercial Communications Satellites—1999–2018," *Space Systems Forecast* (Newtown, Conn.: Forecast International, February 1999), 2–4. **90.** Daniel Gonzales, *The Changing Role of the U.S. Military in Space,* MR-895-AF available at http://www.rand.org/publications/MR/MR895/ (RAND Corporation, 1999), xiv, 13, 14.

91. Paula Shaki, "VSAT Suppliers Pursue Broadband Systems," *Space News,* June 21, 1999, 10. **92.** "Eyes in the Sky," *Wall Street Journal,* May 30, 2000; "U.S. Stops GPS Signal Disruption," *Space and Missile Defense Report,* May 11, 2000, 1. **93.** Gonzales, *The Changing Role of the U.S. Military in Space,* 12. **94.** Bruce A. Smith, "Imaging Satellites," *Aviation Week*

& Space Technology, September 13, 1999, 19. **95.** During the 1999 air bombing campaign in the Balkans, U.S. officials sought to restrict public information of U.S. satellite operations and other military operations that could have aided Belgrade. Bill Gertz and Rowan Scarborough, "News Blackout," *Washington Times*, April 9, 1999, p. 8.

96. See Lambakis, "Space Control in Desert Storm and Beyond," 417–33. **97.** Preston, *Plowshares and Power*, 24, 25. **98.** The United States reportedly operates a secret satellite tracking program that will warn U.S. commanders when foreign satellite spies streak overhead called Project SATRAN (Satellite Reconnaissance Advance Notice). The emergence of commercial remote-sensing satellites was one of the reasons this system came into being. SATRAN will alert installation commanders to the overhead presence of hostile satellites, although there is not likely a program in place to use these same databases to perform active antisatellite operations. "Project SATRAN Warns of Hostile Recon from Space," *Military Space* 14, no. 8 (April 14, 1997): 1. **99.** Cited in Grossman, "Information Operations to Play Key Role," 1. **100.** Vyacheslav Georgiyevich Bezborodov, "Competition in Orbit: It Is Becoming Increasingly Fierce and Creating a Real Threat to Russia's National Interests," *Armeyskiy Sborniki*, December 1995, 20-23, in FBIS-UMA-96-076-S, April 18, 1996, 11.

Chapter 6

1. See B.H. Liddell Hart, *Strategy* (New York: Praeger, 1968). **2.** According to former Senators Gary Hart and Warren Rudman, co-chairmen of an independent commission to study the future national security environment, "America will become increasingly vulnerable to hostile attack on our homeland, and our military superiority will not necessarily protect us." See "Hart-Rudman Commission: U.S. Faces Homeland Attacks," *Aerospace Daily*, October 6, 1999. **3.** To borrow one more quote from Sun Tzu's *The Art of War*, "War is a matter of vital importance to the State; the province of life or death; the road to survival or ruin. It is mandatory that it be thoroughly studied" (1:1). **4.** Ken Booth, *Strategy and Ethnocentrism* (New York: Holmes & Meier, 1979), 26, 27. This book is a wonderful treatment of the security implications of ethnocentrism. **5.** Ibid., 27.

6. Winston S. Churchill, "To End War," in Wolff, *Churchill and War*, 348. **7.** Adda B. Bozeman, "Covert Action and Foreign Policy," in Roy Godsen, ed., *Intelligence Requirements for the 1980s: Covert Action* (New Brunswick: Transaction, 1981), 16. Bozeman writes a little later that "some of the deviations from what we view as trans-nationally valid standards of reasoning and behavior are simply dismissed as 'irrational' or 'mindless' actions. However, in general it continues to be customary to ascribe all their shortcomings to the lack of economic development, a purely temporary condition, so the teaching goes, that will be overcome given material aid and modernization over time." We are in fact "dealing in most instances with very ancient societies and thus with firmly imbedded cultural patterns which prescribe or proscribe thought and action, whether in regard to economics or to statecraft." **8.** Perhaps more of a factor in the U.S. decision not to deploy these nuclear-tipped weapons systems was the protests from Americans who did not want them in their "back yard." **9.** See, for example, the testimony of Robert T. Marsh, cited in Joseph C. Anselmo, "U.S. Seen More Vulnerable to Electromagnetic Attack," *Aviation Week & Space Technology*, July 28, 1997, 67; "Spy Satellite Said to Track U.S. Warships," *South China Morning Post*, October 6, 1999. **10.** Chou Kuan-wu, "China's Reconnaissance Satellites," 36–40.

11. See, for example, "Spy Satellites Said to Track U.S. Warships," *South China Morning Post,* October 6, 1999: military expert Zhou Guanwu claims that China has the capability to destroy optical sensors on satellites using lasers. See also William B. Scott, "Wargames Revival Breaks New Ground," *Aviation Week & Space Technology,* November 2, 1998, 56–58; Brian Berger, "'Space Game II' Findings Show the Pros and Cons of Fighting in Space," *Inside the Army,* March 9, 1998, 2; William B. Scott, "Wargames Underscore Value of Space Assets for Military Ops," *Aviation Week & Space Technology,* April 28, 1997, 60; Sean D. Naylor, "Hidden soft spot in satellite might," *Army Times,* March 10, 1997, 20, 21; Barbara Opall, "Study Pits PLA Nukes Against U.S., Taiwan," *Defense News,* September 23–29, 1996, 10; John Pomfret, "China Plans for a Stronger Air Force," *Washington Post,* November 9, 1999, p. 17. **12.** According to Maj. Gen. Kenneth Hagemann, former Director of the Defense Nuclear Agency, the problem with EMP effects in space will only get worse as electronics are miniaturized, meaning that they require less power and are more susceptible to smaller disruptions. See "Rogue Nations with Only a Few Nuclear Weapons," *Aviation Week & Space Technology,* May 1, 1995, 21; and Joseph Fitchett, "Chinese Nuclear Buildup Predicted," *International Herald Tribune,* November 6–7, 1999, 1. **13.** Consider, for example, Maj. Gen. Wu Guoqing, "Future Trends of Modern Operations," in *Chinese Views of Future Warfare,* ed. Michael Pillsbury (Washington, D.C.: National Defense University Press, 1997), 346. **14.** Even ballistic missile defense systems currently under development, which use ground-based interceptors, would not be in position to kill a missile before it reaches apogee—only space-based interceptors could reliably accomplish such an intercept. Consider also this testimony by Lowell Wood: "How can DoD reliably eliminate the prospect of a single moderate-to-large yield nuclear explosion occurring at high altitude over the U.S.? (Indeed, how can it detect that such an EMP attack is underway, or that it is likely?) How can it robustly attribute the origin of such an attack (noting that at least Russia, China, and India manifestly have the capability to execute such an attack today)? *If reliable defense is not feasible and robust attrition-for-deterrence is not possible, how is eventual attack to be rationally judged to be at all unlikely?*" House, *Prepared Statement of Lowell Wood, Before the House Military Research & Development Subcommittee,* July 16, 1997. **15.** B.H. Liddell-Hart, *Strategy* (New York: Praeger, 1967), 338.

16. See, for example, Christopher Jon Lamb, "The Impact of Information Age Technologies on Operations Other Than War," in Pfaltzgraff and Shultz, *War in the Information Age,* 249–50. **17.** Chairman of the Joint Chiefs of Staff, Gen. Henry "Hugh" Shelton, U.S. Army, noted that the greatest threat to U.S. national security in this current period of relative peace is "the potential that we [the Pentagon] will become too complacent." Cited in Bryan Bender, "Shelton: Greatest Threat Is Pentagon Complacency," *Defense Daily,* December 15, 1997, 6. See also Silverstein, "Buck Rogers Rides Again," 23. **18.** See, for example, Steven Lee Myers, "In Intense but Little-Noticed Fight, Allies Have Bombed Iraq All Year," *New York Times,* August 13, 1999, p. A1. **19.** Carl von Clausewitz, *On War,* Michael Howard and Peter Paret, ed. and trans. (Princeton, N.J.: Princeton University Press, 1984), 77, 577–637. **20.** Clausewitz, "Note of 10 July 1827," *On War,* 69: There are, according to Clausewitz, two natures of war. "First, those in which the object is the *overthrow of the enemy,* whether it be we aim at this political destruction or merely at disarming him and forcing him to conclude peace on our terms; and secondly, those in which our object is merely to make *some conquests on the frontiers of his country,* either for the purpose of retaining them permanently or of turning them to account as a matter of exchange in settling terms of peace."

21. See Julian S. Corbett, *Some Principles of Maritime Strategy* (1911; reprint, Annapolis, Md.: Naval Institute Press, 1972), 13–27. **22.** Liddell-Hart, *Strategy,* 334–35. **23.** Corbett, *Some Principles of Maritime Strategy,* 68. Corbett's intellectual approach and repeated references to historical argument are said to have frustrated Prime Minister David Lloyd George to the point where Lloyd George came to regard "thinking as a form of mutiny." See Williamson Murray, "Corbett, Julian," in Robert Cowley and Geoffrey Parker, eds., *The Reader's Companion to Military History* (Boston: Houghton Mifflin, 1996), 108. **24.** "Strategy has not to overcome resistance, except from nature. *Its purpose is to diminish the possibility of resistance,* and it seeks to fulfill this purpose by exploiting the elements of *movement* and *surprise.*" Liddell-Hart, *Strategy,* 337. **25.** Liddell-Hart, *Strategy,* 25.

26. "Limited wars do not turn upon the armed strength of the belligerents, but upon the amount of that strength which they are able or willing to bring to bear at the decisive point." Corbett, *Some Principles of Maritime Strategy,* 55. **27.** Liu Jiyuan, *Aerospace China* (English) 5, no. 1 (June 1996): 2–6, in FBIS-CST-96-015, June 1, 1996. **28.** Chou Kwan-wu, "China's Reconnaissance Satellites." **29.** George Friedman of Global Forecasting, "Chinese Space Center Can Tap U.S. Navy's Pacific Com," *Navy News and Undersea Technology,* October 27, 1997, 5. **30.** David G. Wiencek, *Dangerous Arsenals: Missile Threats in and from Asia,* Bailrigg Memorandum 22, (Lancaster, UK: Centre for Defence and International Security Studies, 1997, 43.

31. Senior Col. Ping Fan and Capt. Li Qi, "A Theoretical Discussion of Several Matters Involved in the Development of Military Space Forces," *Zhongguo Junshi Kexue* [China Military Science], 20 May 1997, 127–31 in FBIS-CHI-970-302, October 29, 1997. **32.** Ping and Li, "Theoretical Discussion of Several Matters." **33.** Wu, "Future Trends of Modern Operations," 347. **34.** Alastair Iain Johnston, "China's New 'Old Thinking,'" *International Security* (Winter 95/96): 23, 24. **35.** Wang Chunyin, "Characteristics of Strategic Initiative in High-Tech Local Wars," *Hsien-Tai Chun-Shih* 245 (June 11, 1997): 19–22, in FBIS-CHI-97-302, October 29, 1997.

36. Min Zengfu, "Transcend 'Air Domination' Theory," *Jiefangjun Bao,* 12 December 1995, 6 in FBIS-CHI-96-010, 12 December 1995. **37.** Huang Xing and Zuo Quandian, "Holding the Initiative in Our Hands in Conducting Operations, Giving Full Play to Our Own Advantages to Defeat Our Enemy—A Study of the Core Idea of the Operational Doctrine of the People's Liberation Army," *Zhongguo Junshi Kexue [Chinese Military Science]* 4 (November 20, 1996): 49–56, in FBIS-CHI-97-113, November 20, 1996. **38.** Ch'en Huan, "The Third Military Revolution," in Pillsbury, *Chinese Views of Future Warfare,* 394, 95; see also Chong-Pin Lin, "Info Warfare Latecomer," *Defense News,* April 12, 1999, 23. Consider also this analysis by Zhang Dyong and Shang Minghua, and Xu Kejian, "Information Attack," in *Beijing Jiefangjun Bao,* 24 March 1998, 6, in FBIS-CHI-98-104, 14 April 1998: "Arms use is just an outcome of a system operation. Take the 'Patriot' missile, for example: a satellite has to locate a target and then send relevant data to a calculation center in the United States (or Australia). . . . After that, a command center will issue an order to a possible target area. . . . Once an enemy missile enters a control zone, the target area will send a 'Patriot' missile to intercept it. In this process, a whole combat system is activated and put into operation. If something goes wrong with just one link in the system, the whole operation will fail." **39.** Shen Kuiguan, "Dialectics of Defeating the Superior with the Inferior," in Pillsbury, *Chinese Views of Future Warfare,* 213–19. **40.** Wu, "Future Trends of Modern Operations," 348.

41. Ch'en Huan, "Third Military Revolution Will Certainly Have Far-Reaching Effects," *Hsien-Tai Chun-Shih* 230 (March 11, 1996): 8–10, in FBIS-CHI-96-169, March 11, 1996. **42.** Cited in John Pomfret, "China Ponders New Rules of 'Unrestricted War,'" *Washington Post,* August 8, 1999, p. A1. Cols. Qiao Liang and Wang Xiangsui wrote, "We realize that if China's military was to face off against the United States, we would not be sufficient. So we realized that China needs a new strategy to right the balance of power." **43.** Stanislav Yermak, "Training Year Results: Reflections, Analysis, Experience," *Armeyskiy Sbornik* (in Russian) 1 (January 1996): 19–21 (FBIS). At the Military Space Forces commander's directorate command post, an enormous sign hangs: "Russia was, is, and will be a great space power!" Cited in Yevgeniy Moskvin, "The Country's Military Space—Preservation of Priorities Requires Elementary State Attention," *Nezavisimaya Gazeta* 19 (23) (October 10, 1996): 1, 3, in FBIS-UMA-96-216-S, October 10, 1996. **44.** *Moscow RIA,* 1741 GMT, 2 October 1997 in FBIS-SOV-97-275, 2 October 1997. **45.** Cited in Anatoliy Bukharin, "Space Is One of Our Priorities," *Armeyskiy Sbornik* 4 (April 1996): 17–19, in FBIS-UMA-96-129-S, 3 July 1996.

46. Cited in Anatoliy Bukharin, "Space Is One of Our Priorities." See also Dmitrii Paison and Sergei Sokut, "Key to Victory in Space," *Nezavisimoe Voennoe Obozrenie* 5 (1999): 6, translated by Andrei Shoumikhin. **47.** See Mary C. Fitzgerald, *The Soviet Image of Future War: "Through the Prism of the Persian Gulf,"* HI-4145 (Indianapolis: Hudson Institute, May 1991). **48.** Cited in Anatoliy Bukharin, "Space Is One of Our Priorities." **49.** Maj. Gen. I.N. Vorobyev, "Strategy," *Voyennaya Mysl,* March-April 1997, 18–24 in FBIS-UMA-97-097-S, 1 April 1997. **50.** Nikolai Mikhailov, "Weighty Answers to Military Challenges. Sensible Russian Military-Technological Policy in Conditions of Crisis Is Possible," *Nezavisimoe Voennoe Obezrenie,* April 30, 1999, 1, 4, cited in *Russian Arms Control Digest,* trans. Andrei Shoumikin (Fairfax, Va.: National Institute for Public Policy, May 3, 1999).

51. Lt. Gen. Alexsandr Skortsov and Maj. Gen. Nikolay Turko, "Strategic Stability: Key to National Security," *Armeyskiy Sbornik* 1 (January 1996): 4–8, in FBIS-UMA-96-081-S, April 25, 1996, 1–4. **52.** Vorobyev, "Strategy." **53.** Lt. Gen. Anatoliy Nogovitsyn and Maj. Gen. Anatoliy Panchenko, "Not the Sum, but the System," *Armeyskiy Sbornik* 3 (March 1998): 1–10, in FBIS-SOV-98-148, May 28, 1998. **54.** Vorobyev, "Strategy." **55.** Andrei Shoumikin, senior analyst with the National Institute for Public Policy, interview by author, Fairfax, Va., April 29, 1999.

56. Dmitriy Gornostayev, "Strategic Balance in Terms of ABM Systems Maintained. Russia Gets Legal Restrictions from Washington on Development of Existing Arms Systems," *Nezavisimaya Gazeta,* 1 October 1997, 4, in FBIS-TAC-97-274, 1 October 1997. **57.** Sudenshna Banerjee, *Pioneer,* 1 November 1997, 1 in FBIS-NES-97-306, 2 November 1997. **58.** "Indian Air Chief: War Success to Depend on Satellites," *Asian Age,* 8 October 1998, 3, in FBIS, Document ID: FTS19981008000873, October 8, 1998. **59.** Amar Zutshi, "Defensive Deterrence: Miles to Go," *Pioneer,* 12 February 1998, 9, in FBIS-NES-98-044, February 13, 1998. **60.** V.W. Karve, "Anti-Computer Electronic Warfare," *U.S.I. Journal,* July-September 1994, 360, 61.

61. Maj. Gen. V.K. Madhok, "Space as a Battlefield," *U.S.I. Journal* (India), January–

March 1989, 24, 25. **62.** Ibid., 22. **63.** Maj. Gen. Harith Lutfi al-Wafiy, "The Combined Ground-Air War After the Year 2000," *Al-Qadishiyah,* 24 May 1997, 2, in FBIS-NES-97-176, 25 June 1997. **64.** "Mena" (Cairo, broadcast in Arabic), 1355 GMT, 7 March 1998, in FBIS-NES-98-066, March 7, 1998. **65.** Maj. Gen. Husayn Hasan 'Adday and Major General Kan'an Mansur al-'Abbadi, *Alif Ba',* interview by Jawad al-Hattab and Faris al-Katib, 4 February 1998, 13, 14, in FBIS-NES-98-067, March 8, 1998.

66. James Bruce, "Israel's Space and Missile Projects," *Jane's Intelligence Review* 7, no. 8:352–54; and Yosi Melman, report in *Ha'aretz,* March 30, 1998, 1, in FBIS-NES-98-089. **67.** Yosi Melman, "The Iraqi Section Has Grown," *Ha'aretz,* February 2, 1998, B3, in FBIS-TAC-98-033, February 2, 1998. There is some question as to *Offeq 3*'s capabilities. Official Israeli reports on the reconnaissance capabilities of *Offeq 3* tout very high resolution, indeed, that the satellite will "allow identification of license numbers of automobiles on the streets of Baghdad." U.S. analysts, however, believe the satellite's orbit is too high to provide imagery of military significance. Craig Covault, "Israeli Satellite's Military Value Limited," *Aviation Week & Space Technology,* April 10, 1995, 67; "Spy Satellite Sent Aloft by Israel," *Washington Post,* April 6, 1995, p. A27. **68.** "Israeli Air Force Chief Seeks Space Supremacy," *Space News,* March 2–8, 1998, 2; Amnon Barzilay, report in *Ha'aretz,* 13 July 1998, A4, in FBIS-NES-98-194, July 13, 1998; Ze'ev Schiff, "Facing Up to Reality," *Ha'aretz,* 9 January 1998, 7, in FBIS-NES-98-009, January 9, 1998; Ze'ev Schiff, "A Serious Satellite Situation," *Ha'aretz,* 26 January 1998, 6, in FBIS-NES-98-026, January 26, 1998. **69.** Serhiy Chornous, "The Space Eyes of Ukraine—Are Cared for by the Ukrainian Military Without State Oversight, *Narodna Armiya,* 20 December 1995, 2, in FBIS-UMA-96-016-S, December 20, 1997. **70.** Volodymir Shelipov, "Can One See an End to the Arms Race in Space?" *Region,* 12–19 November 1996, 17 in FBIS-TAC-97-001, November 19, 1996.

71. Serhiy Chornous, "Major-General Valeriy Lytvynov: 'Space Is the Way to the Future for Ukraine," *Narodna Armiya,* 12 April 1996, 4–5, in FBIS-UMA-96-109-S, 12 April 1996. **72.** Joint Chiefs of Staff, "Joint Vision 2010: America's Military—Preparing for Tomorrow," *Joint Forces Quarterly,* Summer 1996, 38. According to another Joint Chiefs of Staff publication, *Concept for Future Joint Operations,* some states will rely on asymmetric capabilities, such as ballistic and cruise missiles, man-portable air defenses, weapons of mass destruction, advanced space capabilities, information operations and terrorism as a substitute for, or complement to, conventional forces. Joint Chiefs of Staff, *Concept for Future Joint Operations,* 10. **73.** Joint Chiefs of Staff, *Concept for Future Joint Operation,* 15. **74.** Liddell-Hart, *Strategy,* 343.

Part 3

1. Hamre, *Statement before the Senate Armed Services Committee.* **2.** Cited in William B. Scott, "U.S. Adopts 'Tactical' Space Control Policy," *Aviation Week & Space Technology,* March 29, 1999, 35.

Chapter 7

1. Alexander Hamilton wrote in *Federalist* No. 70, "Energy in the Executive is a leading character of the definition of good government. It is essential to the protection of the community against foreign attacks; it is not less essential to the steady administration of the laws; to the protection of property against those irregular and high-handed combinations which sometimes interrupt the ordinary course of justice; to the security of liberty against enterprises and assaults of ambition, of faction, and of anarchy." Conversely, "a feeble Executive implies a feeble execution in government. A feeble execution is but another phrase for a bad execution; and a government ill executed, whatever it may be in theory, must be, in practice, a bad government." **2.** In the October 1999 issue of *A National Security Strategy for a New Century,* the administration published the following: "Unimpeded access to and use of space is a vital national interest—essential for protecting U.S. national security, promoting our prosperity and ensuring our well-being." **3.** A president typically writes several hundred of them during his term in office. Some very old executive orders, such as those issued by President Theodore Roosevelt concerning the preservation of endangered species, are still on the books. Frank J. Murray, "Clinton's Executive Orders Still Are Packing a Punch," *Washington Times,* August 23, 1999, p. A1. **4.** White House, *Fact Sheet: National Space Policy,* 1, 2. **5.** Cited in Burrows, *This New Ocean,* 167.

6. In a memorandum on disarmament submitted by the United States to the First Committee of the 12th General Assembly of the United Nations on January 12, 1957, it stated that "the United States proposes that the first step toward the objective of assuring that future developments in space would be devoted exclusively to peaceful and scientific purposes." **7.** State Department, *Summary of Foreign Policy Aspects.* **8.** April 2, 1958 message to Congress, reprinted in *Congressional Quarterly Almanac,* 85th Cong., 2d sess., 1958, vol. 14, 599. **9.** Cited in Delbert R. Terrill, Jr., *The Air Force Role in Developing International Outer Space Law* (Maxwell AFB, Ala.: Air University Press, May 1999), 47. **10.** Ibid., 45–53.

11. *Congress and the Nation: 1945–1964; A Review of Government and Politics in the Postwar Years* (Washington, D.C.: Congressional Quarterly Service, 1965), 300. **12.** United States Information Agency, "World Opinion and the Soviet Satellite," 24. **13.** Portree, *NASA's Origins,* 7–9, 45, 46. **14.** In a memorandum to President Eisenhower dated March 5, 1958, Killian wrote, "Because of the importance of the civil interest in space exploration, the long term organization for Federal programs in this area should be under civilian control. Such civilian domination is also suggested by public and foreign relations. However, civilian control does not envisage taking out from military control projects relating to missiles, anti-missile defense, reconnaissance satellites, military communications, and other space technology relating to weapons systems or direct military requirements." See Portree, *NASA's Origins,* 56. The United States continues to look to NASA, in the language of the 1958 act, to "exercise control over aeronautical and space activities sponsored by the United States, except that activities peculiar to or primarily associated with the development of weapons systems, military operations, or the defense of the United States (including research and development necessary to make effective provision for the defense of the United States) shall be the responsibility of, and shall be directed by, the Department of Defense." See 85th Cong., 2d sess., *National Aeronautics and Space Act of 1958,* July 23, 1958, Title 1, Sec. 102(b), H.R. 12575—1. **15.** "Memorandum: Subject: Discussion at the 371st NSC Meeting on Thursday, July 3, 1958," copy, declassified.

16. Report by the President's Science Advisory Committee, released by the White House on March 26, 1958, reprinted in *Congressional Quarterly Almanac,* 85th Cong., 2d sess., 1958, vol. 14, 596. **17.** His stance on space weapons may be fairly summarized by the following excerpts from the President's Science Advisory Committee: "For the most part, even the more sober proposals [for satellite bombers and military bases on the moon, for example] do not hold up well on close examination or appear to be achievable at an early date. . . . There may well be important military applications for space vehicles which we cannot now foresee, and developments in space technology which open up quite novel possibilities. . . . Our road to future strength is the achievement of scientific insight and technical skill by vigorous participation in these new explorations. In this setting, our appropriate military strength will grow naturally and surely." Report by the President's Science Advisory Committee, 598. Bracketed comments were inserted by the author. **18.** Burrows, *This New Ocean,* 319–32. Burrows (332) cites this line used by Kennedy in a speech at Rice University on September 12, 1962: Whether space "will become a force for good or ill depends on us, and only if the United States occupies a position of preeminence can we decide whether this new ocean will be a sea of peace or a new, terrifying theater of war." **19.** Cited in Stares, *Militarization of Space,* 70, 71. **20.** State Department, *Summary of Foreign Policy Aspects,* 27, 28.

21. During the 1964 presidential campaign, Republican contender Barry Goldwater stated that the country was "spending entirely too much money on the manned moon program, when a carefully plotted program using unmanned lunar landing equipment could steadily build up a solid body of scientific knowledge. . . . All manned space research should be directed by the military, with national security and control of the access to space as primary goals." *Congressional Quarterly Almanac,* 88th Cong., 2d sess., vol. 20, 1964, 453. **22.** When he was Senate majority leader and chairman of the Preparedness Subcommittee, Johnson outlined the goal for space at a caucus of Democratic senators: "If, out in space, there is the ultimate position—from which total control of the earth may be exercised—then our national goal and the goal of all free men must be to win and hold that position." He cited testimony of scientists, paraphrasing as follows: "Control of space means control of the world, far more certainly, far more totally than any control that has ever or could ever be achieved by weapons, or by troops of occupation." *Congressional Quarterly Almanac,* 85th Cong., 2d sess., 1958, vol. 14, 669. **23.** "The President's Message to the Congress Transmitting His Annual Report on U.S. Aeronautics and Space Activities for 1967, January 30, 1968," *Weekly Compilation of Presidential Documents* 4, no. 1 (January 8, 1968): 172. **24.** The Space Task Group, chaired by Vice President Agnew, concluded that the cheapest way to keep Americans in space was by designing "low-cost, flexible, long-lived, highly reliable, operational space systems with a high degree of commonality and reusability." Cited in Burrows, *This New Ocean,* 440. **25.** Stares, *Militarization of Space,* 168–79.

26. Ibid., 182. **27.** Announcement of Administrative Review, June 20, 1978, *Weekly Compilation of Presidential Documents,* 1135–37. **28.** The administration underscored in its June 20, 1978, public announcement that "national security related space programs will conduct those activities in space which are necessary to our support of such functions as command and control, communications, navigation, environmental monitoring, warning and surveillance, and space defense . . . and to support the planning for and conduct of military operations." Prior to the release of the new policy, one unnamed administration official summarized the concerns in the area of survivability. National security was the focus of Carter's policy reassessment precisely because space warfare "ripples through everything." The more willing you are, said the official, "to have an uncontrolled or provocative space environment, the more you have to worry about the survivability of the assets in less than full wartime conditions." You also have to worry about the survivability of civilian assets, according to the official, and "we use civilian communications satellites to carry a tremendous diplomatic and military load." Craig Covault, "Unified Policy on Space Readied," *Aviation Week & Space Technology,* January 2, 1978, 16. **29.** Testimony of William J. Perry, Under Secretary of Defense for Research & Development, *Hearings before a Subcommittee of the Committee on Appropriations, House of Representatives, 95th Congress, Second Session, Part 3: FY 1979 Research and Engineering Programs and Systems Acquisition Policy,* February 22, 1978, 726. The June 1978 statement to the press reflected this policy of appeasing both State Department and Defense Department interests: "Survivability of space systems will be preserved commensurate with the planned need in crisis and war, and the availability of other assets to perform the mission. . . . The United States finds itself under increasing pressure to field an antisatellite capability of its own in response to Soviet activities in this area. By exercising mutual restraint, the United States and Soviet Union have an opportunity at this early juncture to stop an unhealthy arms competition in space before the competition develops a momentum of its own." **30.** Stares, *Militarization of Space,* 193–200.

31. Remarks at the Congressional Space Medal of Honor Awards Ceremony, October 1, 1978, *Weekly Compilation of Presidential Documents* 14, no. 39 (October 2, 1978): 1684. **32.** President Johnson referenced the U.S. reconnaissance satellite capability, although his remarks were not intended for public dissemination. He said on March 15, 1967, "I wouldn't want to be quoted on this, but we've spent $35–40 billion on the space program and if nothing else had come out of it except the knowledge we've gained from space photography, it would be worth 10 times what the whole program has cost. Because tonight, we know how many missiles the enemy has." Cited in "Carter Confirms Recon Satellite Capability," *Aviation Week & Space Technology,* October 9, 1978, 22. **33.** "Fact Sheet Outlining the Policy," July 4, 1982, *Weekly Compilation of Presidential Documents* 18, no. 26 (July 5, 1982): 874, 875. **34.** The House Science and Technology Committee's space subcommittee thought the Reagan policy built support for military space programs at the expense of civil programs. See "Space Debate," *Aviation Week & Space Technology,* August 9, 1982, 17; Democratic Representative George Brown believed the Reagan space policy was "a strictly military-inspired" policy. "Space Policy Heat," *Aviation Week & Space Technology,* July 19, 1982, 17. **35.** "Announcement of U.S. Government Support for Commercial Operations by the Private Sector," May 16, 1983, *Weekly Compilation of Presidential Documents* 19, no. 13 (April 4, 1983): 723. For a discussion of incoherence in Reagan's space policy, particularly his launch policy, see John Noble Wilford, "Threat to Nation's Lead in Space Is Seen in Lack of Guiding Policy," *New York Times,* December 30, 1986, p. A1; and Kathy Sawyer, "U.S. Faces Self-Doubt, Competition in Space," *Washington Post,* January 27, 1987, p. A1.

36. "Fiscal 1984 Authorization," *Congressional Quarterly Almanac,* 1983, 188. **37.** Reagan, *Report to Congress on U.S. Policy on ASAT Arms Control,* 15. **38.** R. Jeffrey Smith, "Reagan Plans New ABM Effort," *Science* 220 (April 8, 1983): 170. The vision behind this speech originated in Reagan's deeply held belief that U.S. security should not rest upon the threat of instant nuclear retaliation. According to Keyworth, the objective of moving off of a mutual assured destruction posture "certainly resonated with the President's views. He has felt very strongly about this—in fact I'll go so far as to say that . . . I have never seen him anywhere near as committed to anything as he is to this." For early reaction to Reagan's address to the nation, see "Onward and Upward with Space Defense," *Bulletin of the Atomic Scientists,* June/July 1983, 4–8. **39.** Edgar Ulsamer, "Space Command: Setting the Course for the Future," *Air Force Magazine,* August 1982, 48–55. **40.** White House, *Fact Sheet: Foreign Access to Remote Sensing Space Capabilities* (Washington, D.C.: National Science and Technology Council, March 10, 1994); see also White House, *Fact Sheet: Landsat Remote Sensing Strategy* (Washington, D.C.: National Science and Technology Council, May 10, 1994). Eric Schmitt, "Congress Moves to Reverse Clinton's Satellite-Export Procedure, *New York Times,* September 18, 1998, p. 1; Warren Ferster, "U.S. State Department Takes Helm on Export Licenses," *Defense News,* January 18, 1999, 20; Hirsh, "Great Technology Giveaway?" 2–9; Eric Schmitt, "Panel Finds Harm in China Launchings," *New York Times,* May 7, 1999, p. 1.

41. "US Global Positioning System Policy," *Space Policy,* November 1996, 297, 298; Warren Ferster, "U.S. GPS to Transmit Second Civilian Signal," *Space News,* April 6–12, 1998, 31. **42.** For a discussion of GPS policy issues, see Scott Pace, "The Global Positioning System: Policy Issues for an Information Technology," *Space Policy,* November 1996, 265–75; and John Mintz, "The U.S. Opens Satellites to Civilians," *Washington Post,* March 30, 1996, p. A1. **43.** Paula Shaki Trimble, "U.S. Air Force Study Finds No Commercial 'Pot of Gold,'" *Space News,* November 8, 1999, 4.

Chapter 8

1. For a summary of arguments presented in chapters 8 and 9, see Steven Lambakis, "Space and Security: A US Policy Quandary," *Space Policy* 16 (February 2000): 13–18. See also Steven J. Lambakis and Colin S. Gray, *Political and Legal Restrictions on U.S. Military Space Activities* (Fairfax, Va.: National Institute for Public Policy, December 1997), report produced for the U.S. Air Force, XONP, contract no. HQ0006–95–C-0006. **2.** See, for example, Lambakis, "Exploiting Space Control," 42–46. See also Erin Q. Winograd, "SMDC Chief Says Protecting Space Assets Is an Unassigned Mission," *Inside Missile Defense,* July 26, 2000: 25–27. **3.** Cited in Scott, "Space Ops Threatened," 25–27. **4.** Cited in Scott, "Space Chief Warns of Threats," 51. **5.** Hamre, *Statement before the Senate Armed Services Committee.*

6. See Warren Ferster, "One on One: U.S. Sen Robert Smith," *Defense News,* March 2–8, 1998, 22. When asked whether he intended to resurrect the vetoed programs in the upcoming budget cycle, Smith responded, "Absolutely I will, and there's going to be a debate about it." See also Warren Ferster, "Congress Renews Fight for DoD Space Efforts," *Space News,* February 23–March 1, 1998, 1, and Gigi Whitley, "Pentagon Officials Mull Where to Direct Space Budget Increase," *Inside the Air Force,* October 16, 1998, 2. **7.** Daniel G. Dupont, "Space Control Funds Added to Prompt Congress to 'Deal with' Pentagon," *Inside the Pentagon,* January 28, 1999, 1. Stewart M. Powell, "Bell at the White House," *Air Force Magazine,* February 1999, 40. Robert Bell, a member of the Clinton National Security Council, apparently helped forge this "compromise" with Congress. **8.** Dupont, "Space Control Funds Added to Prompt Congress," and Keith R. Hall, *Presentation to the Committee on Armed Services Subcommittee on Strategic Forces United States,* March 22, 1999. **9.** David Rogers, "Clinton's Defense-Item Vetoes Favor Conventional Pork Over Space Weapons," *Wall Street Journal,* October 15, 1997, 2; John M. Broder, "Clinton Gently Vetoes Items in Military Budget," *New York Times,* October 15, 1997. **10.** Robert Bell, in *Press Briefing: OMB Director Frank Raines, NSC Senior Director for Defense Policy and Arms Control Bob Bell, and Deputy Defense Secretary John Hamre on the Line-Item Veto,* (Washington, D.C.: White House, October 14, 1997), 5, 6; see also Frank Gaffney, "Risky Space Agenda Revealed with Veto," *Washington Times,* October 21, 1997, p. 21.

11. Ferster, "Congress Renews Fight for DoD Space Efforts," 1. **12.** "Air Force May Reprogram FY98 $ for Space Plane," *Military Space,* October 27, 1997, 1. **13.** It should be noted that October 1997 vetoes ran counter to the findings of the president's own Space Control Advisory Board, which endorsed the army ASAT as a deterrent to space-based reconnaissance. Clinton's detractors also pointed out that the president failed to consult properly the Joint Chiefs of Staff, the air force, or the commander of U.S. Space Command before arriving at his decision. Some analysts believed that Clinton's decision against these space weapons may have been bolstered by an arms-control proposal from Russian president Boris Yeltsin, which arrived just days before his decision, to ban all antisatellite weapons. Hackett, "Clinton's Unilateral Disarmament Move," 19. **14.** Broad, "In Era of Satellites," C1; and Bender, "Russia Concerned," 3. **15.** Bender, "Russia Concerned," 3; DoD News Briefing, September 2, 1997.

16. Cited in Scott, "U.S. Adopts 'Tactical' Space Control Policy," 35. **17.** Powell, "Bell at the White House," 40. **18.** Scott, "U.S. Adopts 'Tactical' Space Control Policy," 35. See also George C. Wilson, "No Defense Debate Serves No One," *National Journal,* March 18, 2000. **19.** "Rationality, in fact, dictates reflection on peace despite the uproar of the melee, and on war when weapons are silent. The commerce of nations is continuous; diplomacy and war are only complementary modalities, one or the other dominating in turn, without one never entirely giving way to the other except in the extreme case either of absolute hostility, or of absolute friendship or total federation." Aron, *Peace and War,* 40 (see also 47 and 93). **20.** See, for example, Hafner, "Averting a Brobdignagian Skeep Shoot," 41–60.

21. Hamre, *Statement before the Senate Armed Services Committee.* **22.** Powell, "Bell at the White House," 40. **23.** Steve Weber, "ASAT Proponents Fail to Reverse White House Policy," *Space News,* September 19–25, 1994, 7. **24.** See, for example, Lambakis, "Space Control in Desert Storm and Beyond," 429, 430. **25.** Marcia S. Smith, *U.S. Space Programs* (Washington, D.C.: Congressional Research Service, September 10, 1997), no. IB92011, CRS-7, 8. See also the *National Defense Authorization Act for Fiscal Year 2000,* Senate 1059, where all of DoD's space-control technology programs received a total of $10 million. Pat Cooper, "U.S. Air Force Considers Antisatellite Weapons," *Space News,* February 26–March 3, 1996, 4. According to Marcia Smith, policy analyst for the Congressional Research Service, "The current Congress is certainly more supportive [of the development of ASAT weapons] than the last several congresses." Evidence of this has been congressional support for KEASAT and the removal of the prohibition on using the MIRACL to test the ground-based laser on satellites.

26. Daniel G. Dupont, "Approval of Army Laser Test May Be Beginning, Not End, of ASAT Debate," *Inside the Army,* October 6, 1997, 1. **27.** Daniel G. Dupont and John Liang, "Harkin Implores Clinton to Deny Army Request for Laser Test Permission," *Inside the Army,* September 29, 1997, 1. **28.** Jonathan S. Landay, "Dawn of Laser Weapons Draws Near," *Christian Science Monitor,* October 20, 1997, 1. **29.** Richard Lardner, "New Mexico Lawmakers Blast DoD Plan to Cut Laser Facility Money," *Inside the Army,* March 11, 1996,

5. **30.** Senator Harkin in debate over Amendment No. 2402, in *Department of Defense Appropriations Act 1996* (Senate, August 10, 1995), S12183.

31. Cited in the June 19, 1996, Senate debate on Amendment No. 4058 of the *National Defense Authorization Act for Fiscal Year 1997,* S. 6455; "ASAT Effort in Limbo," *Aviation Week & Space Technology,* May 1, 2000, 64. **32.** Senator Smith argued on June 19, 1996 during Senate debate on Amendment No. 4058 of the *National Defense Authorization Act for Fiscal Year 1997.* The countries named by Smith were Argentina, Australia, Brazil, Canada, China, the Czech Republic, France, Germany, Great Britain, India, Indonesia, Iran, Israel, Italy, Japan, Korea, Luxembourg, Malaysia, Mexico, Norway, Pakistan, Portugal, Russia, Saudi Arabia, South Africa, Spain, Sweden, Thailand, Turkey, and Ukraine. **33.** Amendment 851 to Amendment 829 to the *National Defense Authorization Act for Fiscal Year 1994,* September 10, 1993, S. 11508. **34.** Amendment 851 to Amendment 829 to the *National Defense Authorization Act for Fiscal Year 1994,* September 10, 1993, S. 11508. See also remarks by Senator Robert Smith in the June 19, 1996, Senate debate on Amendment No. 4058 of the *National Defense Authorization Act for Fiscal Year 1997.* **35.** According to Senator Smith, "If the administration does not request it and the policy folks do not want it, if we send it back to the space architect, who is a policy person, to study it, you can pretty well conclude what the results will be. They will not fund it." *National Defense Authorization Act for Fiscal Year 1997,* June 19, 1996, S. 6458.

36. *National Defense Authorization Act for Fiscal Year 1997,* June 19, 1996, S6457. **37.** Amendment No. 2401 to the *DoD Appropriations Act for 1996* (Senate, August 10, 1995). **38.** U.S. Congress, *National Defense Authorization Act for Fiscal Year 1993: Report of the Committee on Armed Services House of Representatives on H.R. 5006 together with Additional and Dissenting Views,* Report 102-527, May 19, 1992, 395. **39.** Senate debate on Amendment No. 799 of the *National Defense Authorization Act for Fiscal Year 1998,* July 11, 1997. See also Ron Dellums, in U.S. Congress, *National Defense Authorization Act for Fiscal Year 1993: Report of the Committee on Armed Services House of Representatives on H.R. 5006 together with Additional and Dissenting Views,* Report 102-527, May 19, 1992, 394. **40.** See, for example, U.S. Congress, *National Defense Authorization Act for Fiscal Year 1993: Report of the Committee on Armed Services House of Representatives on H.R. 5006 together with Additional and Dissenting Views,* 168.

41. Senate debate on Amendment No. 799 of the *National Defense Authorization Act for Fiscal Year 1998,* July 11, 1997. **42.** Amendment No. 2401 to the *DoD Appropriations Act for 1996* (Senate, August 10, 1995). **43.** Senate debate on Amendment No. 799 of the *National Defense Authorization Act for Fiscal Year 1998,* July 11, 1997. **44.** Senate debate on Amendment No. 799 of the *National Defense Authorization Act for Fiscal Year 1998,* July 11, 1997; see also Senator Kyl's remarks during debate on Amendment No. 2401 to the *DoD Appropriations Act for 1996,* Senate, August 10, 1995. **45.** "The arms control community," he maintained, "largely wants to prevent any activity that in their opinion would result in the so-called militarization of space, when in fact there are other countries that are taking steps that could lead them to attempt to obtain a capability in space that we would not be prepared to deal with." Cited in Ferster, "Congress Renews Fight for DoD Space Efforts," 1.

46. Jeff Bingaman, "Remote-Sensing Policy Needs Work," *Space News,* April 3–9, 1995, 29. The Bush policy carried more forceful language regarding space control than the Clinton National Space Policy. **47.** Jeff Bingaman, "Rethink Remote Sensing," *Space News,* February

26–March 3, 1996, 13. See also Jennifer Heronema, "Industry Denounces Senate Imagery Ban," *Space News,* July 1–7, 1996, 1, and Warren Ferster, "U.S. to Restrict Satellite Images of Israel," *Space News,* August 5–11, 1996, 3. The launch of the first commercial one-meter imaging system, Ikonos, highlighted for the editors at *Space News* some gaping holes in the U.S. remote-sensing policy regime. The policy, they noted, gave the U.S. government to place "certain geographic areas and customers off limits to commercial imaging companies." This raises questions: "Who in government will make those decisions? What are the criteria on which those decisions will be based? What are the time limits? To prosper, U.S. industry needs a regulatory environment that makes it possible to predict how long it will take to obtain permission to collect imagery for a particular client of a particular region." "Only the Beginning," *Space News,* October 25, 1999, 14. **48.** Speech of Senator Joe Lieberman before the Association of the United States Army, February 27, 1999, http://www.senate.gov/~lieberman/r030298a.html **49.** See also Ferster, "One on One: U.S. Senator Robert Smith," 22. **50.** "All politics must bend before the right," wrote Immanuel Kant in his essay *On Perpetual Peace.* The "perverseness of human nature" is basically revealed in the "uncontrolled relations between nations." Left unrestrained, war is the only method states have to plead their cause. Peace, he maintained, is naturally desired by all and depends on the establishment of law and progress in History. Only when statesmen subordinate politics to morals, and all men submit themselves to lawful rule under a republican form of government, and all states submit to lawful international organization, will we reach the state of perpetual peace. For Kant, eternal peace was a goal to be forever approached through progress. Through advancing education and the gradual incapacitation of states to make war (as wars become more costly), we can achieve harmony among men. "The homage which each state pays (at least in words) to the concept of law proves that there is slumbering in man an even greater moral disposition to become master of the evil principle in himself (which he cannot disclaim) and to hope for the same from others." Immanuel Kant, "Second Definitive Article for a Perpetual Peace," in Kant, *On Perpetual Peace.*

51. See, for example, DeBlois in "Space Sanctuary," 55. **52.** Isaiah 66:1; see, for example, Revelation 21:1–22 and Psalm 104: "You are clothed with majesty and glory, robed in light as with a cloak. You have spread out the heavens like a tent-cloth." See also interpretation in the hymn "O Worship the King, All Glorious Above," by Johann Michael Hayden and Robert Grant: "Whose robe is the light, Whose canopy is space." **53.** Edward C. Aldridge Jr., "The Myths of Militarization of Space," *International Security* 11, no. 4 (Spring 1987): 151. **54.** Stares, *Space and National Security,* 143. **55.** News Release from Congressman Peter DeFazio, "DeFazio Introduces Bill to Kill Star Wars: Attacks Republican Military Spending Increases," February 16, 1995, mimeo.

56. Amendment No. 2401 to the *DoD Appropriations Act for 1996* (Senate, August 10, 1995). **57.** Landay, "Dawn of Laser Weapons Draws Near," 1. **58.** Lord Zuckerman, "Reagan's Highest Folly," *New York Review of Books,* April 9, 1987, 35. **59.** Philip M. Boffey, "Dark Side of 'Star Wars': System Could Also Attack," *New York Times,* March 7, 1985, p. A24. **60.** Rip Bulkeley and Graham Spinardi, *Space Weapons: Deterrence or Delusion?* (Totawa, N.J.: Barnes & Noble, 1986), 200.

61. Bulkeley and Spinardi, *Space Weapons,* 192. **62.** See, for example, William D. Hartung and Michelle Ciarrocca, "Star Wars II—Here We Go Again," *Nation,* June 19, 2000, 11. **63.** Patricia M. Mische, *Star Wars and the State of our Souls* (Minneapolis: Winston Press, 1985),

2. **64.** Douglas C. Waller, "The Strategic Defense Initiative and Arms Control," in Burns, *Encyclopedia of Arms Control and Disarmament* 2:1122. **65.** McDougall, "How Not to Think about Space Lasers," 554. Well-designed space weapons, according to Senator Malcolm Wallop in 1979, would provide "maximal protection to our society in the event of conflict" and "command the portals of space against the rockets of any other nation." Malcolm Wallop, cited in Bulkeley and Spinardi, *Space Weapons,* 193.

66. DeBlois, "Space Sanctuary," 45. See also David W. Ziegler, "Safe Heavens: Military Strategy and Space Sanctuary," in Bruce M. DeBlois, ed., *Beyond the Paths of Heaven: The Emergence of Space Power Thought* (Maxwell AFB, Ala.: Air University Press, 1999), 185–245. **67.** Lyndon B. Johnson, "The Politics of the Space Age," in *Space: Its Impact on Man and Society,* ed. Lillian Levy (New York: Norton, 1965), 7, 9. **68.** See, for example, Donald R. Baucom, *The Origins of SDI, 1944–1983* (Lawrence: University Press of Kansas, 1993), 39–42. **69.** Senator Robb's recommendation is a shared one. See DeBlois, "Space Sanctuary"; Ziegler, "Safe Heavens"; and Karl Mueller, "Space Weapons and U.S. Security: The Dangers of Fortifying the High Frontier" (paper prepared for the 1998 Annual Meeting of the American Political Science Association, September 3–6 September, 1998). Mueller writes, "The case for banning the deployment of space-to-earth weapons is much stronger, in part because it would be far easier to achieve. No other states are yet on the verge of weaponizing space. . . . A prohibition on placing weapons in orbit would be a straightforward extension of the 1967 Outer Space Treaty's proscription of placing nuclear weapons in space." **70.** Gorove, *Developments in Space Law,* 273.

71. There is "no infallible 'gimmick'—armament or disarmament—which will guarantee definitive peace to a violent and divided humanity." Aron, *Peace and War,* 646. See also pages 643–51. **72.** Winston S. Churchill, *The Gathering Storm* (Boston: Houghton Mifflin, 1948), 14. It is suggested that we may have a more perfect and stable peace by banning weapons in space and signing up to arms-control accords. But strategist Colin Gray pointed out that "if arms control is to attempt important missions, if it is to make a difference, it has to be thoroughly political. War and peace are two sides of a political coin. If arms control could not handle the hostile, or at best nonpermissive, political traffic of the 1930s [a reference to the Washington and London naval arms accords], what value can it have? Political hard times do not excuse the collapse of an arms control process; rather, they demonstrate its irrelevance." Colin S. Gray, *House of Cards: Why Arms Control Must Fail* (Ithaca, N.Y.: Cornell University Press, 1992), 110, 111. **73.** "Only the ABM Treaty also covers non-nuclear weapons for which there is a foreseeable prospect of deployment. For that reason, the possibility of limiting or preventing the overall arms race in space, by means of further treaties covering ASATs and other space weapons, now turns primarily on the effective current status and future prospects of that one agreement." Bulkeley and Spinardi, *Space Weapons,* 202. See also DeBlois, "Space Sanctuary," 46. "The treaty" may die for lack of ideological adherents and failing strategic justification. One of the ABM Treaty's principal architects, Henry Kissinger, came out in May 1999 against pinning hopes for peace and security on an arms-control agreement that "constrains the nation's missile-defense programs to an intolerable degree in the day and age when ballistic missiles are attractive to so many countries." In this world of changed strategic circumstances, Kissinger told the Senate Foreign Relations Committee, the treaty "is reckless." Cited in Jasminka Skrlec, "ABM Pact Outdated, Kissinger Tells Panel," *Washington Times,* May 27, 1999, p. 15.

74. Ziegler, "Safe Heavens," 217, 222. **75.** National Intelligence Council, *Foreign Missile Developments.*

76. McDougall, "How Not to Think about Space Lasers," 554. **77.** See *Federalist* No. 49 (Madison): "If it be true that all governments rest on opinion, it is no less true that the strength of opinion in each individual, and its practical influence on his conduct, depend much on the number which he supposes to have entertained the same opinion. The reason of man, like man himself, is timid and cautious when left alone, and acquires firmness and confidence in proportion to the number with which it is associated. When the examples which fortify opinion are *ancient* as well as numerous, they are known to have a double effect." **78.** Newman, "New Space Race," 30. **79.** Theodore Roosevelt, "Washington's Forgotten Maxim," in *American Ideals and Other Essays, Social and Political* (New York: G.P. Putnam's Sons, 1901), 251. **80.** Thus, it has been argued, "military conflict in space is not a matter for U.S. policy choice today—the choice has already been made." Gray, *American Military Space Policy,* 49.

Chapter 9

1. In some significant respects, the United States' written national security space policy, which recites principles and major goals using very bold and forceful language, is like the country's Bill of Rights—a document that itself is meaningless outside the broader and more fundamental framework of the Constitution and the institutions it established. Founders such as Alexander Hamilton and James Madison argued ardently this very point. A separate Bill of Rights was superfluous in a Constitution that provided *the institutional basis* to protect basic liberties. Or as Hamilton asked, "Why declare that things shall not be done which there is no power to do?" James Madison argued that the real threat to civil liberties was posed by the unruly majority of people, not by the acts of Government. What would a Bill of Rights do, he questioned, if the tyranny of the majority could not be denied? The preferred way to prevent the appearance of a vicious and factious majority was to "control its effects" (the alternative being to destroy liberty altogether and establish a tyranny). Madison's solution was to devise "a republican remedy for the diseases most incident to republican government." The freedoms of Americans would be protected, he maintained, by the institutions of federalism and the separation of powers, not by a "parchment barrier," a bill of rights. This document, he believed, would have its greatest value in the context of public opinion, by declaring political truths in "a solemn manner" and as a ground for appealing "to the sense of the community." But, by itself, it was not the tool for protecting U.S. political rights. See Robert A. Rutland, "Framing and Ratifying the First Ten Amendments," in *The Framing and Ratification of the Constitution,* ed. Leonard W. Levy and Dennis J. Mahoney (New York: Macmillan, 1986, 305–16. See also *Federalist* No. 10 by James Madison, and Paul Peterson, "Separation of Powers and the American Constitution," and Martin Diamond, "The Ends of Federalism," in *Readings in American Government,* ed. Paul Peterson (Dubuque, Iowa: Dendall/Hunt, 1979), 78–85, 87–95. **2.** Consider Aristotle's view of this subject: "For just as man is the best of all animals when completed, when separated from law and adjudication he is the worst of all. For injustice is harshest when it is furnished with arms; and man is born naturally possessing arms for prudence and virtue, which are nevertheless very susceptible to being used for their opposites. This is why, without virtue, he is the most unholy and the most savage [of all animals]." Aristotle, *The Politics,* trans. Carnes Lord (Chicago: University of Chicago Press, 1984), bk. 1, chap. 2, 37, 38. **3.** To act imperialistically is to act with specific objectives in mind, whether a state has substantial capabilities or not. A state that has substantial military capabilities is not necessarily a state that throws its weight around the world (although it *could*) or imposes its will regardless of larger ethical, diplomatic, or strategic considerations at hand. **4.** See, for example, the speech of presidential candidate Governor George W. Bush, "A Period of Consequences," *Citadel,* September 23, 1999: "The American armed forces have an irreplaceable role in the world. They give confidence to our allies; deter the aggression of our enemies; and allow our nation to shape a stable peace. . . . We know that power, in the future, will be projected in different ways." It may well be that U.S. dominance in the future will look something like what former Speaker of the House of Representatives Newt Gingrich has described: "A truly global power—that is, a truly space-based power—may, in fact, be permanently everywhere, by having both overhead access for intelligence and overhead access for delivering firepower in a way that has no resemblance to delivering a World War II-style theater campaign with an expeditionary force that John Wayne would have understood." Cited in Jason Sherman, "Grave New World," *Armed Forces Journal International,* November 1999, 18. See also Tidal W. McCoy, "Seeking American Space Dominance," *Space News,* February 2–8, 1998, 15. **5.** See, for example, Alejandro Alvarez, *The Monroe Doctrine: Its Importance in the International Life of the States of the New World* (New York: Oxford University Press, 1924), 3–25.

6. This, of course, is not to say that existing military departments do not receive formal guidance on the development of their forces or that key principles (i.e., freedom of the seas) do not inform their activities. **7.** See, for example, Gen. Richard B. Myers, "Space Superiority Is Fleeting," *Aviation Week & Space Technology,* January 1, 2000, 54, 55. **8.** The issue of control is a matter of perspective. For example, the U.S. Navy does not have to cover the Atlantic Ocean with its ships in order to control these waters. Its reputation as the most powerful naval force in the world permits it to patrol large bodies of water and at the same time allow other states and private operators to ply the waters in a secure environment. Control, in other words, does not mean exclusive dominance. It does mean the ability to selectively influence, through force or other means, the operations of hostile or neutral states. **9.** See, for example, Peter B. de Selding, "GPS Concerns Spur European Support for Galileo Program," *Space News,* December 20, 1999, 11. **10.** First Lincoln-Douglas debate, Ottawa, Illinois, August 21, 1858, in Roy P. Basler, ed., *The Collected Works of Abraham Lincoln* (New Brunswick, N.J.: Rutgers University Press, 1953), 12–30; see also Glen E. Thurow, "Introduction," in Glen Thurow and Jeffrey D. Wallin, eds., *Rhetoric and American Statesmanship* (Durham, N.C.: Carolina Academic Press, 1984), 3–7.

11. "Stability" may not even be an appropriate political or diplomatic objective in so far as we did not want a stable Nazi Germany. U.S. foreign policy tradition, to the contrary, has striven for the gentle transformation of states by exposing other governments to the activities of open markets, democratic ideals, and the idea that peaceful resolution of disputes is possible. **12.** Bill Gertz, "Helms Charges Clinton Cover-Up of Pilot Injured by Russian Laser," *Washington Times,* September 8, 1999. **13.** Bryan Bender, "US Worried by Coalition 'Technology-Gap,'" *Jane's Defence Weekly,* July 29, 1998, 8; "U.S. Forces' Digital Revolution Threat to Interoperability," *Jane's Defense Weekly,* June 11, 1997, 17; Peter B. de

Selding, "U.S. Implores Allies to Cooperate in Military Space," *Space News*, October 21–27, 1996, 10. **14.** For a taste of this debate, see Gray, *House of Cards;* Thomas C. Schelling and Morton H. Halperin, *Strategy and Arms Control* (New York: Twentieth Century Fund, 1961); Albert Carnesale and Richard N. Haass, eds. *Superpower Arms Control: Setting the Record Straight* (Cambridge, Mass.: Ballinger, 1987). Perhaps the most vigorous period of negotiation between the United States and the Soviet on space arms control (when the focus was on ASAT bans and test moratoriums) occurred in the late 1970s and in the early 1980s, following the resumption of Soviet satellite interceptor tests in 1976. **15.** See, for example, Richard Lardner, "Intel Agencies Urged to Build System for Melding Space Needs, Doctrine," *Inside the Air Force* 10, no. 25 (June 25, 1999): 1.

16. Samuel P. Huntington, "National Policy and the Transoceanic Navy," *U.S. Naval Institute Proceedings* 80, no. 5 (May 1954): 484. Huntington also writes (483–93) that there are human and material as well as organizational structure elements to a military Service. **17.** Schriever interview. **18.** Daniel G. Dupont, "Science Board Says Space Funding, Priority Levels Must Be Increased," *Inside Missile Defense*, April 5, 2000: 15–18. Frank Wolfe, "Lieberman: Pentagon Should Link Spending to New Threats," *Defense Daily*, November 3, 1999, 4; Adam J. Hebert, "Best Use of Space Required Break from Linear Thinking, Scientist Says," *Inside the Air Force* 10, no. 39 (October 1, 1999): 1; Bryan Bender, "DoD Urged to Spend More on Scientific Research," *Jane's Defence Weekly*, November 3, 1999; Thomas E. Ricks and Anne Marie Squeo, "Why the Pentagon Is Often Slow to Pursue Promising New Weapons," *Wall Street Journal*, October 12, 1999, 1. **19.** David Mulholland, "Gansler Maps Future Priorities," *Defense News*, April 19, 1999, 26. **20.** For a fine review of some of the organizational, operational, and management challenges facing the U.S. Intelligence community, see William E. Odom, *Modernizing Intelligence: Structure and Change for the 21st Century* (Fairfax, Va.: National Institute for Public Policy, September 1997).

21. Cited in Wall, "Intelligence Lacking on Satellite Threats," 54. **22.** Mark A. Stokes, *China's Strategic Modernization: Implications for the United States* (Carlisle, Pa.: Strategic Studies Institute, 1999), 195–214; Paul Beaver, "China Develops Anti-Satellite Laser System," *Jane's Defence Weekly*, December 2, 1998, 18; Bill Gertz, "China Makes Upgrades to Island Base, Coastline," *Washington Times*, February 11, 1999, p. 1. **23.** Phillip Clark, "Fact and Fiction: North Korea's Satellite Launch," *Launchspace*, January/February 1999, 39–41. By early 1999, former Ballistic Missile Defense Organization director Lieutenant General Lyles and the Pentagon's chief acquisition's officer, Jacques Gansler, both confessed their belief that, based on the Taepo-Dong incident, a ballistic missile threat to the United States from countries other than Russia or China would emerge much earlier. John Donnelly, "ICBM Threat to U.S. by Next Year, General Predicts," *Defense Week*, March 1, 1999, 1; Cohen, *Annual Report to the President and the Congress*, 1998, 65; William S. Cohen, *Annual Report to the President and the Congress* (Washington, D.C.: GPO, 1999), 74. **24.** Bruce D. Berkowitz, "The CIA Needs to Get Smarter," *Wall Street Journal*, March 1, 1999, A22. **25.** See, for example, George S. Robinson, "Space and Maritime Law: 500 Years of Heritage," *Space News*, November 15, 1999, 21.

Index